MW00737290

The Legal, Engineering, Environmental and Social Perspectives of

Surface Mining Law and Reclamation by Landfilling

The Legal, Engineering, Environmental and Social Perspectives of

Surface Mining Law and Reclamation by Landfilling

Getting Maximum Yield from Surface Mines

Robert Lee Aston, **Ph. D.**
University of Aston, UK & University of Missouri-Rolla, USA

ICP

Imperial College Press

Published by

Imperial College Press
57 Shelton Street
Covent Garden
London WC2H 9HE

Distributed by

World Scientific Publishing Co. Pte. Ltd.
P O Box 128, Farrer Road, Singapore 912805
USA office: Suite 1B, 1060 Main Street, River Edge, NJ 07661
UK office: 57 Shelton Street, Covent Garden, London WC2H 9HE

Library of Congress Cataloging-in-Publication Data
Aston, R. Lee.
The legal, engineering and social perspectives of surface mining law and reclamation by landfilling : getting maximum yield from surface mines / Robert Lee Aston.
p. cm.
Includes bibliographical references and index.
ISBN 1-86094-123-0
1. Strip mining -- Law and legislation -- Great Britain. 2. Abandoned mined lands reclamation -- Law and legislation -- Great Britain. 3. Sanitary landfills -- Law and legislation -- Great Britain. 4. Strip mining -- Law and legislation -- United States. 5. Abandoned mined lands reclamation -- Law and legislation -- United States. 6. Sanitary landfills -- Law and legislation -- United States. 7. Strip mining -- Law and legislation -- Canada. 8. Abandoned mined lands reclamation -- Law and legislation -- Canada. 9. Sanitary landfills -- Law and legislation -- Canada. I. Title.
K3904.A93 1999
346.4104'6765 -- DC21

British Library Cataloguing-in-Publication Data
A catalogue record for this book is available from the British Library.

Printed in Singapore.

ABSTRACT

Surface Mine Reclamation by Landfilling

The subject of this work, although primarily a legal and environmental engineering treatment of the subject, has, in its main, a perspective of great interest to the general public — the social and environmental concerns of restoring disturbed earth from surface mining to its natural, pre-mined condition. Of equal social interest is the locating and using of acceptable depositories for the ever-increasing volumes of human waste. These two goals can be accomplished simultaneously at little or no cost to the public, and at a profit to industry, while conserving land by restoring surface mines to beneficial surface uses.

Although this book focuses on the mineral and critical waste disposal requirements and regulations of the United Kingdom, the United States and Canada, the information applies world-wide. Metallic, non-metallic and coal surface mines, but particularly those of construction minerals, are prime sites for solid waste depositories. Construction mineral pits, i.e. aggregate stone, are emphasised and promoted as the prime sites for landfill consideration because they are most widely and universally located, being located mainly near the population centres of waste generation. Wherever there are humans creating waste, there are also aggregate pits to supply the construction works of humans. Annual surface mining creates sufficient void space to accommodate the annual volume of waste generated.

In the area of international environmental law, this book proposes the formulation of one-step planning and permitting regulation for the integrated utilisation of new surface mines as depositories for municipal solid waste. Additionally, the utilisation of abandoned and currently operated surface mines is proposed as solid waste landfills as an integral step in their reclamation. Existing laws, litigation and issues in the United Kingdom, the U.S. and Canada are discussed because of their common legal system, language and heritage.

The critical shortage of approved space for disposal of solid waste has caused an urgent and growing problem for both the waste disposal industry and society. Surface mining can serve four important environmental and societal functions inuring to the health and welfare of the public:

(1) providing essential minerals for goods and construction;
(2) sequentially, providing critically needed, safe depositories for society's wastes;
(3) fully reclaiming surface mined lands; and
(4) conserving land by dual purpose use and restoring derelict land to beneficial surface use.

Currently, the first two functions are treated environmentally, and in regulation, as two different siting problems, yet they both are earth-disturbing and excavating industries requiring surface restoration. The processes are largely duplicative and should be combined for better efficiency, less earth disturbance, conservation of land, and for fuller and better reclamation of completed surface mines by returning the surfaces to greater utility than present mined land reclamation procedures. Derelict lands need be reclaimed for the use of future generations.While both industries are viewed by a developed society and its communities as "bad neighbours", they remain essential and critical for mankind's existence and welfare.

This work argues and demonstrates that most surface mine openings, if not already safe, can economically and profitably be made environmentally safe through present containment technology for use as solid waste landfills. Simultaneously, the procedure safeguards and monitors protection of ground and surface waters from landfill contamination.Successful examples are given in the three Anglo nations. Fully restored mined lands create increased land values by highly profitable landfilling. Industry and environmentalists should rejoice.

Keywords: Surface mining; Opencast mine reclamation; Landfilling; Land conservation

ACKNOWLEDGEMENTS

The author wishes to express special appreciation, gratitude and thanks to:

Dr. Peter D. Hedges, Senior Lecturer, Aston University, Birmingham, England, for his time, dedication, ever-ready advice, critiquing and for his many trans-oceanic faxes, and for his trips to the colonies;
Dr. C. Dale Elifrits, Professor of Geological Engineering at the University of Missouri-Rolla, for his omni-present enthusiasm and encouragement for the need and purpose of this study; for his generous giving of time, advice, assistance and critiquing of the research. Also, to Dr. Elifrits and his former protégé, Michael Owens, for contributing information on their study of surface coal mine reclamation by landfilling in the Western and Illinois Basins of the U.S.;
Dr. Allen W. Hatheway, Professor of Geological Engineering, University of Missouri-Rolla, for his enthusiastic support, critiquing, and for generously furnishing information from his unpublished chronological history of waste;
Mr. Hugh G. Kent, Planning Manager, Clay Colliery Company, Ltd., Telford, U.K., who contributed technical reclamation and cost information, photographs, and time for visits in the U.K. to Clay Colliery reclamation projects of present and former coal and waste sites.

Appreciation and thanks for assistance in supplying valuable research information is expressed to others, but particularly the following:

Ms. Susan Blackman, lawyer and researcher, formerly with the Canadian Institute of Resource Law, University of Calgary, Alberta, Canada;
Mr. Paul A. Tomes, Regional Director, Northern Regional Office, Greenways Landfill, Judkins Quarry, Tuttle Hill, Nuneaton, Warwickshire, U.K., for

valuable technical information on his company's reclamation projects;
Mr. David Harding, Manager, Public Relations, ARC Central, Leicester,U.K.;
Mr. Marc Ramsey, Manager, Fred Weber Inc, Maryland Heights, Mo., USA, for valuable technical information on his company's landfilling and reclamation project;
Mr. Ronald Plewman, Manager, Walker Inds., Thorold, Ontario, Canada, for valuable information on his company's landfilling and reclamation project;
Dr. Robert W. Slater, Deputy Minister, Environment Canada, Hull, Quebec, Canada, and his staff, particularly Mr. Bob Christensen of the Solid Waste Management Division;
Mr. Bill B. Blakeman, formerly a consultant and later with Environment Canada, Hull, Quebec, Canada, for furnishing a copy of his earlier work and valuable technical information;
to my son, Roger A. Aston, a computer expert and consultant, for his help in the computer work; and last, but most assuredly not least, much gratitude and thanks to my faithful, supportive, loving wife and secretary, Mary Pierce-Aston, whose help over several years at the libraries and office made the task lighter, and for her ever-present, patience, cheerful support and encouragement in completing the work.

TABLE OF U.K. CASES

Reported and Mentioned

Case name and citation | Page

TABLE OF U.K. CASES (continued)
Reported and Mentioned

TABLE OF U.K. CASES (continued)

Reported and Mentioned

TABLE OF U.K. — E.C. CASES (continued)

Reported and Mentioned

EC Cases

TABLE OF U.S. CASES
Reported and Mentioned

TABLE OF U.S. CASES (continued)

Reported and Mentioned

TABLE OF U.S. CASES (continued)
Reported and Mentioned

TABLE OF U.S. CASES (continued)
Reported and Mentioned

Case name — Page

TABLE OF U.S. CASES (continued)

Reported and Mentioned

Case name — Page

TABLE OF U.S. CASES (continued)
Reported and Mentioned

TABLE OF U.S. CASES (continued)

Reported and Mentioned

TABLE OF CANADA CASES

Reported and Mentioned

TABLE OF CANADA CASES (continued)

Reported and Mentioned

TABLE OF CANADA CASES (continued)

Reported and Mentioned

LIST OF ILLUSTRATIONS

Figure **Page**

LIST OF TABLES

ABBREVIATIONS / ACRONYMS COMMONLY ENCOUNTERED

AMD	Acid Mine Drainage
ARD	Acid Rock Drainage
AONB	Areas of Outstanding Natural Beauty (UK)
APA	Administrative Procedure Act (US)
ASCML	Abandoned Surface Coal Mined Land
BAT	Best Available Technology
BLM	Bureau of Land Management (US)
CEQ	Council of Environmental Quality (US)
CFR	Code of Federal Rules (US)
COPA	Control of Pollution Act (UK)
Corps	U.S. Army Corps of Engineers
CWA	Clean Water Act (US)
CMZA	Coastal Zone Management Act (US)
EA	Environmental Assessment (UK, US, Canada)
EC	European Community
EIR	Environmental Impact Report
EIS	Environmental Impact Statement (UK, US, Canada)
EPA	Environmental Protection Agency / Administration / Act
FONSI	Finding of No Significant Impact (US)
HDPE	High Density Polyethylene
LG/LFG	Landfill Gas
MEND	Mine Environment Neutral Drainage
MPG	Mineral Planning Guide (UK)
MSW	Municipal Solid Waste

NEPA	National Environmental Protection Act (US)
NFFO	Non-Fossil Fuel Obligation (UK)
NIMBY	Not In My Back Yard (Syndrome)
NOV	Notice of Violation
NPDES	National Pollutant Discharge Elimination System (US)
NRA	National Rivers Authority (UK)
PIP	Public Inquiry Process / Public Hearing Process
RDF	Refuse Derived Fuel
SMCRA	Surface Mining Control and Reclamation Act (US)
SSSI	Sites of Special Scientific Interest (UK)
TCPA	Town and Country Planning Act (UK)
USEPA	United States Environmental Protection Agency / Administration / Act
VER	Valid Existing Rights
WC	Wildlife and Countryside Act (UK)
WCA	Waste Collection Authority (UK)
WDA	Waste Disposal Authority (UK)
WRA	Waste Regulation Authority (UK)
WML	Waste Management Licence (UK)

PREFACE I

The Social Perspective

A critical shortage of approved space for disposal of municipal solid wastes has caused the developed nations to have an urgent and growing need for landfill space. It is a problem for both the disposal industry and society. Simultaneously, the ever-continuing, ever-growing need exists for minerals which supply society with the basic raw materials for its health, well-being, existence and growth.

In its recent publication, "*Living with Minerals*", by the Confederation of British Industry (CBI), it is emphasised that what is assumed to be common, or layman's knowledge, is actually not; i.e., that man in his ordinary, daily activities constantly uses mineral products, is surrounded by them, and is totally dependent on them, and yet, man remains unknowing, unsuspecting, or unmindful that he "cannot survive without them". To illustrate, CBI cites the use of minerals in the home alone — "Apart from the timber joists and floor-boards, most of the fixtures and fittings in the modern house, ranging from the ceramic basin (clay) in the bathroom to the fiberglass insulation (silica minerals) in the roof (and the gypsum walls) have mineral origins ... that a telephone can contain as many as 40 mineral products, or the family car about 15". (Mining Journal, 1995a).

Thus, treating the problem and seeking an acceptable solution is found to be in a very sensitive area between environmental protection and the life-supporting industries. Therefore, in prefacing this work, it should be understood that the environmentally reprehensible acts of the past committed by the mining and waste disposal industries can only be excused by the historical fact that they were done as the accepted way of life at the time, not with prime intent to cause harm or injury to the earth, the environment, or to fellow men. Special favour to the mining industry herein is not intended. However, it is notably emphasised that mining and waste disposal have a proper and fundamentally essential place in society's cultural order and deserve as much support and favour as does clean air and clean water. Mankind's survival is as equally dependent on the mining industry and proper waste disposal as the individual is dependent on the air he breathes or the water he drinks. However, the air and water man consumes must be clean. So it is in the developed nations, and, hopefully, soon to be in the developing nations, that the mining and waste disposal industries are clean-operating and environmentally-safe under regulation. Thus, this work will be rightfully found to support surface mining in creating earth voids for subsequent waste filling and justifying their place in the social order.

Surface mining, unlike others of man's activities, can serve four important environmental functions: (1) providing minerals for the beneficial well-being, health and welfare of the public; (2) sequentially, to provide critically needed disposal sites for society's wastes which also inures to the health and welfare of the public; (3) fully reclaiming surface mined lands thus restoring derelict land to beneficial surface use; and (4) conserving land and sustainability of land by dual purpose use.

Currently, the two needs for minerals and waste deposit are treated environmentally, and in regulation, as two different problems, yet they are closely related by virtue of both being earth-disturbing and earth-moving industries. Mining digs a hole to obtain minerals, then only smoothes and re-seeds the surrounding surface for deficient, but required land reclamation. Waste disposal, whether mounded on the earth's level or sloped surface, or placed in a ravine or topographic depression, must trench below the surface to emplace a leachate collection system and to make the initial pocket for starting the landfill. Additionally, other pits must be excavated to obtain daily soil cover material for the waste deposited, for clays for use as lining and capping material, or for sand for overlay of the clay capping material. Thus, earth-moving processes are duplicative and should be combined into one location for better efficiency, less earth disturbance, conservation of land, for fuller and better reclamation of completed surface mines, and restoration of mined land to surface use.

Whilst both industries are viewed by modern, developed society and its communities as "bad neighbours", they remain, nevertheless, essential and critical for mankind's existence. By way of analogy in a household, the water closet, the coal bin and woodpile, the furnace, the heat pump and air conditioning, are all essentials for modern-day comfortable living; few modern homes in a developed society would be without them. These necessities of the home are not as aesthetically appealing as the living room, den, or boudoir, still they are a necessity included in the home plan but are ordinarily placed in less conspicuous parts of the home or curtilage, and often given some dressing to be more aesthetically acceptable.

So it is with surface mines and trash or garbage dumps (now, more politely called municipal waste disposal sites). They are absolute necessities of life for communities. Without them, society invites a return to the dark ages, living in filth and disease without such conveniences and amenities as medicines, building materials, machines and electronics, all manufactured from minerals taken from the earth.

Since both industries are earth-disturbing, they incur the greatest attention of both the general public and environmentalists seeking protection to the environment by strict controls through law and regulations. This study reviews and analyses, separately, regulatory controls for both industries, and then proposes the joining of them into one. Mining is done for profit. Similarly, solid waste disposal is a lucrative industry. Land reclamation, when done for aesthetics and the environment alone, is tremendously expensive. Combining the two is economic efficiency. Therefore, the proposal herein is logically to utilise the man-made void openings in the earth, or excavations from mining of minerals for backfilling with man's wastes. The logic of this solution seems indicated in a quotation from the Bible, "... for dust thou art, and unto dust shalt thou return". (Genesis 3: 19, King James version).

R. Lee Aston

PREFACE II

Instructions for the Non-Legally Trained Reader in the Use of Case Citations

A case citation is the reference to the parties, the date, the court and the legal publication where the full text of the case may be found.

A Table of Cases for those, either reported on, or mentioned by reference, is given in Appendix A. Cases listed are separated for Great Britain (England / Scotland) (U.K.), the European Community (EC), the United States, and Canada.

Parties: The style of the case refers to the names of the parties, i.e., the plaintiff party, bringing the suit, is listed first; then, following *v.*, for versus, the Defendant's name follows, or the party defending the charge. In this work, the case style, or parties names, are *italicised.*

Date of Decision of Case: In English and Canadian cases, the date of the hearing court's decision immediately follows the litigating parties' names. The date of the decision in United States cases is given at the end of the citation.

The Court rendering the decision: In older English cases, a court is not usually cited. The date is followed by the Case Recorder's name which indicates his series of Reports. In modern English case reporting, no court is given in the citation, but may be found in the publication cited, e.g., P. & C.R., which stands for Property and Compensation Reports, or A.E.R. All England Reports, or W.W.R., Western Weekly Reports.

In U.S. citations, the name of the court is generally given after the publication citation, e.g., 297 S.W. 184 (Mo.App. 1927); the court was the Missouri Court of Appeals found in Volume 297, 1st series of SouthWestern Reporter, on page 184.

In Canadian citations, the name of the court is not always given. It may appear after the parties' names, or after the date.

Publication volumes and pages: The volume number of the publication precedes the initials of the publication. The number after indicates the page number on which the case begins. If a second number follows the first number, it is given as a reference where a particular quotation just given from the case is found. In the U.S. citation example given above, the number 297 indicates the volume; S.W. indicates it is a case in the Southwestern Series (by West Publishing Co., St. Paul, Minnesota); Volume 297 is in the first series, since there is no other number immediately following S.W., as in S.W.2d for the second series; the case will be found starting on page 184.

U.S. Case Reporters: West Publishing Company covers all the United States Courts, state and federal, in its series of publications. For the federal system of published cases, West publishes weekly case decisions for the U.S. Supreme Court in the Supreme Court Reports; for the U.S. Courts of Appeal in the Federal Reporter (F); and for the U.S. District Court decisions in the Federal Supplement (F.Supp.).

In West's regional system of state case reporting system, cases from the lower trial courts, the Circuit and Superior Courts, are not published. West publishes appealed decisions of the lower, trial courts made to state courts of appeal and to the state supreme courts in a series of Reporters grouped according to geographical location of the states. West publishes seven state case reporters, viz., Atlantic, Northeastern, Southeastern, Southern, Southwestern, Northwestern and Pacific Reporters.

For example, Nevada and Oklahoma state decisions are found in the Pacific Reporter; the Pacific Reporter covers most of the Rocky Mountain states; Missouri, Arkansas, Texas decisions are in the Southwest Reporter; the New England states and southward through Pennsylvania are in the Atlantic Reporter;

below the Mason and Dixon Line, the Southern states, Maryland, West Virginia, Virginia, southward to South Carolina are in the Southeastern Reporter; the Southern Reporter covers the Deep South, Georgia, Florida, and westward to Louisiana. Michigan is included in the Northwest Reporter; Illinois in the Northeastern Reporter along with Indiana. West publishes state cases for California and New York in separate reporters. All states also have their own decisions reported and published in bound volumes, usually by another publisher, not West.

Canadian Case Reports: Provincial cases from the provincial court systems are found in the various Provincial case reporters. For example, *Casamiro Resources Ltd. v. British Columbia,* [1991] 55 B.C.L.R. (2d) 346, is a 1991 decision found in Volume 55 of the second series of the British Columbia Law Reports starting on page 346. And, *Tener v. The Queen* [1985] 1 S.C.R.533, 3 W.W.R. 673, is found by either of two citations, viz., Volume 1 of the Supreme Court (of Canada) Reports starting on page 533, or may be found in Volume 3, page 673 of Western Weekly Reports. Few other Canadian case sources are used, e.g., *Regina v. United Keno Hill Mines Ltd.* [1980], 10 CELR 43, indicates that it is found in Volume 10 of the Canadian Environmental Law Reports, starting on page 43. Various publishers are used for Canadian case reporting.

United Kingdom Case Reports: The system of reporting for British case law is more complicated than in either the U.S. or Canada. This is due to there being more levels of courts and more jurisdictions, with each having its own reporting publication. More than that, it is not the purpose of this work to give an extensive discourse on British case reporting.

Explanations of the abbreviations of legal publications is not offered here as the sources are so numerous for the three countries included in this work. If the non-lawyer needs to have a legal citation reference deciphered, a law librarian will be of help.

Statute / Act Citations: Where an act is cited, it is usually followed by letters, e.g., "R.S.C." (Revised Statutes of Canada), the date of the statute and the civil code number; or, "Stat. Ann.", meaning Statutes Annotated.

below the Mason and Dixon Line, the Southern states, Maryland, West Virginia, Virginia, southward to South Carolina are in the Southeastern Reporter, the Southern Reporter covers the Deep South, Georgia, Florida, and westward to Louisiana. Michigan is included in the Northwest Reporter, Illinois in the Northeastern Reporter, along with Indiana. West publishes state cases for California and New York in separate reporters. All states also have their own decisions reported and published in bound volumes, usually by another publisher, not West.

Canadian Case Reports: Provincial cases from the provincial court systems are found in the various provincial case reporters. For example, *Greenpeace Resources Ltd. v. Britannia Mining* (1991) 55 B.C.L.R. (2d) 346 is a 1991 decision found in volume 55 of the second series of the British Columbia Law Reports starting on page 346. And *Torres v. The Queen* (1955), 1 S.C.R. 153, 3 W.W.R. 672, is found in either of two citations, viz., Volume 1 of the Supreme Court (of Canada) Reports starting on page 153, or may be found in Volume 3, page 672 of Western Weekly Reports. A few other Canadian case sources are used, e.g., *Kitimat v. Grand Lake Ltd. Alberta* et al. (1990), 10 C.E.L.R. 43, indicates that it is found in Volume 10 of the Canadian Environmental Law Reports, starting on page 43. Various publications are used for Canadian case reporting.

United Kingdom Case Reports: The system of reporting for British case law is more complex than in either the U.S. or Canada. This is due to there being more levels of courts and their jurisdictions, with each having its own reporting publication. More than that, it is not the purpose of this work to give an extensive discourse on British case reporting.

Explanations of the abbreviations of legal publications are not offered here as the sources are so numerous for the three countries included in this work. If the non-lawyer needs to have a legal citation reference deciphered, a law librarian will be of help.

Statute (Act) Citations: When an act is cited, it is usually followed by letters, e.g., "R.S.C." (Revised Statutes of Canada), the date of the statute, and the division or chapter number, or "Stat. Ann.", meaning Statutes Annotated.

CONTENTS

SECTION II AN HISTORICAL REVIEW — EARLY MAN TO RECENT REGULATIONS 25

Section I

CHAPTER 1

INTRODUCTION

1.1 Opening Statement of the Case

Because this study is quasi-legal in nature, combining subject matters of law, geology, mining, environmental and geological engineering, Counsel will outline the general case to be presented. The Opening Statement is simply to advise of the issues involved, of the forthcoming facts to be relied on and entered into evidence. The anticipated proof of facts will be brought into evidence and presented at the proper time as the claims are made later when the points are presented and the issues argued.

Statements of judicial notice are also made, particularly in the Opening Statement, which are those facts given as true without formal pleading and proof. The essential elements for judicial notice are that the fact stated must be one of common knowledge which persons of average intelligence and knowledge of things about him can be presumed to know, and of which fact is certain. Judicially known facts are those covering matters so notorious that a production of evidence is unnecessary.

The overall aim of this research and study is to develop a best uniform practice model for the Anglo nations for conjunctive regulation to enable all surface mines, but especially non-fossil fuel open pit mining, reclamation through landfilling of municipal solid wastes.

The aim of the study will be achieved through ten main objectives, visually:

(i) an historical review of mineral lands and regulation in the U.K., the U.S, and Canada up to the environmental consciousness of the 1960's;
(ii) an historical review of the disposal of wastes by earth burial;
(iii) an historical review of litigated mining and refuse disposal pollution claims in the three Anglo nations;

(iv) a review of the environmental era regulatory actions taken for surface mining and landfilling with litigated interpretations;
(v) a review of later legislative environmental responses to update the initial regulations;
(vi) a report on today's weaknesses of environmental regulations concerning surface mining and waste disposal;
(vii) a report of present and future mineral and waste disposal trends in the U.K. and North America;
(viii) an updated report on landfill technology making open pit landfilling environmentally feasible;
(ix) a report on expected future legislative problems;
(x) the proposed solution through a best practice regulatory model for land conservation, sustainable land use, reclamation of surface mining by subsequent landfilling, and restoration of derelict mined lands to beneficial surface uses.

As observed in the preceding objectives and the List of Contents, the work is a progression, from past to present to future; from whence surface mining and waste disposal came, existing for centuries without concern for the environment or regulatory controls, then passing rapidly into late-twentieth century's rigid regulation within a relatively short period, and where it must go for the future good of mankind, as well as his environment.

Before presenting the hard issues of the case, the historical review of mining, litigated damage claims from early mining and early disposal methods of man's wastes by burial in the three concerned Anglo nations is presented in Section II in the following three chapters, i.e., 2, 3 and 4. Histories are traced from the earliest written recordings in Great Britain, the United States and Canada to the advent of the 1960s, when rumblings of environmental concern were loudly heard.

In tracing the separate histories of mining and waste disposal, litigated claims have been researched over ensuing and parallel periods of time. Such study enables us to compare the development of Anglo law for meeting the complaints over the many decades leading up to recent environmental concerns and the subsequent formation and enactment of stringent environmental regulations.

The purpose of studying the separate development courses for legal treatment of mining and waste disposal, from earlier time to the recent past, is to reveal the former earth-disturbing practices and the environmentally damaging claims made against each. Historical nuisance claims from mining and garbage dumping have become today's regulated environmental concerns. It is found in reviewing the past that there were environmental concerns for both the so-called "bad-neighbour" industries, i.e., mining and waste disposal, long before the advent of the 1960s' green movement. Both industries are earth-disturbing operations. They share yesteryears' common environmental concerns for noise, dust, traffic, odours, air and water pollution, along with the more recent addition of today's society's concerns for disturbance to wildlife habitats and aesthetic objections of intrusion into and destruction of natural beauty areas.

The historical background chapters are followed by three chapters, 5, 6, and 7, which bring the study up to the present. They look into the early environmental regulatory period beginning in the 1970s with its unprecedented, pervasive environmental controls for all industry and the public. As might be expected, with any newly and intensely regulated social reformation, the era was fraught with political and legal contention in deciding to what extent the parameters of regulatory control for the improvement of the environment could and might go.

Towards this end, selected, recent and current litigated cases are examined to illustrate the conflicts over regulations which arose and were challenged as they affected surface mining, waste disposal and water pollution. Resolution of these challenges by the judiciary in some cases have clarified the regulations, whilst in a few, a regulation has been found *ultra vires,* or beyond the delegated powers authorised by the governing and enacting bodies, or lastly, as found in a majority of cases that the regulation was upheld. The purpose of examination of litigation in this era is to accomplish an understanding of the extent of regulation that had been undertaken by Anglo governments, the effects on industry and society, its degree of acceptance by industry and society, and its to date realised successes and weaknesses in cleaning up the environment.

Section III, Ch. 8, follows with an inspection and critiquing of the weaknesses of certain existing environmental regulations and actions.

Conclusions are drawn and made to improve and expedite the regulatory processes for permitting and licencing new operating sites, particularly to combat the NIMBY (Not in My Backyard) syndrome largely caused by the hangover distrust from a former age's disrepute for surface mining and refuse disposal sites.

The focus of the case then turns to the future in Section IV, Chs. 9, 10, 11 and 12. First examined are the indicated trends for society's future mineral requirements. As more products are produced from increased mineral production, the resulting increase in waste will, in turn, cause an even more critical, increased demand and urgency for more landfill space in the future. Strong evidence of environmental landfill safety is presented in Ch. 11 with a short review of current landfill technology and the latest methods, treatment innovations, and potential for more remote disposal sites follows in support of the argument that waters around most any former or current surface mine can be made environmentally safe when a pit is used for depositing municipal solid waste (MSW). A discussion of hydrology, landfill technology, and ground water protection is included in Ch. 11. The proof of the case is presented with successful examples of current, combined surface mines being backfilled with MSW. This section on the trends culminates in Ch. 12 treating possible future legislative problems. Concerns are expressed for mounting limitations on the available land base for future surface mining in the name of land conservation which could seriously hamper mining's ability to adequately and economically furnish the mineral requirements for society's future needs to maintain standards of quality living.

In the final Section V, closing arguments are made in Ch. 13, in support of the use of surface mines to relieve the present and predicted future critical need of space for landfills. The combined present and future requirements are supported by the argument of the steady population growth and concentration in urban masses, accompanied by the repetitive process or cyclical nature of increased societal demands for goods and construction, requiring more construction materials for ever-growing population centres, which in turn increases the size and number of aggregate surface mines near the population masses. In turn, construction materials from mining provide more potential municipal solid waste disposal sites near the population centres where the greatest volumes of waste are generated. The cycle is continuously repetitive.

With the case presented, accompanied by supportive arguments and evidence, the final Ch. 14 concludes with a model formula, or law, proposed to alleviate and resolve the critical need for waste disposal space through utilisation of surface mine sites.

Having stated the case to be presented, the evidence will show that the name "Bad Neighbours" no longer properly describes the surface mining and solid waste disposal industries, that in fact, they are a new generation and "breed" of operators with environmental awareness, and should no longer be equated to their predecessors, the miners as "rapers of the land", and refuse collectors as "spreaders of pestilence". They deserve a new, greatly improved image in the public's eye, truly pictured as citizens equally conscientious and environmentally responsible.

In finality, the last chapter formally proposes the solution to the problem of urgency for landfill space through the utilisation of surface mines by backfilling them with MSW, making for fuller and better reclamation of the mined land, and simultaneously conserving land use. Additionally, a proposal is made for alteration and improvement of the present, general planning, public inquiry and permission process, or systems, in place.

The first proposal concerns reduction of the vexatious, prolonged, litigious, time-consuming and costly delays caused by the public inquiry process (PIP) which confronts projects, and at times defeats essential projects vital for the general public's good and benefits. At times projects are defeated by the public hearing process for no other reason that they are undesirable and unwanted by a group in the local area, i.e., "not in my back yard".

The second, and main proposal is made in the form of a best practice model law for the planning, licencing and mandatory use of such dual purpose surface sites in a one-step approval process as opposed to the present individual, separate, two-step process for individual operations at different locations.

In pleading this case for surface mining and the utilisation of open cast mining pits as depositories for municipal solid waste (MSW), it is felt necessary that, at least, an attempt should be made to combat and assuage, if not refute, their age-old "bad-neighbour" reputations. In this attempt for modern surface miners and MSW collectors of the 1990s to cast off the disrepute to which former surface mining and trash dumping had fallen, and in defense of their present, essential existence, inuring to society's benefit, the greatly needed facility of licencing such operations without impediments by an uninformed, or perhaps more appropriately, a misinformed public is pleaded. The pleading for this case might be analogised to a fictional civil action, as the case of *Modern Surface Mining vs. The General Public.*

The parties in such a fictional class-action case would be: Plaintiff, the modern surface mining industry, joined by the solid waste disposal industry as co-Plaintiff. Defendants would be the General Public.

Plaintiffs might bring the action against the General Public for slander, libel and defamation of character, complaining that they have been maliciously and wrongfully maligned, falsely accused of perpetrating certain injurious acts against the environment, the well-being, health and welfare of the General Public.

Defendant General Public might answer the Plaintiffs' complaint stating that it is a well-established fact, perhaps one of judicial note and common knowledge, that both Plaintiffs are generally known as "bad neighbours", having earned and established their reputations over past centuries. Plaintiffs are in disrepute because of their past harmful acts to the Public, to Nature and the environment. Public opinion has been justified. Truth is always a defence to slander, libel and defamation, and the past accusations made against the Plaintiffs are true. Therefore, the reputations of the "bad neighbours" are truthful and not slanderous.

Plaintiffs, Modern Surface Mining and Solid Waste Disposal Industry respond to Defendants' Answer by way of Amendment to its Complaint saying that they are a new generation, being neither culpable, nor responsible for the acts of their antecedents and forebearers. Both new generation industries are responsible, conscientious, law-abiding citizens conforming to environmental regulatory controls. They are very aware and concerned with preservation and conservation of the earth's environment, and that their acts in performing its essential earth-disturbing job are carried out with care and concern for the protection of the public good and the environment. Plaintiff Modern Surface Mining rejects the appellation of "Bad Neighbour" as a wrongful and misapplied character description.

Modern Surface Mining pleads, as good citizens of many international and local communities, in a demonstration of good faith, and in view of its improved and law-abiding environmental conduct in recent years, that it be accepted as a "good neighbour" and as a responsible member of society in the local communities where it is so critically needed.

To demonstrate Modern Surface Mining and the Waste Disposal Industry's public spirit, it proposes to come to the aid of society, the General Public, at a time of critical need in solving the municipal solid waste space-deficiency problem by offering its mining void spaces for waste disposal, to enhance reclamation of its disturbed land by total restoration of the original surface whilst conserving land use by combining worked-out mines as solid waste landfill depositories as opposed to requiring a separate location for depositing municipal solid wastes. Future land-use planning and permitting is proposed to be accomplished under a pre-planned, one-step licencing regulatory process rather than the present haphazard, by chance, double licencing method. This concludes the Opening Statement by Counsel for Plaintiffs.

1.2 A Brief Outline Of The Problem

The basic problem is to find an environmentally acceptable solution for the critical shortage of approved space for disposal of solid waste materials, particularly in the areas of population concentrations.

A correlative problem is to find better utilisation and fuller restoration for mined-out, non-fuel, open pits than under the present land reclamation regulations.

A third problem is to find a less contentious, less controversial and less combative way for projects to be approved, permitted and licenced than through the extended PIP / public inquiry (hearing) process. It is pleaded and urged that more public trust and confidence be placed in the scientific management of the in-place regulatory agencies already charged with the protection of the public and the earth from environmentally damaging works. The constant contentious litigation and antagonism by environmental groups demonstrates distrust in the established environmental protection regulations and the agencies to carry out their duties.

The proposed solution for the first two problems is to utilise existing surface mining voids wherever possible for landfills, and in the future to join the making of new surface mines and landfills into a sequentially, one-step procedure from the initial planning and permitting stage. Thus, in a world currently conscious of re-cycling, the worked-out surface mines should be "re-cycled" to further serve humanity, and finally to fully restore the mined land to beneficial surface uses.

The essence of this work is to stress and prove the importance of the very desirable joining in regulation for the planning, permitting and much needed co-management of two major earth-disturbing, earth-working, essential-to-man industries contributing environmental problems in the present world, viz., mining and solid waste disposal. Though waste disposal may not be thought of as an earth-working industry making large excavations as does mining, it does in fact require excavations to prepare, grading to slope, and daily fill-cover to operate the waste disposal site. Landfill sites are commonly deposited on top of the earth's surface, or in a topographic depression as a ravine or canyon, as opposed to burial in some previous man-made excavational void. However, waste landfilling, whether mounded on the earth's level or sloped surface, or placed in a topographic depression, must trench below the surface to emplace a leachate collection system and to create and grade the initial pocket with a drainage gradient for starting the landfill. It must, additionally, excavate other pits to obtain daily cover material for the waste deposited, and for clays for use as lining and capping material, and for sand or other vegetative-supporting material for overlay of the clay capping material. Consequently, waste disposal / burial by landfilling is indeed very much earth disturbing, earth-moving and excavating in nature as is mining. For support that landfilling is earth-disturbing, evidence is cited in "A case of excavating three clay pits to cover one landfill", *Ontario (Joint Bd.) v. Metropolitan Toronto,* infra, Ch. 6 §4 (ii). Also, it is noted that sand and gravel was used as periodic cover material in the 1968 Ontario case of *Plater v. Town of Collingwood et al*, Ontario High Court of Justice [1968], 1 O.R. 81, infra (Ch. 6 §4). The sand and gravel obviously had to be mined from some location. In a California landfill, the volume of cover material is reported to be 20% of the whole. The cover material is dirt / clay that had to be excavated from some other site.

Presently, the regulated attitude of environmental and land-use planners and controllers treat them as two distinct and separate problems and industries.

In reality, the two "problems" still co-exist in the general public's mind as "bad neighbour", earth-disturbing industries. Both should be regulated to coordinate and reduce disturbed area excavations to compliment each other for a far simpler and more satisfactory solution. New regulatory coordination, as proposed herein, should pre-plan for utilisation of mined-out pits as waste disposal sites. This would yield not only more efficient and beneficial results in greatly relieving the critical space problem for waste disposal, but reduce by nearly a whole the required land for MSW sites and greatly increase land conservation. The periodic and frustrating search by municipal and county governments for new waste disposal sites can be greatly reduced by simply using available area mining pits. Nearly every shire or county has an aggregates (crushed stone) surface mine within its boundaries.

1.3 Environmental Overview

The Anglo nations of the world, particularly the United Kingdom, the United States and Canada, are well advanced in establishing environmental controls for industry through legislation, laws and regulations to conserve and preserve natural resources, to minimise or eliminate pollution of water, soil and air, and to protect mankind, the health, safety, and welfare of his environment. Illustrative of this statement is the fact that the U.S. was the earliest (1969) of nations to install a fully comprehensive, all-inclusive environmental protective act, *per se*. It was followed closely in time by Canada, the U.K. (see Ch. 7 §2.6, infra), and other northern European nations also enacting similar environmentally protective laws. That is not to say that other northern European nations had not made prior, initial in roads with limited acts protecting certain areas of their environment. Nevertheless, the Environmental Policy Act of the U.S. (NEPA) is the most internationally copied law in the history of the U.S. According to Nicholas C. Yost, a United Nations Conference on Environment and Development (UNCED) delegate from the U.S. to the 1992 Earth Summit conference in Rio de Janerio, 84 nations have borrowed, copied, or adopted the Environmental Impact Statement (EIS) feature of the NEPA. In the last two and a half decades the mining and waste disposal industries in these industrially-developed nations have severely felt the impact of a multitude of environmental constraints and controls over their development, operation and land reclamation.

By their very nature, mining and solid waste operations necessarily disturb the surface of the earth. During exploitation of the mineral deposit and during the solid waste covering, or burial, good earth tilling and care practices are temporarily in direct conflict with industrial operational practices. Still, neither can miners, as "gardeners" of the earth, reap the fruit that has been placed in the earth's crust for mankind's benefit and well-being without disturbing the matrix that holds them, nor can society's solid wastes be disposed in earth sites without disturbing the earth's surface.

By virtue of the distribution and varying depths of the occurrence of the "fruits", or minerals, that lie within the earth's crust, mining falls into two categories, viz., surface mining (quarrying, open pit or open cast), and subsurface, or underground mining. Both extractive operations disturb and affect the earth's environment. It is debatable which has a greater effect on the environment, particularly the aqueous resources, but certainly, surface mining has the more noticeable effect because of its greater public exposure. Surface mining alone produces overburden, that overlying waste dirt and rock that must be stripped from the surface to reach the mineral(s) to be mined. The overburden is stored on the surface for possible, later use, but is generally not replaced in mine reclamation due to prohibitive costs. Both types of mining produce "waste" rock in their excavations and mill tailings in their mineral beneficiation processes. Disposal and dispersal of waste rock and mill tailings become a problem for surface reclamation and for surface water pollution prevention.

Because of a significant physical difference between underground and surface mining methods, as well as the former's lack of effect on the surface environment, and its different potential effects on ground and surface waters, underground mining will not be treated in this study. Open-pit coal mining is largely omitted from this study as separate regulations are usually made for the surface coal mining industry which presents different problems requiring different treatment. Thus, this study has been partially confined to surface, open cast mining of non-fuel minerals, with particular emphasis on the more numerous, the more conveniently located, construction materials mines, e.g., stone, gravel, sand, and clay, etc, and to a lesser extent, of metallic minerals that are mined by open pit or surface mining methods, as gold, copper and iron. This not to say that surface coal mine pits should not be seriously considered as disposal sites for placement of solid wastes. In fact, there are successful examples of utilising coal pits for MSW (Bowmans Harbour, Wolverhampton, in the U.K., infra, Ch. 10 §7.2, and a number of currently permitted landfills in former surface coal mines in Kentucky, Kansas, Illinois, Indiana, et al, in the U.S. See infra, Ch. 10 §7.6).

In prior decades before reclamation regulations for mining were enacted, disposal of the waste rock from both types of mining operations was simply by piling it on the earth's surface where it was left to eventually become, once again, part of the earth's crust (for an example, see the 1990 case of *Crowell Constructors v. N.C. State,*. infra, Ch. 5, §4.9.2). Natural processes take over and Mother Nature, in many cases, heals the earth's surface's wounds without help from man. Generally, no re-seeding, anti-erosional, anti-leaching, or anti-contamination controls were taken by the mine operators for the abandoned surface. With subsequent reclamation regulations in developed countries, the rock waste piles must be stabilised for erosion control by wind and water, and to abate or lessen the pollution of surface and ground waters from any naturally existing contaminants in the waste piles, e.g., lead, arsenic.

Various claims are made that mining's waste rock disposal on the earth's surface causes contamination and pollution of surface and subsurface waters. Some claims are founded on truth; others on part-truths, whilst still others are unsubstantiated, dubious to unfounded, being merely caution or misinformation. Waste rock from much of surface mining is chemically inert and has no different effect on water resources than were it in its original location. It should be borne in mind that under the USEPA's standards, even inert rock sediments placed into navigable streams are considered water pollutants, whether chemically neutral, or not. Such rigid standards were upheld in 1990 in *Rybachek v. U.S. EPA,* where the EPA interpreted dredged soil and inert rock as pollutants and required settling pond treatment before discharge. (See *Rybachek* case, infra, Ch. 7, §3.1.1).

An historical investigation made herein, infra (see Ch. 3), for environmental damages against non-fuel surface mining pits indicates that claims for contamination and pollution of underground water were from nil to rare. It is the author's theory and conviction that, in general, non-fuel and non-metal surface mining has a negligible to no-effect in contaminating subsurface waters, whether mining occurs below the water table, or not.

It is conceded, and the historical investigation made herein of litigated pollution claims bears out, that at some sites containing mine-mill tailings waste and ponds, chemical additives and minute amounts of naturally occurring hazardous metals and compounds from the milling process remain and can produce a contaminative effect by its surface runoff waters into nearby surface and ground waters. This concession applies mainly to ore milling operations that were seeking to remove valuable metallic compounds from a rock matrix that was wasted. In virtually all cases where the mining operation is for physical values or attributes of the rock, such as stone aggregates and dimension stone, occurring in massive, relatively homogeneous lithologic bodies as limestone, marble, shales, granites, et al, there will be virtually no chemical additives involved in the plant-upgrading or milling process. Investigation of historical litigation also bears out that claims against surface stone quarries and sand pits for pollution of underground waters has been virtually non-existent.

The mining industry and its operating techniques are international in nature. Thus, the problems of waste rock dispersal and disposal from open pit mining, and ensuing problems of mined-land reclamation, prevention of erosion, leaching and contamination from the resultant open pit and waste piles, and water discharges are also internationally common.

Since mineral development and exploitation is for profit, the cost of opening and closing a mine in the present era of environmental controls has increased the costs of mining dramatically. Many discovered metallic mineral deposits in the developed nations are marginal, or less, as to development costs. With the added cost of environmental safeguards, the demarcation between marginal and sub-marginal deposits has been raised thereby making previously marginal deposits sub-marginal and unmineable. In the course of determining what environmental measures are sufficient for protection of the environment during mining operations, a struggle has developed between the mining industry and those that would over-regulate. At present, ever-tightening environmental controls, whether too stringent or not, appear to be the promise of the future for mining operations.

On a worldwide basis for environmental concern, prospective countries for development by the international mining industry shows aspects of having a divided camp, falling into two groups, the developed or industrialised countries, and the Third World or developing countries. An incisive, speculative treatment of this environmental mining issue is made by an Editor of the International Bar Association's *Journal of Energy and Natural Resources Law*, in an article, "Environmental Policies Towards Mining in Developing Countries (Walde, 1992). Professor Walde points out that in the developed countries, "... mining is no longer automatically assigned precedence over other land-uses, and environmentally (socially or culturally) important land is no longer available for mining". In contrast, in developing nations mineral development has become a priority.

Industrial exploration and operation mineral planners are well aware of the extreme environmental sensitivity of land use in developed countries. Environmental risk assessments and risk modeling are becoming standard procedure for prospective mineral operation evaluations. In prior decades, the question of whether a mineral deposit would be developed by surface or underground mining methods was largely determined by the economics of the depth of overburden to be stripped. With the addition of high environmental and land reclamation costs in the developed nations, the choice of mining methods is no longer principally determined by overburden thickness and stripping economics. An additional environmental risk consideration for surface mining involves its exposure to the public. The obvious, temporary disposal of large volumes of overburden on the surface from surface openings in the earth frequently creates an anti-mining feeling in the host-community. As a consequence, frequently costly, and at times basically unnecessary, speculative and precautionary protective environmental measures must be incorporated into the permitting process to assuage the local citizenry's fears.

Professor Walde's article quotes an estimate of environmental costs from the *Financial Times*, June 5, 1991, in industrialised countries as "likely to raise mining costs by 20%." Overall, in the opinion of this researcher, the figure appears to be an underestimate. For example, in the case of the Kennecott Corporation's new copper sulphide Flambeau open-pit mine in Wisconsin, U.S., "75% of the capital investment was tied up in environmental protection." (Engineering & Mining Journal, Feb 1992).

Consequently, the increasingly high costs of environmental controls of mining in developed nations have caused the mining industry to look elsewhere for lower mining costs. (For support, see the article in Ch. 7, entitled "Canadians Abroad: Canadian Mining Money moves overseas".)

The underdeveloped and the developing nations are striving to overcome their lower living standards and attain a richer lifestyle, similar to that of the developed nations which have already largely developed their mineral resources. These Third World nations have openly solicited foreign mining industry to invest in the development of their wealth of minerals without the costly and stringent environmental and mine permitting regulations of the developed nations. They reason, to raise their standard of living they must exploit their mineral wealth first as did the developed nations; environmental concerns will follow. Members of the mining industry of the developed nations have responded to their invitation with very large expenditures. For support, see foreign advertising in Ch. 12, infra. The following news item of 24 February 1995 from *Mining Journal, London*, illustrates the mining movement away from the developed nations and was headlined,

U.S.-CANADIAN MINING COMPANY ECHO BAY MOVING ABROAD

> "Echo Bay is set to change the geographical focus of its operations. ... all four of the company's gold mines ... are in the U.S. and Canada. However, the Denver-based company is now setting its focus for growth on the international scene. Like most other North American mining companies, Echo Bay is increasing its foreign exploration expenditure (North America's share of an unchanged total exploration budget falling from 60% in 1991 to 40% this year) and has stepped the level of overseas strategic alliances.
> ***Echo Bay would be concentrating its search on countries where mining companies were welcomed.
> ***The executives also noted that Echo Bay would apply the same environmental criteria abroad as at home". (The Mining Journal, 1995b)

Perhaps ironical, yet laudable and beneficial to the world environment, is the fact that some mine operators, while seeking lower mining costs in countries where environmental expenditures are negligible to non-existent, are taking some of their indoctrinated environmentally protective mining management principles with them to the essentially environmentally-unconcerned developing nations. As an example, in early 1994 the Bangladesh Government approved BHP Corporation's plans to explore and exploit coal reserves by open cast mining "**incorporating full environmental safeguards**". (The Mining Journal, 1994a, emphasis added).

Since the Anglo nations — the United Kingdom, the United States, and Canada — are well advanced in dealing with and treating environmental problems, it is fortunate that they have not only a common language, but a common legal system evolving from England. All three nations have recognised an environmental need for regulation of surface mining and waste disposal, and all three have reacted with controls to aid in the prevention of further contamination and pollution of air and water resources. However, the formulation of controls for surface mining has taken place in the past on an individual national basis with little or no coordination between the nations to resolve common problems. That is not to say that international mining symposia have not stressed and ultimately effected mining environmental awareness amongst the various nations.

The advanced state of development and higher standard of living of the industrial nations have brought with it other serious environmental problems. Deterioration of water and air quality have steadily mounted in intensity and volume until society would no longer tolerate uncontrolled pollution by industry. High living standards in a deteriorated environment triggered society to cry out for quick relief, which in turn brought stern corrective measures for the culpable industrial world. The higher standard of living, with its often superfluous amenities for an easier lifestyle, has also brought with it massive volumes of waste resulting in the problem of how and where to dispose of it without further deterioration of the environment that fostered and gave birth to the higher lifestyle. Advanced technology of the developed nations, in response to the public's demand and craving to make a utopian dream of luxury a reality for every individual, has resulted in increased production to meet the continuous demand of higher living standards for all persons without regard to societal costs. Consequently, the pursuit of such luxury and abundance ideals has at the same time generated tremendous volumes of municipal, solid, hazardous and toxic wastes. Simultaneously, there has been a demand for immense amounts of energy supplied by fossil fuels, or alternatively by radioactive minerals, to meet the power requirements for the increased production. The wastes from increased amounts of fuels have only compounded the environmental problems. The developed societies, having lived in the lap of luxury for several decades since the end of World War I, without great concern for the deterioration of the environment, have come to realise that the time has come "to pay the piper" and make amends to the environment.

Mining, as a basic industry, produces the essential raw materials necessary for the manufacturing of societies' necessities and luxuries for its well-being. Lest we forget, and to illustrate with some typical examples of the many minerals provided by the mining industry to maintain the standard of living, in the U.S. every day 18 million tonnes (16 M metric tonnes) of raw materials must be mined, cut or harvested ("if it can't be grown, it must be mined") to meet the individual, daily demands and consumption of U.S. citizens for goods, ("things and stuff"), which amounts to about 150 pounds (68 kg/d) for every man, woman and child. (Similar figures would hold true for the U.K. and Canada.)

Daily mineral production required by the American public for products consumed daily:

3,000 new homes and 650 mobile homes are completed daily, requiring limestone, clay and iron slag for cement, and crushed stone for concrete for foundations, blocks, and driveways; asphalt for waterproofing foundations; iron and zinc nails, copper and aluminium for wiring, guttering and window frames; window glass (silica sand); steel beams and window frames; clay, feldspar, talc and silica in porcelain tiles, sinks, bathtubs and toilets; iron, brass, steel, aluminium, copper in plumbing fixtures; iron, zinc, brass light fixtures and bulbs; clays in brick and tiles; counter tops, fireplace and hearth stones of marble and granite; clay putty and joint mortars with limestone and gypsum; gypsum for wallboard; asphalt roofing with stone grains embedded, mineral wool insulation from silica and silicate minerals; etc, etc.
carpeting: one square mile — 640 acres per day is woven daily and backed with barite and calcium carbonate;

plate and window glass: 9.7 million square feet (902,000 m^2) — about 223 acres (90 ha) per day — are used— enough to cover 200 football fields — made from silica sand and trona (hydrous sodium carbonate);
pavement: 2,750 acres (1,100 ha) of concrete and asphalt paving are laid daily — highways, roads, re-paving of broken highway surfaces, parking lots, driveways, etc, using concrete (limestone) crushed stone, sand and gravel, asphalt;
auto batteries: 150,000 lead-acid auto batteries are replaced each day (lead);
medical x-rays: 650,000 X-ray pictures are taken daily — each requires lead shielding for patients and technicians (silver, iodide, lead);
dental care - toothpaste: 550,000 pounds of tooth paste are used daily — 2.5 million tubes — (calcium carbonate/limestone, zeolites, trona, clays, silica)
dental care- cavity fillings: 80 pounds of gold are used to fill 500,000 dental cavities daily;
light bulbs: 3.6 million light bulbs purchased daily — made from tungsten, trona, silica sand, copper, aluminium;
glass bottle and jars: 120 million are used daily — made from silica sands, trona;
telephone wiring: 150,000 miles of copper wiring is added daily to handle the 80 million telephone calls made daily;
paints: 3 million gallons of paint used daily for "sprucing up" 200,000 homes — made with titanium, iron, silica, wulfenite (lead molybdenate), mica;
mixed cement: daily ready-mixes make 187,000 tons (170,000 mt) daily, enough to construct a 4-foot (1.2 m) wide sidewalk from coast to coast — from limestone, sand and gravel, crushed stone;
photography: 21 million photos are taken daily — equal to more than 29 acres of wallet-sized pictures — made with silver and iodide; (Soc. Mining Engineers, 1995)

The list for the public's daily mineral consumption goes on and on. There are the many other large, obvious volumes of minerals consumed for manufacturing vehicles, autos, trucks, trains, aircraft, and ships; for personal and industrial electric and electronic equipment; for many medicines and hospital equipment; ad infinitum, all benefiting the public's welfare and well-being.

Mining, the procurement of essential materials from the earth, is as necessary to mankind's health and well-being as clean water and air. It is not suggested herein that all new proposed mining projects and solid waste disposal sites be given *carte blanche* approval, but does suggest that for one to oppose new waste disposal sites and new mining pits at public hearing is to oppose one's own continued well-being and high level of living. Unfortunately, for society to continue in the manner of life to which it has become accustomed, it is caught between the demands of cleaning up its deteriorating environment and having a continuing supply of basic raw materials for its mode of "good living". An advanced society must have "clean living" conditions at the same time with "good living", neither of which is expendable. The price tag for this combination must be paid by society. The price is, indeed high, and becoming higher every year with increasing environmental constraints placed on the mining and manufacturing industries. The caveat for society is to avoid extremism and over-regulation in environmental concerns.

As indicated, a part of the problem of restoring the environment to an improved state of cleaner living has resulted from the wastes generated in the manufacturing of the raw materials taken from the earth. In addition, society itself generates tremendous volumes of disposable waste from its consumption of products which have a short-term life, such as containers, paper and plastic products. The resulting problem is how to cope with increasing waste disposal without degrading the environment, or at least, minimising the degradation, and particularly to protect the water quality and supply.

Unclassified and unrestricted landfills, dumping in the oceans, rivers and large lakes have been former solutions for disposing of society's waste. In a few passing decades those methods boomeranged to haunt society and the environment, particularly in the more highly populated areas. Even attenuated landfills of a few years ago are revealing injury to the ground water sources. The current trends are toward incineration of bulk waste, controlled, classified, and totally contained sanitary landfills, and recycling of reusable materials.

Although incineration greatly reduces the bulk of waste, incineration has not been readily accepted because of its offensive odors and polluting emissions into the atmosphere. However, highly-utilitarian re-cycling of MSW is reviewed in Ch. 10 §1. Recycling does not adequately resolve the necessary volume reduction for disposal of society's waste. A major problem for worldwide recycling advocates is that the daily volume of MSW is generating faster than the rate of recycling. Thus, large amounts of waste for disposal continues to be a problem. Undoubtedly, recycling is helpful in reducing the volume of MSW, but disposal of the remaining bulk in controlled landfills is the method still utilised and preferred. The waste disposal industry has mushroomed as a result. This is supported by a statement in a paper given by Paul A. Tomes, FIQ, M. Inst. W.M., Company Landfill Manager-ARC Aggregates, 2d October 1989, and presented to the Institute of Quarrying, Annual Conference Symposium at Bristol, "The waste management business in Britain has a turnover of £5 billion (annually)". The industry has continued to grow since 1989 with ever-safer landfilling technology. Hughes has also stated, "Certainly waste and its disposal is now a major industry". (Hughes, 1992, p. 247).

New excavations for surface mounding-type landfills, i.e., either for initial grading of pockets for the waste and emplacement of the leachate collection systems, or for excavating cover materials for surface mounding of waste, must be made. With the enormous volumes of waste, more landfills are needed in an increasing amount of geographical locations. Along with this need comes the problem of siting, approval and permitting for landfills. The shortage of and need for more landfill sites has become severely critical. The legislative mandating for re-cycling of worked-out open mine pits for landfilling with solid wastes, before otherwise inferior reclamation methods are employed, is the proposed solution investigated in this research. Mandatory re-cycling of surface mines, thus, relieves the critical shortage of disposal sites; offers full restoration of the land to its original surface, thereby making fuller and better use of land, resulting in greater conservation of land.

1.4 A Proposed Solution

This study recognises that there are many existing worked-out and abandoned surface mining pits in the industrialised nations left in an unreclaimed condition from before the era of enforced reclamation laws; also, many pits have been left in a "reclaimed" condition that have had only minor surface grading and allowed to fill with water. Special attention and emphasis herein is given to surface mining for construction materials, e.g., stone and rock for aggregate (crushed stone), clay, etc. As old pits are worked-out, new pits are being started every year. These new surface mines for crushed rock are normally located and opened in the proximity of the periphery of new and growing metropolitan areas, close to the new metropolitan-suburban area construction. This is largely true because crushed stone is a low-priced, high-volume, low-profit per ton mined material that generally cannot tolerate high-cost haulage or freight cost for distance in addition.

Many of the older, abandoned pits are located near urban areas where the generation of municipal solid waste is in the largest volumes because of high population densities. These older pits were abandoned, either because they were worked out, or the operation was forced to move further away from the metropolitan-urban area by the encroaching population. Still, they remain unrestored, partially filled with water. They are landfill-usable because no backfilling has never been done. (These observations and comments are based on a lifetime of geological work in the construction minerals industry and whilst exploring for new aggregate and materials source-sites, either as replacement for older pits to be abandoned, or for new sites to meet expanding construction for new growth areas.)

Logic indicates that such worked-out mine pits in population centres are fortuitous locations for solid waste landfills. The critical-space deficiency of the municipal solid waste (MSW) disposal problem has provided an opportunity to make a further and maximum use of the once-used natural resource locations. In a proper sense, the abandoned and worked-out mining pits may be "recycled" as solid waste landfill sites. Currently operated pits and unmined pits of the future, may, in turn, as they are worked-out for their mineral content, or closed for rapid encroachment by population growth, also be recycled to fill the omnipresent, critical demand for disposal sites of society's solid refuse. Of even greater logic is their common bond by virtue that one excavates a hole in the earth, and the other requires an earth site to deposit wastes. Mining of construction materials is concentrated in proximity to the population centres where major construction is also concentrated. Those same populations centres generate society's larger volumes of solid waste. The location compatibility of the two industries are made for unification of purpose from beginning to end. This relationship is referred to, infra, as the "urban quarry-landfill cycle".

The excavation of new holes in the earth and further earth disturbance simply to deposit solid wastes is misused, unnecessary environmental disturbance, wasted land use, and expense when ready-made excavations for waste disposal are provided by the construction mineral industry. In addition, excavating holes in the earth separately for two purposes, i.e., to mine construction minerals, and secondly for depositories of solid waste, only serves to exacerbate the crisis of land-shortage in metropolitan areas by taking twice as much land than needed to serve the two purposes. Furthermore, the spoil piles present at abandoned and working pits provide a ready-made and suitable source of daily landfill cover material. It replaces the overburden from whence it came.

Research included here investigates the possibility of promulgating uniform laws for mandatory recycling of the worked-out mine pits as landfill sites. Under the present applicable regulations in all three Anglo nations, the U.K., the U.S., Canada, open pit and solid waste landfill site permitting are separate processes often involving difficult, arduous, and exhaustive environmental assessment investigations. There are open pits in the U.K. that are already simultaneously being filled with municipal waste whilst being mined. The U.K. has already successfully proven to a satisfactory degree the practicality and utility of reclaiming worked-out open pits by infilling with solid refuse. This study gives support to that British practice, and proposes that the practice become universal, and in addition, the study proposes that the permitting processes be combined and coordinated from the beginning of the planning stage with the end purpose that before completion of mining, the open pit site, upon pre-planned safe, pre-studied and pre-determined conditions, automatically becomes an environmentally approved disposal site for solid wastes. The economic, political and practical advantages to simultaneous environmental investigations for a dual purpose opening of the earth's surface should be obvious.

With the conveniences of a common language and a common legal system, it is urged that the Anglo nations make concerted and cooperative efforts to adopt the best practice model law proposal herein; to develop uniformly beneficial environmental regulations for the mandatory recycling of worked-out surface mine pits for solid waste disposal sites in the land reclamation and land conservation process. Beyond adoption by the Anglo nations, the prospect may be entertained that environmental laws and practice regulating the dual use of open pit mining with landfilling will be eventually copied, emulated and transferred from the developed and industrialised nations to the developing nations.

In finality, offered inducement and appeal is made to the mining industry to give serious consideration to full and complete reclamation of its mined lands, restoring it to beneficial surface re-use by the solid waste landfilling option. The restored, reusable mined land will be greatly appreciated in value, particularly those sites occurring in newly urbanised and growing areas. As additional incentive to industry for this type of mined land reclamation, landfilling can be, and actually is, a profit-making business. Further, governments can be persuaded to create mined land reclamation tax deductions as incentives for fully restoring the derelict land by the proposed landfilling process.

Section II

AN HISTORICAL REVIEW — FROM EARLY MAN TO RECENT REGULATIONS

CHAPTER 2

A BRIEF HISTORY OF MINERAL LANDS AND REGULATION

2.1 Introduction

Mining in the Anglo nations [Anglo nations or countries hereafter in this work refers to the United Kingdom (U.K.), the United States (U.S.) and Canada] falls into two land control groups, viz., (1) privately owned lands, and (2) government-owned lands, e.g., Crown lands for the U.K., the public domain or federal lands for the U.S., and Crown / federal lands for Canada.

A brief history follows from the birth of the nation until the period just prior to the "green revolution" for each nation's development of its interest in and control of minerals, mining and mineral-bearing lands along with the manner of division into privately owned and governmentally owned mineral lands.

2.2 Mineral Land Interests

2.2.1 *United Kingdom*

Knowledge of mining in the British Isles antedating recorded history of the Roman occupation is scant to non-extant. The long history of mining in the Isles is evidenced by the discovered underground workings for flint made during the Paleolithic period in East Anglia. A recent announcement of an archaeological discovery was made of prehistoric tribesmen mining for copper, and possibly lead, in 2,200 B.C. in County Kerry, Ireland. "Among the debris were found hundreds of stone hammers. To extract the copper ore, the prehistoric miners drove tunnels at least 15 m. into the hillside. Rock was removed by heating it with fire (decrepitation) and then shattering it with stone hammers. Rubble was removed with shovels made from the shoulder blades of oxen". (Mining Environmental Management, 1993).

Six hundred years before the birth of Christ, the Phoenicians referred to the British Isles as the Cassiterides, meaning the Tin Islands. Herodotus (c. 485–425 B.C.), sometimes called the "father of history", in writing about Europe, said "that the Greeks obtained their tin from the furthest part of Europe". The tin country alluded to was supposedly the Cornwall district. And Caesar, in his writings after his invasion of Britain (c. 54 B.C.), reported that tin was produced in the midland, and iron from the maritime area. Discouragingly, Cicero (c. 50 B.C.) wrote, "*In Britannia nihil esse audio neque auri neque argenti*". [In Britain there is neither gold nor silver] (Rogers, 1876, Chapt. VI).

By right of conquest and according to Roman law, the ownership proper of all lands was vested in the state (*dominium strictum,* i.e., strict domain), and while it might have claimed the beneficial use (*dominium utile*, i.e., useful domain), its possessory ownership and enjoyment was vested in the grantees or tenants of the state subject to rents or royalties payable to the state. (Bainbridge, 1900, pp. 106–107). Thus, the broad distinction between the titled ownership of the state and the equitable or beneficial ownership by tenants through their tenure and occupation was formed. Consequently, under the Roman occupation and rule of *dominium strictum*, all mines, minerals and quarries belonged to the state. Though Emperor Tiberius (c. 14–26 A.D.) reportedly claimed absolute dominion over all mineral lands and minerals, that doctrine was abandoned later, and in accordance with *dominium utile*, a demand was made for royalties on minerals produced by private persons who worked mines. Under Emperor Gratianus of the West (c. 367–383 A.D.), all gold and silver was reserved to the State; for all other, baser metals and minerals the general right to mine from private lands was granted to the beneficial landowner on a royalty payment of 1/10th to the state. If the mining was performed by another under agreement with the beneficial owner, the mine-worker paid 1/10th to the state and 1/10th to the property owner. Such doctrine of mineral royalties was known as *canon metallicus* (metal, or "mining" law). (Bainbridge, 1900, p. 108) That mineral land policy was adopted and continued by succeeding Roman rulers over Britain, e.g., Valentinian II and

Theodosius I (c. 346–395 A.D.), Theodosius II (408–450 A.D.) and Valentinian III (425–455 A.D.). According to Rogers, "In some of the Roman-ruled provinces, on the allotment of the land to private individuals, the conquerors reserved the rights to the mines, minerals and quarries, for the benefit of the state; in other provinces mineral rights were assigned to the allottees of the soil. Hence, the property in minerals became, not infrequently, distinct from the property in the soil". (Rogers, 1875, p. 20).

An innovative feature for the promotion of mineral development was created under a Valentinian II rescript decreeing that state mineral lands could be worked by private persons for gold for their own advantage by paying a royalty to the state. The royalty was set at "eight scruples (*i.e., 1/3 oz.*) in gold dust for each worker in the mines". However, a right of pre-emption was included for the state to take over the mine when the gold discovered reached a certain quantity (Rogers, 1876, p. 21).

With further Roman rule, the doctrine of absolute ownership of property in the state, and later, in the sovereign or Crown, became firmly established; also, the principles that mines and minerals on state lands belonged exclusively to the state; and, on private lands where there had been a grant of property from the state, the concession became entrenched that ownership of the surface was *prima facie* (at first sight) title to ownership of the non-royal minerals therein and thereon. It also became established that all mines and mineral lands were subject to certain servitudes and became a source of revenue for the sovereign.

Thus, it was the hypotheses of Rogers' and Bainbridge's works that the civil law of the Romans, as applied to mines and mineral lands in all continental lands under Roman rule and in Britain as well for several centuries, was perpetuated by succeeding sovereigns in Britain, viz., that royal mines, or gold- and silver-bearing lands, were the exclusive property of the Crown; and all other mines, referred to as baser metals or substances, were conceded in *prima facie* as belonging to the surface owner, *ab inferis usque ad coelum* (from below all the way to the sky). (Bainbridge, 1900, p. 108).

In the intervening six centuries, between the end of Roman rule (c. 449 A.D.) and the beginning of Norman rule (1066 A.D.), the invasions, occupation, settlement and rule of Britain by the Jutes, Angles, Saxons and Danes occurred.

There appears to be a hiatus in recorded history of mining and mineral land activities during that time. With the subsequent Norman occupation and rule, the Normans, having adopted the strong Roman influence of making written records under centuries-long occupation, recorded once again English history of mining, mineral lands and mining law which have been left for study.

The Norman rulers gained title *in posse* (in possibility; not in actual existence) to all lands of Britain by virtue of their conquest. William I, The Conqueror, demanded fealty of the defeated Anglo-Saxon barons, and the lands of those not giving allegiance were confiscated, thus becoming sovereign lands *in esse* (in being, or actuality) as well as *in posse*. The earliest Norman grants of royal lands, which included the minerals thereon, were made by William I to the earldom of Cornwall, a half-brother of William I, and to other Norman royalty in Devon which included, even in those times, the well-known tin mines.

In the reign of Richard I, *the Charta Stannararium Domini Regis,* 1198 A.D., concerned and protected certain rights of the Crown in the stannaries (tin deposits and mines) of Cornwall and Devon. In 1201 A.D., King John granted "the celebrated charter of liberties to the tinners of Cornwall and Devon which may justly be considered the foundation of their rights". (Rogers, 1876, p. 164). However, the ownership *in esse* of the tin mines of Cornwall in particular, and those in Devon, was the subject of dispute between the King and lesser lords for several centuries. The question of whether the tin mines were 'royal mines', was at constant issue and unsettled. The answer, in turn, depended on whether the attendant occurrence of the 'royal' minerals of gold and silver along with the occurrence of tin in the stanneries, or with any other minerals, as in the mines of copper, lead, et al, made the occurrences 'royal' mines.

England, seemingly, was not blessed with an abundance of the royal minerals. Gold was discovered in the Dolgellau area of North Wales as early as the 12th century. However, gold mining in the area did not have any import until 1863 when the Gwynfyndd mine was started and continued until 1916. (The mine was reopened in 1981, and new zones of mineralisation were found in 1992.)

It was not until 1568, after the great *Case of Mines*, [1567] 1 Plowden 310, 75 E.R. 472 (Exch.), between Queen Elizabeth and the Earl of Northumberland heard, that an attempt was made to settle the question of when mines are to be considered 'royal' and in the domain of the Crown. The Queen claimed as the royal prerogative entitlement to all "mines and ores of gold and silver, and of all other metals whatsoever containing in them gold or silver, with the appurtenants, which might or could be found in any lands within the realm of England or other the dominions thereof, *** in the soil of the Queen and in the lands of any of her subjects; ***" In the case at bar, gold and silver occurring in the copper mines on the lands of the Earl of Northumberland were at issue. Prior to the action, the Queen had ordered mining to be done on the Earl's lands for the winning of the royal metals. In so doing, the copper content of the ores was taken along with the gold and silver since the royal metal could not be separated without smelting. The Earl "hindered and disturbed" the mining by the Queen causing the action and claim by the Queen, and for damages of £1,000. The Court of the Exchequer held that the copper mines containing gold and silver, and also the ores therefrom, belonged to the Crown (Bainbridge, 1900, pp. 110–11). Much later, in the case of *Lyddal v. Weston*, [1739] 2 Atk. 20, it was decided that the *Case of Mines* upheld the right of the Crown for the working of royal mines to enter private lands and to use and disturb the surface, i.e. surface rights, including the right to cut timber (Bainbridge, 1900, p. 113).

The uncertainty of the status of royal mines continued for mines where baser minerals (lead, tin, copper) were contained with gold and silver. Contention lay between the argument that if any gold or silver were present, it was sufficient to be a 'royal' mine, and the argument that to be deemed a 'royal' mine, the value of the gold and silver contained therein must exceed in value the other mineral with which it was found. According to Heton's Account of Mines, p. 21, it was laterally found from the *Case of Mines* decision that "if the gold or silver, although of less value than the other mineral, was yet sufficient to bear the charge of refining it, or was of more worth than the base metal spent in refining it, then it was a mine-royal, and as well the base metal as the gold and silver in it belonged to the Crown". (Bainbridge, 1900, pp. 113–114). Nevertheless, the difference of opinion and dissension continued until William and Mary's Act I, c. 30, was passed declaring that "no mine of tin, copper, iron, or lead should thereafter be taken to be a royal mine, although gold or silver might be extracted out of the same". It was further enacted,

Act 5 of William and Mary, c.6, entitled "An Act to prevent Disputes and Controversies concerning Royal Mines", that "all owners or proprietors of any mines in England or Wales, wherein any ore then was, or thereafter should be discovered or wrought, and in which there was copper, tin, iron, or lead, should hold and enjoy the same mines", notwithstanding claims of the Crown to be royal mines, the Crown was given the right to purchase the ore of any such mines, other than the tin ore in Devon and Cornwall counties, within thirty-days after the ore was "raised, washed, and made merchantable". Act 5 also set prices for the ores of copper, tin, iron and lead for purchase by the Crown. If the Crown failed to make the purchase, the miner was free to sell the ore to other buyers. (Bainbridge, 1900, pp. 114–115).

As to Crown claims on other minerals, in 1607, under the reign of James I, in the *Case of Saltpetre,* 12 Rep. 1, the court unanimously found that the Crown had no claim to the mine, but in the interest of the "defence of the realm" could grant licences for the working of the saltpetre (potassium nitrate) for gunpowder on private lands, and finally, that the owners of saltpeter-bearing lands could not be restrained from mining it. Similar holdings were found for alum-bearing lands [potassium aluminium sulphate] (Bainbridge, 1900, p. 115). The more common minerals and rocks, used mainly as construction materials and obtained by surface mining, were in far less contention as to rights to mine, quarry or dig and that title was in the surface owner. It may be noted here, the Judkins Quarry-Landfill example, located in North Warwickshire, and discussed in Ch. 11, §81, infra, may be traced back to 1581 when quarrying of stone on Hartshill Ridge for construction materials was first mentioned.

Over the ensuing centuries, it became evident that British mining law could trace its roots to doctrines and principles established by Roman civil law, was perpetuated by the Normans and, subsequently, was adopted by the British. However, the developing English common law varied from the Roman civil law in that it incorporated and firmly developed the Anglo-Saxon's ideals of the importance of rights of the individual in opposition to, and to the exclusion of, oppressive property rights of the sovereign. With regard to mineral lands, mineral ownership and mining rights, Crown ownership was gradually reduced over time from a claim to 'all minerals whatsoever' to a claim only over gold and silver mines. The claim to the "royal" minerals was extended to the British colonies and dominions overseas.

2.2.2 *United States*

During the colonial period, as the Crown lands of the American colonies were disposed of for settlement, the sole concern for minerals was in the Crown retaining royalty rights from discovered and operated deposits. Even then, only interest in the "royal" metals was reserved. Most of the colonial charters contained royal-mineral royalty reservations. The 1584 Carolina charter from Queen Elizabeth to Sir Walter Raleigh reserved one-fifth of the gold and silver that might be discovered or mined. Similar reservations to the Crown were contained in the charters of several American colonies; one of the Virginia charters reserved "a one-fifteenth of all copper"; other charters for Massachusetts Bay Colony (1629); New Hampshire (1629); Maryland (1632), Maine (1639), Rhode Island and Providence Plantations (1643); Connecticut (1632); and, Carolina (1663 and 1665) reserved one-fourth of all gold and silver ores (Lindley, 1914, Sect. 31).

After 13 of the American colonies severed ties with Great Britain, large areas of mineral-bearing lands went into private ownership to encourage western settlement without the state governments succeeding to the royal interests therein. An exception was in New York, where the state asserted a right to mines of gold and silver at an early date. "The New York statute also declared that 'mines of other minerals on lands owned by persons not citizens of any of the United States' are also claimed by the State of New York". (Kent, 1896).

Along with other English cultural traits deeply embedded in the American colonies was its judicial system of the English common law, inherited from its mother country and retained after its independence. Under the common law, the property owner was the *prima facie* owner of the minerals in the land, with three exceptions, viz., (i) "royal mines" which contained gold and silver; (ii) particular customs to the contrary, such as practiced in the tin mines in Cornwall and Devon, or the lead mines in Derbyshire (Note: The influence of "particular customs to the contrary" in the formulation of early U.S. mining law is discussed later in (2) Government-owned lands (a) United States, infra); and, (iii) cases where the minerals had become severed from the surface and were held in different, private ownership. [Am. Law of Mining, 1986, Volume I, Ch. 4, Sect. 4.03, citing *Lindley on Mines*, Sect. 2 (3d ed.1914)].

The Continental Congress enacted the Land Ordinance of 1785 calling for a method of surveying the public lands of the new nation. Included, were directions for the surveyors to list "... all mines ... that shall come to his knowledge" and also provided for a royalty reservation of "one third part of all gold, silver, lead and copper mines, to be sold, or otherwise disposed of as Congress shall hereafter direct". [Ordinance of May 20, 1785, 28/ Continental Congress 375, 376, 378 (Fitzpatrick ed.1930)] After the end of rule by the Continental Congress, the reservation ceased to be in force.

The first acquisition of public lands for the new nation took place after the end of the hostilities with Britain, gaining the area westward of the coastal colonies to the Mississippi River by the Treaty of Paris, 1783, and with the cession of claimed lands from the former colonies to the federal government during the period of 1781 to 1802. With further rapid acquisitions and areal expansion of the U.S. through the Louisiana Purchase in 1803, the cession from Spain in 1819, the Oregon Compromise with Great Britain in 1846, the Texas acquisition in 1845–1850 (lands outside of the state of Texas), the Mexican Cession of 1848, the Gadsden Purchase of 1853, and the purchase of Alaska in 1867, the public domain outgrew the pace of settlement (see Figure 1, p. 35, Growth Map of the U.S., 1776 to 1867.)

In the early part of the nineteenth century, large areas of copper-bearing lands of Michigan in the Northwest Territory (NW of the Ohio River, see Figure 1) went to the highest bidder. Between 1829 and 1847, mineral resources in the central U.S. and the region of the Great Lakes went into private ownership as a result of various Congressional Acts offering mineral lands for sale.

In the following decades of frenzied development of the Western U.S., wholesale land grants in fee simple were made by the federal government to the public at large. Nearly 52,812,626.5 ha (130.5 million acres) of the public domain were granted to railroad companies for construction of lines to encourage the westward building of them and for settlement of the acquired western lands. In some cases of railroad grants, reservations of the minerals were made by the federal government, although with frequent exceptions for "coal, iron, and other minerals". Thus, up until the first half of the nineteenth century, public lands had been regulated by the federal government with concern mainly for revenue and disposal for settlement. Under the common law, the property owner was the *prima facie* owner of the minerals in the land.

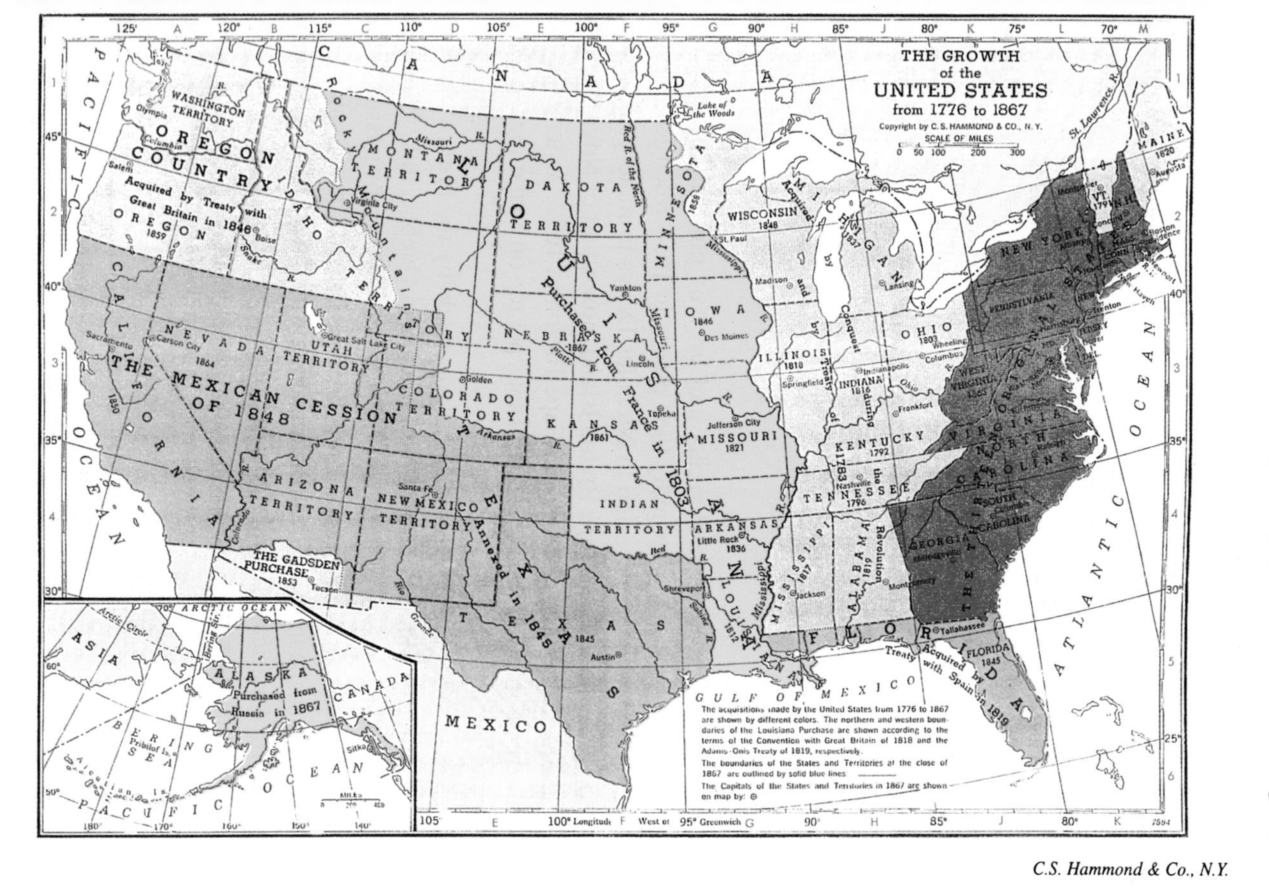

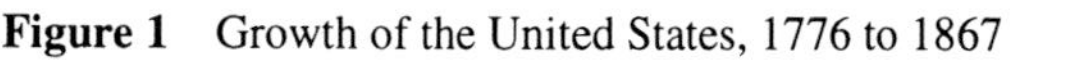

Figure 1 Growth of the United States, 1776 to 1867

Consequently, a large amount of mineral-bearing lands in the U.S. became privately owned. The development of federal land law by which mineral interests were retained in public lands did not start taking place until the enactment of the Mining Laws of 1866 and 1870.

A threefold distinction results when referring to mineral land ownership in the U.S., viz., (1) those lands which came under private ownership in the 13 original colonies; (2) those lands in which private ownership has since been granted by the federal government under sale or patents; and (3) those lands which still belong to the federal government. As a practical matter, and as in England, the owner of originally private lands is the owner of everything *ab inferis usque ad coelum* (from the interior to the sky).

2.2.3 *Canada*

The earliest disposal of mineral rights and levy of mineral royalties on government lands in Canada reputedly took place "in 1654 when King Louis XIV of France granted a concession to Nicholas Denys to mine gold, silver, copper and other minerals on Cape Breton Island, now a part of Nova Scotia". Denys subsequently discovered coal on Cape Breton in 1672. As a result, the French government levied "a royalty of 20 sous per ton on coal mines in Cape Breton". (Hodgson, 1966, p. 1).

Although the first settlement in Canada was made by the French in 1604, the British militarily gained control of New France, or Quebec, in 1759, which was confirmed in the Treaty of Paris of 1763. Ownership thereby passed to the British Crown. Former New France lands became British Crown lands with minerals subject to the laws of Great Britain along with other separate colonies in British North America. Mineral rights in Quebec, prior to the Confederation of 1867, had been granted under French seigniorial tenure or English tenure. In the case of lands granted under the former, mineral rights remained reserved to the Crown unless expressly granted; under the latter, lands granted for agriculture or colonisation purposes, the mineral rights, except for gold and silver, formed part of the grant unless expressly reserved to the Crown. Private owned mineral rights in Quebec are, however, subject to provisions of the Civil Code (Hodgson, 1966, p. 125)

Canada suffered various periods of political strife in its early history which undoubtedly served to slow the process of settlement and development of its mineral lands, e.g. (i) external pressures of border disputes with the United States in the central, Great Lakes area as well as at its eastern (Maine) and western (Oregon) extremities; and, (ii) internal disputes and problems between the Anglo Protestants and the French Canadian Catholics. With reference to the latter issue, after the acquisition of New France by the Treaty of Paris 1763, Britain faced the problem of fitting an old colony with alien laws, religion and institutions into the existing British imperial structure. After failing in its early attempt to govern New France as if it were an ordinary English-peopled colony, Britain attempted in 1774, by enactment of the Quebec Act, to assuage the Canadian French with measures of self-autonomy and acceptance. The Quebec Act provided for a governor and council with limited legislative powers, opened public offices to Roman Catholics, and gave statutory protection to the Church. The Act added to the grievances of the English colonists but the British felt that it had kept the French Canadians neutral during the struggle with the rebelling English colonists of the 13 American Coastal Colonies to the south.

After the Colonial Revolution, the regular British representative system continued to be acceptable to the English of Nova Scotia, but the Quebec Act was no longer satisfactory to the peoples of the St. Lawrence River Valley area (later, Ontario). In 1791, the old province of Quebec was divided into Lower Canada, largely French and Catholic, and Upper Canada, predominantly English, loyalist and Protestant. Each had a governor, an appointive council, and an elective assembly (American Peoples Encyclopedia). In succeeding decades, with rapid growth of the area of the Great Lakes, Upper Canada became Ontario and Lower Canada became Quebec. (see Figure 2, p. 39, Growth of Canada, maps of 1791 and 1873.) Both were founding provinces in the Confederation of 1867.

A vast area of western and northern Canada, known as Rupert's Land, was under the jurisdiction of and governed by the Hudson's Bay Company by a charter granted in 1670 by King Charles II until the British North American Act of 1867, the Act of Confederation, which made Canada a federal state. Mineral interests under the Hudson's Bay Company were subject to Crown reservations. At the time of the 1867 Confederation, Rupert's Land was divided between the provinces of Ontario and Quebec. (see Figure 2, p. 39, Canada 1791 map.)

Although Canada's rich mineral wealth compares favourably with that of the U.S., the development of its mineral lands grew at a less frenzied pace. This is likely due to several factors; (i) the colder climate of Canada was less conducive to large waves of immigration for settlement and mineral development in the nineteenth century; (ii) the internal strife and disunified status of the various British colonies delayed land development; (iii) later colonial status comprising British North America until past the middle of the nineteenth century kept mineral lands solidly in control of the English Crown minimised disposal to private ownership; (iv) the confederation of British North America did not occur until 1867, well past the United State's earlier period of attractive-to-immigrants, wholesale development of its public lands for mining and settlement; (v) upon confederation of the provinces, ownership of minerals was continued in the provincial governments [Note: At the time of confederation, July 1, 1867, only four provinces were included in the federal union, viz., Ontario, Quebec, New Brunswick and Nova Scotia. Manitoba entered in 1870; British Columbia in 1871; Prince Edward Island in 1873; Alberta and Saskatchewan in 1905; and, Newfoundland in 1949. As provinces were created and joined the Confederation, they acquired ownership of the Crown mineral resources located within their boundaries, except on reserved federal lands (see Figure 2, p. 39, Canada growth map.)]; (vi) Canadian Provinces did not enact programs of wholesale disposal of fee simple and mineral rights on its lands to entice development as did the U.S.; and (vii) government participation in mining.

2.3 Privately-Owned Lands

Generally, in all three Anglo countries, where purchasing or leasing of surface mining rights or on private lands was concerned, it was, and remains, a contractual matter between the landowner and the mine operator. In leasing, or by *profit a prendre*, the essential terms for mining of the minerals as to area, duration of the right, royalties to be paid for minerals removed, surface rights, restrictions, damages and liabilities, were expressed in the contract. Except in

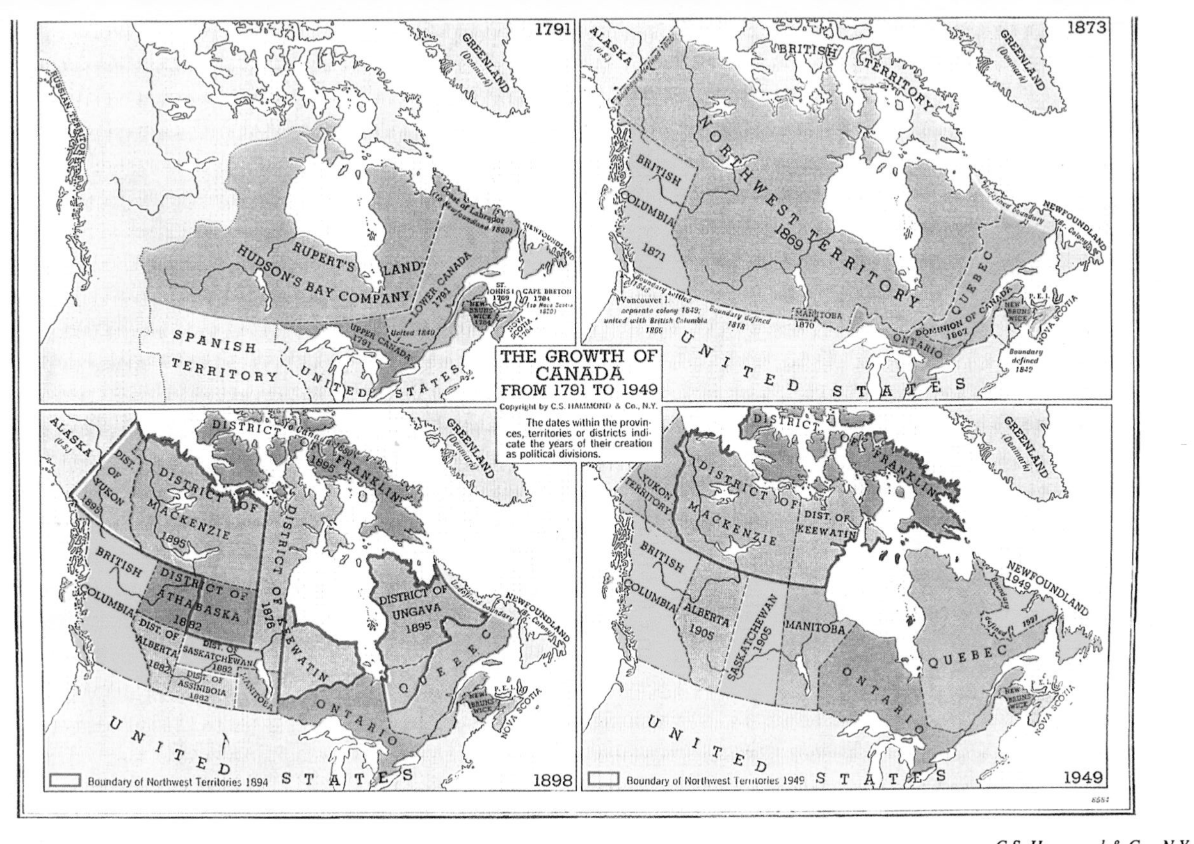

C.S. Hammond & Co., N.Y.

Figure 2 Map of Canada Growth, 1791 to 1949

the U.K., little control was exercised over surface mining until recent decades of safety and environmental regulations.

Whether by ownership or leasing on private lands, where the working of minerals by surface mining was involved, the owner's or lessor's expectations for the life of a surface mine was generally long enough that any thought for reclamation of the land on termination of mining was seldom given consideration. Further, it was the established custom of surface mining to simply pile the stripped overburden into a spoil-bank on the surface of the property leaving it for natural reseeding and Nature to take over. If reclamation of the mined land was agreed to between lessor and lessee, it was a contractual matter between the parties and enforceable under contract law or a common law action. Furthermore, properties with open-pit operations often had histories of being re-opened at a later date after being closed for some reason other than being worked-out. Reclamation even under private contractual agreement with the mining operator would only have hindered and made re-opening less possible or inviting. This is still true today, and a point of contention between pit operators and governmental regulations that mandate reclamation of the land after a regulatory-stipulated short period of inoperation (example: 2 years). At times, mineral economics may cause temporary periods of enforced closures. Present U.S. governmental reclamation regulations are quite intolerant of mineral economics that govern open-pit operations' forced temporary closings. Mandated, premature reclamation before a pit is worked-out serves as an unnecessary cost and a deterrent to its re-opening and extended life for maximum extraction of the natural resource.

Again, Britain was the early history exception for the reclamation of surface-mined lands, and for maintaining a liberal time period before legally finding that a surface mine has been "abandoned" (see "abandoned mine" discussion in Ch. 2 §3.1, ¶3, United Kingdom, infra).

2.3.1 *United Kingdom*

The right to work mines and minerals on private lands as developed in the history of the U.K. in II, 1(a), supra, has been and remains to the present, a private contractual matter between the fee or mineral owner and the prospective mine operator. However, the operations of all underground and surface mines in the U.K. (inclusive of Scotland and Ireland) have been subject to government inspections and enforcement of comprehensive mining regulations at a considerably earlier date than in the U.S. and Canada. The Metalliferous Mines Regulations Acts, 1872 and 1875, governed "all mineral operations whatsoever except those that the Coal Mines Regulation Act, 1872, applied to", viz., coal, stratified iron, shale and fire clay (Rogers, 1876, p. 708).

Early mine case law in the U.K. defined an "abandoned" surface mine and the need for reclamation, whilst the U.S. and Canada were unconcerned about unreclaimed, abandoned and worked-out mines. Although under the English common law of all three Anglo nations, actions for damages to the surface of the land due to surface mining could be made, the U.K. exhibited an earlier concern in its history for surface protection from mining. As early as 1718 the case of *Bishop of London v. Web* [1718] 31(2) L&T; involved a lease for the surface mining of brick-clay. The court stated that where a lease permitted the lessee to exhaust the "brick-earth" (i.e., to mine it out completely), it was "subject to his afterwards leveling the field". And, in *Rosse (Earl) v. Wainman* [1845] 11 Comns, which involved a stone quarry, the court found that the stone belonged to the lord, who may quarry it, but the Inclosure Act (giving use of the surface to the commoners for cultivation), expressly provided that the lord should preserve the top-layer of the 'soil' for respreading on the surface. The court's interpretation of the Act showed a clear intent that where the winning of stone by quarrying was contemplated, the surface soil shall be kept separate and the surface subsequently restored. Lastly, in *Boileau v. Heath* [1898] 34 Mines, where pit waste had been piled on the surface in a spoil-bank, the surface miner was "under a duty, at the end of his lease, to fill up the pit, and to remove the spoil-bank, and to restore the surface land to its original agricultural condition". (Bainbridge, 1900, pp. 3;19; 262).

With regard to the argument against premature reclamation under present U.S. regulations of the so-called "abandoned" mine, or quarry, referred to in §2.3, supra, a more fitting nomenclature in some cases might be "dormant". As argued, supra, a quarry closure by the operator might be for market or economic reasons at the time without permanent closure or abandonment intended. This more appropriate description of "dormant" for a temporary closing was found in older case decisions in the U.K. holding that a mine closure may be either "abandoned" or "dormant". According to the decision in *Bagot v. Bagot* [1863] 2 Agric, it is "a question of degree, that is to say, depending on enquiry, whether the mine may yet be considered an 'open mine' or not; after one hundred years' abandonment, it would probably not be considered an open mine, but after twenty or thirty years' abandonment, it might be otherwise; and (if any rise in the price of the mineral produced *(sic, coal),* justified the expense of re-opening) it might, in such case, be regarded as an open mine". Similar holdings were found in *Stoughton v. Leigh* [1808] 34 Mines; *Bartlett v. Phillips* [1859] 19 Eccl (Bainbridge, 1900, p. 18).

2.3.2 *United States*

Prior to World War II, the U.S. federal government had not possessed any powers over minerals and their development on private lands. They had been governed exclusively by the state law in which they are located. Generally, state governments have never exercised any sovereign rights over the minerals. Up until World War II, and for approximately two decades thereafter, mining on privately owned or leased lands had simply been a matter of starting excavating on one's own land, or as a contractual matter between the fee / or mineral owner, as the lessor, and the mining operator as the lessee, or in the case of a severed mineral estate, by agreement with the mineral estate owner. Not until 1891 did any mining operations first become subject to very limited federal mining regulations by an Act for the Protection of the Lives of Miners in the Territories. This Act provided for one, annual mine inspection of underground coal mines in the territories and did not apply to the states, or to non-coal surface mines.

From its inception, regulation of mining in the states had been considered beyond federal constitutional concern and jurisdiction, being strictly within the jurisdiction of the individual states according to the constitutional powers reserved to the states. Hence, the 1891 Act, above, was limited to coal mines in the federally-controlled territories. The U.S. Supreme Court's position on this jurisdictional issue of control over mining operations continued to be maintained for almost 50 years after the 1891 Act. The Court's position in preserving for the states the power of regulation over mines was based on its interpretation of the federal Constitution's Commerce Clause (U.S. Const., Article I, Section 8). Federal control of interstate commerce under the commerce clause could only be found and affirmed for goods "**in** interstate commerce", i.e., between the states. Mine ores, concentrates and unfinished mineral products were considered as produced "**for** interstate commerce", consequently, mineral products and the mines producing them were not subject to federal authority. In 1936, in the case of *National Labor Relations Board (NLRB) v. Jones & Laughlin Steel Corp.,* 301 U.S. 1 (1936), the U.S. Supreme Court changed its former position holding that Congress had the power to regulate production of goods *for* commerce as well as goods *in* commerce. However, no further federal legislation for mine regulation was enacted until 1941 with the Coal Safety Act. Although coal mine inspections were authorised by this Act, there were no penalties authorised (American Law of Mining, 1986, Sect. 201.01[4]).

During this long period, surface mines for metals and non-metals remained without federal or even state regulation. In more settled, urbanised areas, zoning restrictions for/against allowing heavy industry might be the only controls encountered when opening a surface mine. Even then, a re-zoning might be arranged, e.g., agricultural land to heavy industry, allowing location of a surface mine without incurring the protests of an irate community over environmental concerns. In fact, environmental concerns over surface mining operations were rarely expressed in the first half of the twentieth century. The first federal mine regulation that affected non-coal surface mining was the federal Metal and Non-Metallic Mine Safety Act, 1966, but not made effective until July 31, 1970. This Act provided for primacy of state control whereby a state might enact its own version of the regulations but wherein equal stringency was required.

Thus, in the U.S. until 1969 when the great, popular tide of the "green movement" overcame the nation with mounting national environmental concerns causing its federal congress to respond with enactment of the National Environmental Protection Act (NEPA) of 1969, surface mining operations were virtually unrestricted by governmental regulations. In fact, it would be a few more years before the concerns of NEPA filtered down to all 50 state levels where surface mining would be affected by regulations. Initially, before NEPA, state and federal mining regulations were concerned only with safety practices. After NEPA and its ensuing ancillary legislative acts, earth reclamation measures for surface mining followed, and lastly by preventative air, water and waste pollution and contamination regulations.

The Surface Mining Control and Reclamation Act of 1977 (SMCRA), applicable only to surface coal mining operations, mandated a federal study for the Congress' consideration of whether to enact similar legislation for the regulation of non-coal surface mining. A report prepared by the Committee on Surface Mining and Reclamation is commonly referred to as the COSMAR Study (1979). COSMAR found that non-coal surface mining can be categorised in two types, viz., (i) construction materials mines, and (ii) regional mineral deposits as iron, porphyry copper, sedimentary uranium, phosphate, and oil shale. The study concluded that the mining of such near-surface deposits has limited reclamation potential due to the complete removal of the ore body, thereby leaving little to no overburden material to be replaced, and that the impacts from mining such deposits are localised. In addition, such mineral deposits were less frequent than coal, giving them lesser national significance, and there were more variations in mining methods than in coal. As a consequence of the COSMAR Study, the federal government left the regulation of non-coal surface pit and quarry mining to control by state law (Am. Law Mining, 2d Ed., Sect. 173.02).

2.3.3 *Canada*

It is noted here that, as in the U.S., Canada's legal system is firmly rooted in the English common law system. In many areas of the law, Canadian courts receive both English and American authority in law. However, Canada's legal system has the additional influence of the French civil law, although largely confined to the province of Quebec. By the Treaty of Paris in 1763, England allowed the province of Quebec to retain its civil law system, which is in effect today. Some mineral rights are owned privately in all provinces of Canada. The acquisition of those rights took place in title grants from the Crown, from previous fee simple grants, or were acquired under mineral laws granting mineral titles in fee simple with work and fee requirements to be met, similar to the patenting process of the U.S. In some provinces, title to fee simple mineral holdings may be reclaimed by the Crown where no development work has been done for a statutorily specified number of years.

Mining agreements with private property owners, where the owner held title in fee simple and without mineral reservations by the Crown, and until the green movement overflowed the northern border of the U.S., were also largely a contractual matter between the owner and the mineral operator. In 1976, "approximately 80% of mineral aggregate producers were operating on lands having privately owned mineral rights. (Blakeman, 1976, p. 41). Where the mining operator owned the minerals in fee simple, the operator was free to start excavations without governmental restraints. Until recent decades, the largest concern for beginning a mining operation was for foreign individuals and corporations. Prior to the environmental movement, as in the U.S., for surface or open cast mines in Canada, overburden was piled on the earth's surface to become a part of the earth's surface and left for Nature to reclaim.

2.4 Government-Owned Lands

Historically, there has been a notable difference between the amount of government-owned lands and the consequential treatment of mineral interests in Great Britain and those of the U.S. and Canada. The difference is certainly accountable by two pertinent facts regarding the population and area of each, viz.: (i) that the former is ancient in time and fully settled having a present population of approximately 62 million people, while the latter two are relatively newer lands with populations of approximately 250 million (U.S.) and 25 million (Canada), and (ii) the land area of the former being 244,109 square kilometres (94,226 square miles), with a population density of approximately 253 per square kilometre (660 per square mile) (England's density is 362 per square kilometre or 940 per square mile), and the latter two with areas and population densities, respectively of: U.S., 9,169,277 square kilometres (3,539,341 square miles) and 27 persons per square kilometre (71 persons per square mile); and Canada, 9,978,780 square kilometres (3,851,809 square miles) and a population density of 2.5 persons per square kilometre (6 persons per square mile). (National Geographic Society, 1990) In addition, large areas of Canada's far North are ice-laden much of the year, or subject to permafrost all of the year, making them relatively uninhabitable or the surface difficult to work.

Such figures make obvious the facts that the U.K. is population-density saturated, with virtually no unexplored land areas under government ownership for mineral development and settlement, whilst the U.S. and Canada are sparsely settled with vast amounts of government-owned lands for mineral development and settlement. England was well-developed and settled centuries before and whilst the American colonies and Canada were just being discovered. As a consequence, there are a limited amount of Crown lands in the U.K. where the opening of new mines may be encouraged, or on which new minerals may be discovered.

2.4.1 *United Kingdom*

As a result of the decision of the *Case of Mines* (1585) certain areas were excepted from its application to Crown rights, e.g., in Devon and Cornwall as to tin, in Derbyshire as to lead, in the Forest of Dean and the Hundred of St. Briavel's in Gloucester County, and the Isle of Man as to the baser minerals, generally. These Crown mineral rights remained intact for several hundred years. Under the Crown Lands Act, 1873, the power of leasing was created for the Commissioners of Forest and Woods to grant leases of any mines, metallic or non-metallic minerals, for mining and quarrying in any of the foregoing areas, gold and silver being excepted. (Bainbridge, 1900, p. 123).

The following provisions for mining on Crown lands of the seashore and seabed demonstrate the early concern for disruption of the local environment and as established under the Crown Lands Act, 1866, whereby:

> "*** one moiety (a half) of the net annual income of the land revenue of the crown received in respect of any coal, ironstone, or mineral, stone, slate, clay, gravel, sand, or chalk or of any substance obtained by mining, quarrying or excavating, shall be treated as capital, and the residue thereof as income. … And all persons for the time being entitled, in right of or under the crown, to, or to the management of any beds, seams, veins,

> mines, or quarries in or under the foreshore, or in or under any lands immediately adjacent thereto, and their respective tenants, may (subject to the provisions of the Act) enter into possession of, and use or pass over or under, any portion of the foreshore under the management of the Board of Trade, in order to make or sink any pits, shafts, adits, etc.; or to erect and repair steam and other engines, buildings, works and machinery; or generally, to do any such other acts as are for the time being necessary or convenient for working, searching for, digging, raising, carrying away, dressing, or making merchantable the coal, stone, or other minerals, ... giving at least two months' previous notice in writing of the intention to exercise the powers of the Act, and doing as little damage as may be in the exercise of those powers, and making full compensation (to all persons interested) for all damage sustained by them by reason or in consequence of the exercise of such powers; ... nothing in the Act is to authorise any person to sink, drive, or make any pit, shaft, adit, etc., which will injure, weaken, or endanger, or be likely to injure, weaken, or endanger, any pier or other structure on or near the foreshore; ... and the person for the time being, exercising the powers conferred by the Act, is required to make and maintain all works and conveniences necessary or proper for the safety and accommodation of the public". (Bainbridge, 1900, pp. 124–125).

By virtue of its longevity as a civilised and settled nation, the United Kingdom had also greatly developed its laws to regulate mines and mineral lands more than a century before its offsprings (i.e., the United States and Canada) were first considering such controls. By the time the U.S. had enacted its first meaningful law (1872) to register mining claims on public lands, and by the date that Canada was first forming its unification, the Confederation in 1867, various Acts had been passed in England for the inspection and regulation of mines. In 1872, all former acts regulating mines in the U.K. were repealed and two new comprehensive Acts were enacted, viz., the Coal Mines Regulation Act and the Metalliferous Mines Regulation Act (MMRA). All surface mines were included in the Acts.

The U.K.'s Mines Regulation Acts, 1872 and amended 1875, covered regulatory ground in one act that took the U.S. a period of 75 to 85 years to similarly cover. The MMRA provided for inspection of all mines in the U.K.;

for the powers and duties of inspectors; for duties of owners and agents; age and sex restrictions for employees in mines and surface plants; for provision of wages and hours; for storage and use of explosives; for mine maps and plans; for reporting mine accidents and deaths and investigations of causes thereof; provisions for opening and closing mines; for fencing and enclosing dangerous abandoned mine openings [also under the Quarry (Fencing) Act, 1887]; provisions for liabilities and damages, civil and criminal penalties for violations of all regulations of the Act. The extensiveness of the Act is illustrated by its provision authorising the Commissioners of Public Works "to lend and advance money out of the Consolidated Fund, to any person or company, for the support of any mine or colliery". (Rogers, 1876, p. 746).

As commented by barrister John R. Pickering in his 1957 review of the Mines and Quarries Act, 1954, "It is perhaps remarkable that in such industries as mining and quarrying, in which new techniques are being constantly developed and new knowledge acquired, the legislation which has governed their working, namely the Metalliferous Mines Regulation Act, 1872, and ... the Quarries Act, 1894, should have remained in force without substantial revision for so long a time". The 1954 Act still dealt mainly with health and safety measures in mining. Reputed to be a more comprehensive and stringent Act than its predecessor, it was actually only an updated post-war version. It provided for greater responsibilities and liabilities of the mine operator, raising the duty of the owner from one of reasonable care to conform to that of absolute compliance (Pickering, 1957, pp. 1–3). Claims for environmental damages remained under the common law actions of nuisance and trespass.

Beginning in the early 1960s, the general public's concerns for the environment mounted, including those resulting from the dereliction of land by the surface mineral extractive industry. By the mid-1970s, surface mining was taking about 5,059 ha (12,500 acres) per annum with very little acreage being restored after completion. Approximations at that time calculated that 60,704 ha (150,000 acres) in Britain had accumulated in waste lands from former mining operations, and by 1974 some 3,237.56 ha (8,000 acres) had been affected by mining, mine plant, mine buildings, and lagoons in National Parks and Areas of Outstanding Natural Beauty (AONB) (Hughes, 1986, p. 240). The nine British National Parks in 1986 comprised 9% of the land in

England and Wales, whilst 33 AONB's comprised 37,546.6 square kilometres (14,493 square miles) in 1986, or 9.6%. (Hughes, 1986, p. 187). Total AONB area increased to 44,259 square kilometres (17,084 square miles) by 1992. However, David Hughes, Senior Lecturer in the Faculty of Law at the University of Leicester notes that "There seems also to be little consistency with regard to mineral applications in National Parks. In 1979 permission was centrally refused for potash mining in the North York Moors National Park, but in 1978–80 permission to quarry limestone on the Peak Park was given ministerially". (Hughes, 1986, p. 245).

Over the decades of the twentieth century, planning law has played an increasingly important role in affecting and for permitting surface mining operations. Barrister David Hughes traces the history of planning law back to Edwin Chadwick, who, in 1873 "called ... for a new class of 'town surveyors', men trained in science and engineering, able to build healthy towns with adequate protection for health". And, the emergence of the term "town planning" appears to have originated in a report of the City of Birmingham Housing Committee in 1906. The term was first given legal recognition and status in the Housing, Town Planning &c Act 1909. The making of plans was made compulsory in the amended Act 1919. Further planning powers were incorporated into the Town and Country Planning Act 1932. Various Royal Commissions in the early 1940s reported on studies made, and a new pattern of thinking and principles was founded for land development. The Town and Country Planning Act 1947 declared that "all land is subject to the jurisdiction of planning authorities". (Hughes, 1986, pp. 16–17). Consequently, permission for new-start surface mining has required approval under the Act. However, under the 1971 Regulations, it was stated that "*** *generally,* 'use' in relation to the development of land does not include use of land by carrying out mining operations". (Hughes, 1986, p. 246). Nevertheless, various litigated cases over permitting for new mining operations have involved the Town and Country Planning Act.

English legislative acts and the common law had early on shown concerns for protecting the rights of others from injurious acts to the environment, persons and property. Common law actions in nuisance, defilement of waters, and personal injuries covered all the ground that modern environmental and mining regulations attempt to resolve. The improvement in law for the general public through mining and environmental regulations is that the problem, potential or actual, is sought to be diminished or abated at the source by enforced compliance rather than letting each harmed individual bring his own action for damages after the injurious or tortious act has occurred.

2.4.2 United States

Environmental controls and regulations for mining operations on private or public lands did not make an appearance until after NEPA (1969). Until NEPA, public lands had only been regulated by the federal government with concern to allocation for revenue, disposal for settlement, and resource conservation. The nation has since moved toward a period of retention of resources and management thereof. While the federal government has been the owner of the surface and minerals of public lands, unlike Canada, it has never made any attempt to work the minerals itself, e.g., the phosphates deposits of Saskatchewan.

Federal leasing of minerals on a royalty basis for a concession found its beginnings around 1807. Shortly after the discovery of lead in Missouri, Congress passed an act establishing a leasing policy. The mineral leasing policy was extended to lead and copper in the Lake Superior region. The program deteriorated from pressures to sell the lands for agricultural purposes, various abuses and confusion. Congress terminated the lead leasing in Missouri in 1829 with the sale of the mines, and eventually abandoned the entire mineral leasing program by 1847 (James, 1898).

No meaningful federal mining law had been established in the U.S. until the enactments of the Lode Law of 1866, the Placer Law of 1870 and the General Mining Law of 1872. Their formulation had taken into consideration the mining laws established in the earlier western districts following the California gold rush of 1849. Similarities are noted between the laws and customs established in the California gold districts and those of the lead miners of Derbyshire, England. Notable were the making of rules by the miners, diligence in working and marking their claims, reference of disputes to miner's courts, emphasis upon the vein as opposed to a surface area, and customs with reference to crossing veins and extralateral rights. (Lindley, 1914, Sec. 8) Miners' courts in the California gold district had precedent in the Stannary Courts of Cornwall and Devonshire for the administration of justice among the miners and tinners. These courts were held by virtue of a privilege granted to the tin miners to sue and be sued in their own courts only, in order that they might not be drawn away from their business by having to attend law suits in distant courts (Black, 1968, p. 1577, "stannary courts"). In Derbyshire, the customs were limited to those of the lead mines. In Australia today, Wardens Courts, and in the U.S. administrative law courts, serve a similar purpose in hearing complaints of miners before being passed on to the superior or circuit courts when an appeal is sought.

In early California, mining districts were organised with a camp recorder to keep record of the claims and transfers. By-laws were adopted for claim registers, conflict resolutions, and the settling of other claims, transgressions and complaints between miners (Gates, 1968, p. 709). From 1849 to 1866, more than a thousand mining districts had been organised throughout the new West (Shinn, 1884). The major elements adopted for the mining laws of 1866, 1870 and 1872 that came from various mining districts were, e.g., that the first in time is first in right; rules dealing with the amount of land that may be claimed; discovery as an essential element of a possessory right, with a work requirement to show good faith to hold a claim; and, extralateral rights for a lode deposit [Am. Law of Mining, 1986, Sect. 4.09 fn. (citing Lindley on Mines, Sects. 68–74, "Some of the miners' regulations appear to have been copied in part from customs or laws in the Galena (Illinois) district, from Cornish miners in England and from Mexican law traditions.", id., Sects. 5,13.)

The General Mining Law of 1872 in effect rewrote the mining laws of 1866 and 1870. The 1872 Law established the first of three current policies governing the exploration, development and production of minerals on public lands. Under it, where public lands are open to mineral location, i.e., not having been withdrawn for specific allocations as for national parks, forests, wildlife reserves, locators are able to initiate rights to mineral deposits merely by discovery and without prior administrative approval. The locator may acquire legal title to the land claimed where the deposits are proven "valuable", by applying for a form of deed known as a patent. Even without a patent, the miner may produce minerals without any payment in any form of royalty, fee or tax. Although the 1872 law itself does not define the meaning of "valuable mineral deposits" [Note: 30 U.S.C. Sect. 22 (1976) simply states "valuable mineral deposits"; Sect 23 states "gold, silver, cinnebar, lead, tin, copper, or other valuable deposits ..."], subsequent legislative changes, court and administrative agency decisions over the years have narrowed the application down to only hardrock metallic minerals which may be located and mined. (For an overview on current debate over the U.S. General Mining Law of 1872, see the article from *Mining Magazine*, London, October 1994, Appendix C).

The loophole, whereby unreserved minerals passed to homesteading patentees under agricultural entry statutes, was closed in 1916 with the enactment of the Stock Raising Homestead Act. Under that Act, the grantee's patent for homesteading land contained a reservation by the federal government reserving "all coal and other minerals in the land so entered and patented". However, under the same Act, a prospector may enter agriculturally granted lands and locate mineral deposits and receive a mineral patent on them, subject only for liability and to the rights of the surface owner (43 U.S.C. Sect. 299). The Act was amended in 1982 and makes "leasable" minerals subject to the Mineral Lands Leasing Act of 1920 (see infra).

Nearly one-third of the U.S.'s land area is publicly owned, and roughly 55% of that was open to mining in 1991. The Bureau of Land Management (BLM) manages the largest part, approximately 190,206,394 ha (470 million acres), lying in 11 western states and Alaska. The National Forest Service (NFS) manages about 76,891,947 ha) 190 million acres in 40 states. NFS claims the presence of over 40 industrial (non-metallic) minerals on NF lands. Aggregate stone production from open pits on NFS lands has made it the one of the larger crushed stone producers in the U.S.

It is reported that during the entire history of mining on public lands in the U.S., less than one-quarter of 1% of the available public lands have been touched by mining. As stated above, there are three different federal policy systems governing the exploration, development and production of minerals on U.S. unappropriated public lands, viz., (1) hardrock (generally, metallic) claims under the General Mining Law of 1872; (2) mineral leasing under The Mineral Leasing

Act of 1920; and (3) by the minerals disposal system under the Materials Act of 1947. The latter two acts, in general, affect minerals that are usually mined by open pit, or surface mining methods.

Under the Leasing Act of 1920, specific minerals (e.g., phosphate, sodium, potassium, zeolites, et al) were removed from coverage by the General Law and placed under the Leasing Act. [Aston, 1992, p. 33; Note: "Zeolites, as double salts, become problematic as to whether they are locatable or leasable minerals. A two-pronged test to determine whether a zeolite is locatable or leasable was developed in a 1977 case, GFS (MIN)34(1977). Under the test, if a zeolite is classified as a silicate of sodium, it is locatable, provided the sodium present is in sufficient quantity to be commercially valuable; or, if the presence of sodium or any other material listed in the Mineral Leasing Act is essential to the existence of the mineral.] Under the 1920 Leasing Act, deposits of borates, carbonates, chlorides, nitrates of potassium, sulphates were authorised to be leased. [Aston, 1992, p. 33 Note: "In addition, a potassium lease may include magnesium, aluminium or calcium that is associated with the potassium. However, in the case of *Foote Mineral Co. v. U.S.*, 654 F.2d 81 (Cl.Ct. 1981), it was determined that lithium is not a deposit associated with potassium".]. The leasing method requires annual rentals until production, with royalties paid thereafter. The responsible, supervising government agency has complete discretion to accept or reject offers.

The Materials Act makes available common minerals such as sand, gravel, pumice, volcanic cinder, stone, and clay, at a market price usually determined by competitive bidding (contract sale). The non-metallic minerals are acquired by mining claim or by a sale contract which is based on their chemical composition, physical properties and prospective use. Uncommon varieties of sand, gravel, stone, pumicite, clay, etc., are locatable under the Act of 1955. [Aston, 1992, p. 34; Note: "Common quartz varieties as jasper, obsidian, opal, etc,. are not locatable unless they exhibit uncommon qualities. Semi-precious minerals as amethyst, garnets, topaz, geodes, etc., and precious stones (emerald, ruby, sapphire) are locatable".] Geothermal steam and associated geothermal resources on BLM lands are also leasable under the Geothermal Steam Act of 1970. (30 U.S.C. 1001–1025, 1976, Suppl. V, 1981). Under the Mineral Leasing Act and the Materials Act, agency permits are required for explorational activity for the minerals thereunder.

2.4.2.1 *The U.S. conservation movement and increased regulation*

Beginning around 1864, the progenitors of conservationism sowed the seeds of concern which eventually lead to modern resources environmentalism. The seeds grew at a slower pace than did development of the public lands. One of the progenitors was a lawyer and diplomat by the name of George Perkins Marsh. In his travels over Europe and the middle East, he noted massive deforestation with resultant soil erosion, and man's interference with the balance of nature. Marsh published a book in 1864 entitled *Man and Nature,* in which he not only described his observations of results of man's poor stewardship of Nature through wasteful practices in forestry and agriculture, but also advocated restorative measures.

The creation of Yellowstone National Park in 1872 is regarded as the initial act of the U.S. government's concern for conservation of the nation's natural resources. This environmental concern and advocacy of measures to conserve natural resources did not take a serious hold until the 1890s when the U.S. Congress authorised the President, by the General Revision Act of 1891, to establish forest reserves on the public lands. In 1891, large amounts of the public domain were withdrawn for national forests. Several milestones of natural resources conservation followed in the early twentieth century, e.g.; 1902, the Reclamation Act for water resources in the West; 1905, the Forest Service was created as part of the Department of Agriculture to manage the forests; 1906, Congress's authorised mineral lands were withdrawn from entry until valuations could be made by the Department of Interior; 1906, Land-Grant colleges were established to promote improved farming practices in the states; 1908, Theodore Roosevelt "fathered" a conservation spirit for planning of natural resources and national parks under a conference producing a "Declaration of Principles on Conservation"; 1909, Theodore Roosevelt hosted a North American conference for advocating natural resources conservation; the National Park Service was created in 1916, although 16 national parks and about 18 national monument sites had already been created by that time.

From the early decades of the twentieth century, natural resources concern has grown and many federal acts have been passed in an effort to conserve and preserve resources. The advent of the environmental "green" movement of the 1960s eventually lead to Congressional response by passage of the National

Environmental Policy Act of 1969 (NEPA, effective January 1, 1970), and establishment of the first observed Earth Day, April 22, 1970. With the subsequent plethora of environmental laws created for industry, including mining, whole volumes are now being devoted to mineral and environmental regulation. Special attention and stress affecting the mining industry is being placed in the areas of air and water pollution and reclamation of mined land.

2.4.3 *Canada*

Before confederation of the Canadian Crown colonies, mining legislation in British Columbia and Ontario influenced regulations of the later-founded prairie provinces. Ontario's mining statute of 1846 established regulations for licences to explore for and discover minerals on Crown lands. A licence gave the right to explore on a 8.05 km x 3.22 km (5-mile by 2-mile) area, and upon discovery of a mineral, entitled the discoverer to purchase the lands in fee simple at 4 shillings per acre. The area was limited in 1853 to 161.9 ha (400 acres). Ontario's Gold Mining Act, 1864, introduced the staking of a "mining claim" and conferred a right to mine, but curiously did not require an actual discovery of minerals (Bartlett, 1984).

Under the British North American Act (B.N.A. Act) of 1867 (the Confederation of Canada), authority to make laws relating to mining and mineral-bearing lands was divided between the federal and provincial governments, and the provincial legislatures were given jurisdiction over management of Crown/public lands [Note: The B.N. A. Act is also known as the Canadian Constitution Act (1867); Section 109 states that all "Lands, Mines, Minerals and Royalties" shall belong to the provinces.] Subsequent judicial decisions have redefined the allocation of various powers. Provincial legislatures now have the power to enact laws over mineral titles, the exploration and development of minerals, and for mine operations [Am. Law of Mining, 1986, 210.03(2)]. Present provincial powers also include conservation of mineral resources and environmental protection. Current Dominion, or federal power and control of mineral lands, essentially, exists only over the Northwest and Yukon Territories.

Early Canadian mining and mineral land legislation was a reaction to the gold rushes of the times. During the Cariboo gold rush in British Columbia, the "free miner" concept was adopted from Australian models and employed in the Goldfields Proclamation, 1859. Prospectors could obtain a free miner's certificate enabling them to stake a claim on any waste lands of the Crown not already lawfully occupied, and by which the exclusive right was granted to take the minerals (Crommelin, 1974).

The Ontario Mining Act, 1869, set precedent for the later incoming provinces to the Confederation by establishing the right to a staked mineral land claim and conferring fee simple title upon payment of $1.00 per acre (.4 ha). The area of a mining claim, or location, was generally fixed at 129.5 ha (320 acres), with no limit as to the number of mining locations that could be purchased. Until 1963, under the Ontario statutes, it was still possible to obtain title in fee simple by performance of an annual assessment work on a mineral claim. Since 1963, mineral land claimants may lease lands for a period of 21 years upon performance of assessment work. Entitlement to a patent may still be obtained by the holder of a mining lease who satisfies the Minister as to continuous production of substantial quantities of minerals (Bartlett, 1984).

Similar to the U.S., ownership of non-granted minerals and mineral-bearing public lands is vested in either the provincial or federal government and titled in the Sovereign / Crown. Minerals in Indian reservations, national parks, and other federally-owned lands are owned by the federal government [Canadian Constitution, Const. Act, Sects. 91(24), 108 (1867)]. Title to minerals in Canada's two territories, the Yukon and the Northwest, are in the Crown. Minerals under territorial seas and the continental shelf are in the federal government. The practice of reserving the mineral interest from fee simple Crown grants was established near the end of the 19th century. Later legislation insured that all minerals were reserved from Crown land grants and became subject to leasing policies. The free miner concept still prevailed in the controlling legislation, but present day claim staking usually leads to a lease from the Crown rather than a fee title or patent. Mineral claims have at times still received Crown mineral grants in fee simple, e.g., as under the Mineral Act, R.S.B.C. 1936, c.181 (repealed 1977, c.54) and mineral claims that have been leased may obtain patents in certain provinces, e.g., in Ontario, under the

Mining Act, Ont. Rev. Stats., Ch. 268, Sect. 96 (1980) which provides that leases can be taken to patent if in continuous production for more than a year. Unfortunately, under an interim British Columbia Mineral Act, 1973, the entitlement of a mineral claimant to lease was removed; the claimant could only apply for a lease. The application was evaluated by the Minister and, reputedly in his discretion, to insure that the best method of development of the minerals was to be used. "Mining was no longer clearly a right ... instead this right was replaced by a permission granted by the Minister. The security of the right to mine was lost ... assurance was removed that development of a deposit was possible after a discovery was made". A great outcry of public protest brought a change by the Mineral Act, 1977, and reversion was made for the claimant to mine the land claimed without the discretionary approval of the Minister (Bartlett, 1984).

The Dominion Mining Regulations of 1889 provided for the disposition of minerals, other than coal, on vacant Crown/Dominion lands, and required actual discovery. In 1917, the entitlement of a claim holder to a fee simple grant from the Crown was rescinded, however, under the Quartz Mining Regulations, 1917, the holder of a claim was entitled to a lease for 21 years for a fee of $50, and renewable for another 21 years for $200. Such leases were not assignable without the consent of the Minister. The government's attitude toward mineral land leasing was greatly altered by 1961, as expressed in the enactment of the Mineral Dispositions Regulations. The concept of a mineral exploration licence was introduced, which did not confer a right to mine minerals discovered, but provided for mining of minerals only under a leasing arrangement. As in the British Columbia experience, above, the uncertainty of miner's rights was increased. Bartlett stated in 1984 that "The evolution of mining legislation in Canada suggests a continuing movement towards retention of a larger interest and discretion by the Crown (Bartlett, 1984, pp. 22, 27).

The right to prospect, explore and exploit minerals owned by the Crown required meeting permitting, licencing and leasing procedures under the governing provincial laws. Crown-land minerals are separate from surface estate ownership and rights to mine are granted independently from the surface rights in all parts of Canada. Permission to use the surface, timber, water and access reserved to the Crown were usually granted to the extent necessary for mineral development. National park lands are not open to mineral prospecting or development, except where valid mineral claims and title existed before creation of the park. And, as in the case of *Casamiro Resources Corp. v. British Columbia, 1991,* infra, those reserved grants of mining rights in British Columbia have been encroached upon by subsequent legislation until finally lost.

Several prominent acts principally govern the disposition of Canadian mineral rights and are briefly discussed.

(i) **The Public Lands Grants Act 1990:** regulates and governs the sale and leases of public lands, including mines and minerals.

(ii) **The Territorial Lands Act 1978**, revised 1985, regulates and governs the disposition of public lands, including mines and minerals, in the Yukon, and Northwest Territories. The Northwest Territories covers one-third of the land mass of Canada. The Territories have been divided into mining districts (four in the Yukon, viz., Dawson, Mayo, Whitehorse and Watson Lake; three in the Northwest, viz., Nahanni, Mackenzie, and Arctic and Hudson Bay). The Act is administered by an official, the Mining Recorder, located in each of the district headquarters. The Mining Recorder acts in a quasi-judicial position, recording claims, issuing permits and licences, and interpreting regulations under the Act.

Under the Act, "mineral" is defined as "all deposits of gold, silver, and other naturally occurring substances that can be mined, excepting soil, limestone, gravel, peat, coal, oil, helium, natural gas, or other related hydrocarbons". Territorial dredging regulations allow dredging for gold and silver and other natural substance, but not for clay, sand and gravel. Territorial quarrying regulations apply to limestone, granite, slate, marble, gypsum, marl gravel, loam, sand, clay, volcanic ash and stone. Mineral claims are for an initial term of 10 years and for as much as 64.75 ha (160 acres) for the specified quarrying materials, except for loam, which is limited to 8.09 ha (20 acres). (Hodgson, 1966, pp. 9, 14–15).

Under the 1985 revision, a preliminary prospecting licence is required. Without it, a claim may not be recorded. A prospecting licence entitles a prospector to stake a claim up to a maximum area of 1,045.12 ha (2,582.5-acres). Representation work requirements are required for both the permit and the staked claim. If production of over C$100,000 in minerals per annum is expected, a lease is required, and may be obtained by the claim holder for 21 years (C.R.C. 1978, c.1516 under Territorial Lands Act, R.S.C. 1985.).

(iii) **Construction Materials acts:** These lesser mining acts, although of no great import to the more glamorous segments of the mining industry, are important to this work as they provide the applicable laws for surface mining pits for which surface voids are made as potential landfill sites. Non-metallic minerals and common rocks, or construction minerals, as sand and gravel, clay, and stone, for use primarily in aggregate and brick manufacturing, are ordinarily excepted from provincial mining acts and in some provinces are placed under separate acts, e.g., Saskatchewan's Sand and Gravel Act, R.S.S. 1978, places ownership of sand and gravel with the surface owner; Ontario regulates construction minerals under their Aggregate Resources Act, R.S.O. 1990, with sand and gravel being specifically excluded from its Mining Act. Certain other non-metallic minerals on Crown lands are regulated under both Acts. Under Quebec's Mining Act 1987, S.Q. c.64, §§5,6, separate provisions are made for surface minerals, e.g., building stone, limestone clay, and marl. British Columbia has bandied about its treatment and inclusion, or exclusion, as the case may be, of construction materials between its Land Act, S.B.C. 1988, c 5, §65, and its Mineral Tenure Act, R.S.B.C. 1979, c.214. The latter now includes limestone, dolomite, marble, shale, clay, volcanic ash and diatomaceous earth.

(iv) **The Yukon Quartz Mining Act 1952**, 1970, revised 1985: Although only applicable to the Yukon, this act is important because of the great mineral wealth of the territory. Minerals are defined by the Act as including a list of some 23 metallic and sub-metallic elements, and additionally, asbestos, emery, mica, mineral pigments, corundum and diamonds. It also defines and lists certain earth materials as not being considered minerals and therefore, are part of the agricultural surface of the land: viz., "limestone, marble, clay, gypsum, or any building stone when mined for building purposes, earth, ash, marl, gravel, sand, and any element forming a part of the surface". Location of claims for mining iron and mica are regulated separately. (Hodgson, 1966, p. 16)

Under the Act, a staked mineral claim, not more than 1,500 square, not exceeding 20.9 ha (51.65 acres), is valid on an annual basis, providing initial working requirements are met, e.g., claim has been surveyed, $500-worth of work done, and a discovery of a deposit made. The claimant may get a certificate of improvements, which entitles the claim holder to extract minerals without a lease, or to keep the claim without further work. However, a 21-year lease is obtainable after receiving the certificate.

(v) **Yukon Placer Mining Act 1906, rev. 1985:** Placer mining is defined "to include every method of working whereby earth, soil, gravel or cement may be removed, washed, sifted or refined for the purpose of obtaining gold or such other minerals or stones, but does not include working rock in place".

Water rights are defined and regulated under the Act. "The owner of a claim is entitled to use that part of the water naturally flowing past his claim, which is not already lawfully appropriated, as may be necessary to the working of the claim. A water grant may be made by a mining recorder which gives the applicant the right to take water from any stream or lake. It also conveys the right of entry for constructing or repairing ditches and flumes, provided that the applicant has posted a bond securing payment for damage caused by such construction".

The Act also stipulates compliance by the holder of mining rights with acts and regulations governing Indian Mining Regulations and Indian Reserves (Hodgson, 1966, pp. 18–19).

2.4.4 *Provincial Acts And Regulations Governing Minerals And Lands*

In each province, adaptations of the various prominent federal acts affecting mineral lands and mining rights were made. The acts frequently defined and specified which minerals may be claimed and worked, and which were not minerals but were included in the surface estate ownership. These regulations are still important in determining for surface mining whether the prospective pit operation has to deal with a private surface owner, a private mineral owner, or the Crown. Royalties for Crown mineral rights under the provincial acts were established for many of the mineral lands.

As indicated, authority over the Crown mineral lands and disposal of minerals and rights within the provinces had been delegated to the provincial governments. A brief historical summary of provincial mineral land regulations follows.

2.4.4.1 *British Columbia*

In 1859, the Crown declared ownership of all mines and minerals in the colony. British Columbia was the first province to join the Canada Confederation after its formation in 1867. It retained full control of its Crown lands and mineral rights. Rights to minerals were included in Crown grants under the Land Act until 1891, and surface rights were included in Crown grants under the Mineral Act until 1893. Thereafter, mineral and surface rights were separated in most Crown grants. Crown grants of freehold mineral rights have been both in fee simple, i.e., inclusive of surface, but have not necessarily included all minerals. Such involved and restrictive grants require careful examination of the granting terms themselves. In 1957, the Mineral Act was amended whereby fee simple Crown grants were replaced by a mineral leasing system. It is of interest that British Columbia offered bounties in 1960 and 1961 for the production of iron ore, and for blister and refined copper, respectively, mined and smelted in the province (Hodgson, 1966, pp. 38–39,42,46,48).

2.4.4.2 *Alberta*

By the Alberta Act of 1905, enacted upon entry of Alberta as a province, the control of mineral resources remained under federal control. In 1930, under the passage of the Alberta Natural Resources Act, full control of its minerals and Crown lands was transferred to the provincial government, except for National Park and Indian Reserve lands. Prior to 1930, some 19% of provincial mineral rights were granted to homesteaders, the railways and Hudson's Bay Company. The remaining 81% of Alberta's mineral lands and rights were placed under a leasing system in 1930.

Special acts in Alberta declared clay, marl, sand gravel belong to the surface owners and were not minerals. However, those materials may be leased for mining under the Public Lands Act (Hodgson, 1966, pp. 52–68).

2.4.4.3 *Saskatchewan*

Similar to Alberta, Saskatchewan was taken out of the Northwest Territories and made into a province by Act in 1905, and control of its natural resources remained in the federal government until 1930. Prior to 1930, about 50% of its mineral rights had been alienated. After 1930, the province assumed control of mineral rights of its lands.

Under the provincial Mineral Resources Act 1961, surface rights were protected from mining dispositions. A mining rights holder was conveyed no automatic right to use the surface and was required to compensate the surface owner for any surface area that might be required or damaged as a result of a mining operation. Also, the Act's Subsurface Mineral Regulations (1960), made applicable particularly to potash deposits, contained an interesting protective measure for the surface owner, presumably against destruction of the surface by open pit mining to reach shallow mineral deposits. Crown mineral rights ("including all mineral salts of boron, calcium, lithium, magnesium, potassium, sodium, bromine, chlorine, fluorine, iodine, nitrogen, phosphorus, sulphur, and their compounds … or such other minerals as may be designated from time to time …") could only be disposed of for deposits "lying more than 200 feet (60.96 meters) below the agricultural surface of the land …".

The Saskatchewan mining regulation raises two interesting questions concerning the ownership of minerals above the 200-foot (60.96 m) level; whether they belong to the surface owner, or to the Crown?; and, who has the right to mine them? The U.S. State of Texas has litigated a similar question in which the Supreme Court of Texas developed the "surface destruction" test to determine whether mineral estate owners had the right to surface mine minerals (uranium and iron) from shallow mineral deposits occurring less than 200 feet (60.96 meters) below the surface [*Crewes v. Plainsman Trading Co.*, 827 S.W.2d 455 (Tex.App.-San Antonio 1992)].

In 1957, with regard to surface mining, Saskatchewan enacted The Quarrying Regulations in which " 'quarriable substance' was defined to include bentonite, building stone, clay, granite, gravel, gypsum, limestone, marble, marl, sand, slate, volcanic ash and any other substance that may be declared a quarriable substance by the Lt. Governor in Council". It should be noted that under Saskatchewan's Sand and Gravel Act, the surface owner is "*** entitled to all sands and gravel found on the surface of the land and that which is obtainable by stripping off the over-burden or other surface operation and, the owner of the mineral rights is also entitled to all volcanic ash, marl, bentonite, ceramic clays and other industrial clay except any clay required for the construction of an earthen dam or road grade". (Hodgson, 1966, pp. 69–87).

The Mineral Disposition Regulations 1961 greatly altered the mineral tenement. No longer was the right to mine conferred, but in lieu, provided for mining by leasing from the Province.

2.4.4.4 *Manitoba*

Manitoba was carved from the Northwest Territories in 1870 by the Manitoba Act. Provincial mineral resources were placed under the control of the federal government until 1930, at which time full control of its mineral resources were transferred to Manitoba. Prior to 1930, Crown-owned mineral rights had been granted in substantial areas making up about 80% of the total. The remaining 20% of Crown mineral rights land have been held under a leasing system since 1930 (Hodgson, 1966, pp. 92–109).

2.4.4.5 *Ontario*

As Ontario was one of the original four founding provinces at the time of the Confederation, it has had control of Crown mineral lands within its borders since 1867. About 88% of the land in Ontario is presently held by the Provincial Crown, 1% by the federal Crown, and 11% is privately owned (Blakeman, 1976, p. 32).

Modification of its mineral land granting system in 1963, whereon the mineral leasing system was substituted. Mineral rights may be granted under a leasehold patent from the Crown upon fulfillment of assessment work and other requirements. A freehold patent could be acquired after 1963 by achieving specified quantities of mineral production.

Ontario's Mining Act 1960, Part vii, dealt with Quarry Permits, requiring that a permit must be obtained for mining on Crown lands for "any stone or rock quarried for any industrial or commercial purpose; limestone, marble, granite, quartzite, feldspar, fluorspar, gypsum, diatomaceous earth, clay, marl, peat, sand or gravel". (Hodgson, 1966, pp. 110–116).

The Mining Act was revised in 1970 and amended several times in subsequent years. The Mining Act was augmented by the Pits and Quarries Control Act 1971 and as amended. These two acts were the two basic statutes regulating the mining of mineral aggregates and industrial minerals in Ontario. The Mining Act applied only to operations on Crown lands, whereas the P & Q Control Act was applicable to operations on private land and to those on Crown land if it was situated within a designated township. However, the new Aggregates Resources Act 1990 replaced former provisions of the two acts governing aggregate operations. Ontario's Mining Act is claimed to be the most comprehensive mining statute in all of Canada.

2.4.4.6 *Quebec*

Quebec is the largest province in Canada and as one of the four founding provinces joining the Confederation in 1867, Quebec maintained control of its Crown mineral resources from that date. Prior to 1880, the time of its first mining act, mineral rights had been granted under French seigniorial tenure or English tenure [See §2.1.3 Canada, ¶2, supra]. The Mining Act of 1880 separated mining rights from surface rights and reserved the rights on all minerals to the Crown. Grantees of land for agricultural purposes, or for colonisation, no longer received land in fee simple, i.e., with the mineral rights. Under the 1880 Act, mining concessions could be purchased on mineral lands which gave the right to letters patent and absolute title. However, mining concessions sold after 1880 were made revocable by the Mining Act of 1965 under certain conditions of prolonged dormancy (mining inactivity) or failure to pay taxes. (Note: here, a similarity to U.S. dormant mineral and mineral lapse statutes is noted.)

In Quebec, the Provincial Crown owns by far the highest proportion of land with 92.5%, whilst only 7.3% is privately owned, and the Federal Crown owning the very small balance of less than 1%.

As to stone, sand and gravel in Quebec, the rights to mining of it were no longer reserved to the surface owner after 1966. The Lt. Governor in Council was authorised to dispose of working sand and gravel deposits after 1966 (Hodgson, 1966, p. 125, 130).

Quebec's Mining Act is the main regulating mining act. The Quebec Urban Community Act 1969 and as amended, and the Cities and Town Act 1972, and as amended, has empowered local and regional governments to have some control of zoning and by-law control over surface mining operations, and may exclude pits and quarries if thought necessary.

2.4.4.7 *New Brunswick*

As another of the founding provinces, New Brunswick maintained full control over its Crown mineral lands from the time of the 1867 Confederation. Grants of land made prior to 1784 when New Brunswick was part of the British colony of Nova Scotia contained a reservation to the Crown of "All mines of gold, silver, precious stones, lapis lazuli, lead, copper and coals". All minerals since 1805, with minor exceptions, have been reserved to the Crown. Crown-owned minerals since have been disposed of by mining claims, licence and leasing. The provincial Mining Act of 1962 applied to "all minerals" and "included salt, infusorial (diatomaceous) earth, ochres, or paints the base of which is found in the soil, fire clays, carbonate of lime, sulphate of lime gypsum, coal, bituminous shale and albertite and such others as may, from time to time, be declared mineral by the order of the Lt. Governor in Council".

Under New Brunswick's Crown Lands Act 1952, leases were "granted for the quarrying of building stone, for the taking of sand and gravel for construction purposes, and for the taking of ceramic substances, mineral waters, soapstone and peat". And, under the Sand Removal Act 1954, "the amount of sand and gravel that could be removed from a shore line was limited to 1/2 cubic yard (.382 cu.m.) per day unless a lease had been granted under the Act". (Hodgson, 1966, p. 137,138,141).

New Brunswick was the first of two provinces in eastern Canada to make statutory distinctions between bedrock minerals and surficial minerals such as sand, gravel, clay and peat. The provincial Mining Act applies to the extraction, under licence or lease, of minerals from privately owned lands and those in which the mineral rights are vested in the Crown. Minerals in this act are defined "as any natural, solid, inorganic or fossilised organic substance, and other materials declared to be minerals by the Lieutenant Governor in Council, but excluding 'ordinary stone', building or construction stone, sand, gravel, peat, ordinary soil, oil and natural gas".

The Quarriable Substances Act applies to extraction under permit or lease on all Crown lands and designated shore areas which might lie outside Crown lands. Quarriable substances is defined as "ordinary" stone, building or construction stone, sand, gravel, peat and peat moss. Similar to acts in the other provinces, local governments are empowered to regulate surface mines and pits within their jurisdiction, e.g., The Municipalities Act and the Community Planning Act 1972, amended.

Presently, the Provincial Crown owns about 43.1% of land, the Federal Crown owns 2.8%, whilst the balance of 54.1% is privately owned.

2.4.4.8 *Nova Scotia*

In 1861, before the formation of the Canadian Confederation, control of minerals and lands in Nova Scotia was under the Chief Gold Commissioner. As with the other original founding provinces of the Canadian Confederation of 1867, it maintained full control over the Crown mineral lands within its borders under the British North American Act. Under its first mining act of 1885 exceptions to the Crown minerals were stated to be for limestone, gypsum and building materials which belonged to the surface owners unless decreed otherwise by order in Council. Again, that power of decree excepted gypsum which, by regulation of the public interest, could never be declared a mineral. Under the Lands and Forest Act of 1954 exceptions to reserved Crown minerals continued to be for limestone, gypsum and building materials which were the property of the surface owner unless decreed otherwise by order in Council. The Mines Act 1954 apparently closed "loopholes" by stating that this Act applied to all minerals including limestone and building materials that have been declared to be minerals by order in Council.

Hodgson's Mineral Report 13 of 1966 declared that there are no fee simple mineral rights held in Nova Scotia. Also, under the province's Water Authority Act 1964, all waters are the property of the Crown. A permit is required by a mining operator for use of water, and discharged mine and mill processing waters are monitored and controlled by the Water Authority (Hodgson, 1966, pp. 144–145, 152). However, the largest percentage, 68.5%, of surface rights of provincial lands is owned privately, whilst the Provincial Crown owns 28.7% and federal Crown 2.9%.

By the Mineral Resources Act 1975, the rights to all minerals, defined by the Act as any "natural, solid, inorganic substance or fossilised organic substance, except 'ordinary' stone, building or construction stone, sand, gravel, peat, limestone, gypsum, oil and natural gas, belong to the Crown. Further exceptions are those excepted surficial materials that were vested in the owners of the land prior to 1910 under previous surface right grants.

The principal Nova Scotian acts applying to minerals are the Mineral Resources Act and the Metalliferrous and Quarries Regulation Act 1967, as amended. Under the former, the Minister of Mines has authority to require operators to eliminate or reduce practises which are injurious to the environment and to reclaim their lands. As with the other provinces, a Municipal Act 1967 and Planning act 1969 empower local governments with zoning controls for surface mining.

2.4.4.9 *Prince Edward Island*

Prince Edward Island joined the Canada Confederacy in 1873, but retained full control over its Crown mineral lands and minerals on its entry. All minerals were reserved to the Crown. Under its Oil, Natural Gas and Minerals Act of 1957, minerals are defined as "all naturally occurring minerals or any combination of them with themselves or with any other element including oil and gas, coal, salt, sulphur and potash". The mineral regulations of the province were almost totally directed to the production of oil and gas (Hodgson, 1966, pp. 154–155). This is attributed to the scarcity of mineral deposits on the island.

Although 100% of the minerals are owned by the Crown, 94% of the surface land area is privately owned, with the balance being in the Provincial Crown. Like New Brunswick and Nova Scotia, Prince Edward Island does not consider surface materials to be minerals. Therefore the right to extract and use sand and gravel is vested in the owner of the surface. The Environmental Protection Act requires a permit for extracting sand, gravel and shale, and rehabilitation of closed pits.

2.4.4.10 *Newfoundland*

Newfoundland was the last to join Canada as a province, in 1949, having been a royal British Colony until then. Before provincial status, its first mining law of 1860 empowered the Governor General to issue mining leases for terms of 99 years with options for grants of fee simple. The Crown Lands Act 1930 offered the same mining leases terms convertible to fee simple grants. In 1966, 2.7% of the land area of the province was held in fee simple mineral rights. In 1951, the Mines and Quarries Act provided for a leasing system of mineral rights, and did not include surface rights. Under the Crown Lands (Mines and Quarries) Act 1961, minerals were defined as "any naturally occurring inorganic substance, not including quarry materials, coals, oil, natural gas or salt. Quarry materials are limestone, granite, slate, marble, gypsum, peat, marl, clay, sand, gravel, any building stone and volcanic ash". (Hodgson, 1966, pp. 160, 162).

Although the provincial Crown owns title to 95% of the land, various subdivisions of provincial government made numerous land grant and development concessions to pulp and paper, construction, railway, mining and development companies prior to joining the Confederation in 1949. "Consequently, the land surface of the island is now subdivided into an interminable number of 'lots' of various sizes and shapes. These lots are classified as being 'fee simple', 'fee simple mining grants', grounds under development licence or mining leases and concessions". (Blakeman, 1976, p. 61).

The Crown Lands (Mines and Quarries) Act, 1961, above, was replaced in 1970 by a same titled act, and again replaced in 1977 by the Minerals Act and the Quarry Materials Act. The latter act regulates the extraction of quarriable construction materials, requiring permits for mining on government lands. Operations on private lands at that time did not require permits for mining. Permits for mining leases are available for 20-years, renewable for two more 20-year periods. Environmental controls for mining are exercised under the Act. Again, local and municipal controls for surface aggregate mines are found in the Urban and Rural Planning Act, 1970.

2.4.4.11 *Canadian conservation movement and increased regulation*

By the late 1960s and early 1970s confrontations between citizens groups, government agencies, and members of the mining industry were becoming commonplace. This was particularly so in the more densely populated areas where mining's visual intrusion into the landscape was much more apparent than in remote and rural areas of Canada. "In Southern Ontario, particularly, the conflicts centred on issues involving 'rape of the landscape' and degradation of the visual environment, disturbance of water tables, destruction of agricultural lands, damage to recreational lands / parks, in addition to noise, dust, truck traffic and public nuisance, all of which are normally associated with rock quarry and gravel pit operations. *** Due to an apparent lack of environmental concern exhibited by many operators, it is easy to understand why the industry acquired a poor public image, and why the sentiment is often expressed — "Yes, we realise that mineral aggregates are vital to the economy, but the pits and quarries don't have to be here"! (the Canadian NIMBY syndrome).

"Consequently, measures were taken to protect the land, water, and air environments through provincial statutes and regulations, *** and through municipal / regional zoning by-laws which, in many instances, essentially prohibited sand, gravel and rock aggregate operations within local boundaries". (Blakeman 1977, p. 1, 2, 66.).

2.5 Conclusions and Comments

The principles of state-owned minerals and privately-owned minerals in the Anglo nations are based on Roman law. Other mineral rights that developed from Roman law were: the right of the landowner to possess or work the minerals in his land; royal minerals, that is auriferous and agentiferous, or gold- and silver-bearing, belonged to the State; state-owned minerals on state-owned lands could be worked by publicans on a royalty-basis reserved to the State; the ownership of "royal" minerals on all lands has been preserved by the U.K. and Canada, whilst the U.S. has not retained its title to them on privately-owned lands; title and the right to mine all baser metals and non-metallic minerals on private lands belongs to the surface owner, except where severed from the surface ownership.

U.K. mining law, regulating mining operations, developed fully about 1875, 75 to 85 years before the U.S. reached the same degree of regulation. Many of the modern U.S. protective regulations governing mines, mining health and safety had long before been incorporated into those of the U.K. Similarly, the

U.K. led the U.S. in developing laws governing zoning and land use for mining operations.

In the younger U.S., a government policy, more or less, of *laissez-faire* has been in place for minerals to encourage free-enterprise development of mineral lands, mining, and the industry in order to bring about settlement of the vast new country. In slightly over a century, since the General Mining Law of 1872 regulating minerals on public lands, the nation has been slowly moving towards a period of retention of resources and management thereof. Since 1992, a battle has been taking place between the environmentalist and mineral industry proponents to strongly revise, if not scrap the 1872 Mining Law, or to keep it intact, and which has embroiled the U.S. Congress.

In view of the stastistical fact that during the entire history of mining on public lands in the U.S., less than one-quarter of 1% of the available public lands have been touched by mining, a reasonable question seeks a reasonable answer for all the environmental furor clamouring for revision of the 1872 General Mining Law.

In Canada, also a younger Anglo-nation, the historical development of its mineral lands grew at a less frenzied pace than in its neighbouring cousin's land. The retention of large areas of Crown lands were reserved upon which minerals might be located. Further, title to the minerals located on Crown lands were often retained and the right of free entry (free mining) was limited. Since the formation of the Canadian Confederation, control over mineral lands and mining has been the bailiwick of the provinces, with little control by the federal government except in the Northwest and Yukon Territories. Present day claim staking usually leads to a lease from the Crown rather than a fee title or patent.

Dissimilar to the U.S. has been Canada's governmental participation in the mining of its minerals. Participation in mining by both the federal and provincial governments has been in several forms, e.g., direct equity investments, loans or loan guarantees, or provision of infrastructure as roads, railways or electricity. The federal and provincial governments have the legal authority to undertake direct investment in the mining industry through the exercise of their general spending powers. Special government-owned corporations as mining investment vehicles may be established through legislation at either level of government. A more recent example was the provincial Saskatchewan government playing a direct part in the development of potash deposits through the formation of a Crown corporation, Potash Corporation of Saskatchewan (Am. Law of Mining, 1986, 210.04, and fn. 6). Direct federal involvement has also been justified in the mining development of uranium under the Atomic Energy Control Act, R.S.C. 1970, C.A.-19.

By the 1960s, public concern had mounted and materialised in the form of environmental activism before the legislative tools had been provided by government to respond to public opinion and pressures. Within a very few years, the Canadian provincial and federal governments were responding with several and various environmental acts, many of which affected surface mining. In the following chapter, the early-history claims of pollution and damages caused by open-pit mining operations are reviewed.

CHAPTER 3

A BRIEF HISTORY OF ENVIRONMENTAL DAMAGE AND LITIGATED POLLUTION CLAIMS FROM NON-FUEL SURFACE MINING WITH EMPHASIS ON WATER RESOURCES

3.1 Introduction

A German scientist, Georg Bauer, reviewed the effects of surface mining on the environment and the charges made by environmentalists opposed to surface mining's effects, and wrote:

> "*** The strongest arguments of the detractors is that the fields are devastated by mining operations. Also, they argue that the woods and groves are cut down, but there is need for an endless amount of wood for timbers, machines and the smelting of metals. And when the woods and groves are felled, then are exterminated the beasts and birds, very many of which furnish an agreeable and pleasant food for man. Further, when the ores are washed, the water which has been used poisons the brooks and streams, and either destroys the fish or drives them away. Therefore, the inhabitants of these regions, on account of the devastation of their fields, woods and groves, brooks, and rivers find great difficulty in procuring the necessaries of life, and by reason of the destruction of the timber they are forced to greater expense in erecting buildings. Thus it is said, it is clear to all that there is greater detriment from mining than the value of the metals which the mining produces".

Though sounding in modern day environmental complaint, this quotation was written approximately in the year 1550 by Georg Bauer, better known by his Latinised name as Georgius Agricola, in his treatise De Re Metallica, translated from Latin to English by Herbert Hoover (1950, p. 8).

Agricola, writing to defend the cause of mining, stated, "Without doubt, none of the arts is older than agriculture, but that of the metals (mining) is not less ancient, in fact they are at least equal and coeval for no mortal man ever tilled a field without implements … when an art is so poor that it lacks metals, it is not of much importance, for nothing is made without tools". He considered that mining should not be minimised, or "neglected", since it was in his opinion, "one of the most ancient, the most necessary, and the most profitable to mankind …".

In his defense of the "detractors" charges of devastation by mining to the environment, Agricola wrote:

> "If we remove metals from the service of man, all methods of protecting and sustaining health and more carefully preserving the course of life are done away with. If there were no metals, men would pass a horrible and wretched existence in the midst of wild beasts; they would return to the acorns and fruits and berries of the forest. They would feed upon the herbs and roots which they plucked up with their nails. They would dig out caves in which to lie down at night, and by day rove in the woods and plains at random like beasts, and inasmuch as this condition is utterly unworthy of humanity, with its splendid and glorious endowment, will anyone be so foolish or obstinate as not to allow that metals are necessary for food and clothing and that they tend to preserve life. (Agricola, 1556, (Hoover) p. 14).

Thus, nearly five hundred years ago, the same contentions, arguments and concerns between surface mining and environmentalist existed.

3.2 A Review of Historical Mining Pollution Claims

Historically, prior to the new era of intensive environmental regulation for surface mining in the Anglo nations, surface mining was lightly regulated in England, but virtually unregulated in the U.S. and Canada. Where existing statutes fell short in providing penalties for private or public "environmental" damages caused by mining operations, the English common law system filled the statutory shortfall. Common law liability for damages to the environment are strictly of local and personal concern, i.e., for damaging the environment of one's neighbour, or neighbours, or the neighbourhood.

Environmental regulatory violations and damage claims, allegedly resulting from surface mines of the present, are essentially pollution and contamination of air, water, public health and property caused by various mining and mineral processing activities. In reality, they are the same type of claims made before the advent of the environmental regulatory revolution. In the past and recent history prior to the 1960s, private grievances and relief from such environmental damage claims were left to individuals to pursue through the common law and courts of the Anglo countries. "Environmental" grievances and relief for the general public were made as claims of a "public nuisance" and pursued in law by the local government.

Grounds for grievances and complaints for "environmental" damages from mining pollution and contamination were found in the legal tort doctrines of nuisance, both private and public, trespass, non-trespassory invasion of property, negligence, and employing the equitable doctrine of injunctive relief. Except for claims brought by government entities in the name of the public, the burden of seeking damages and relief for private claims fell upon the individual. Reclamation of private surface mined land was a matter of contract law between the landowner and the miner. No concern was given to reclamation on public lands.

3.3 Claim Definitions

The legal doctrines under which early "environmental" damage claims were made in the Anglo countries with their common English law are defined as follows:

3.3.1 *Nuisance*

Generally defined in law, a nuisance is an unreasonable interference with the use and enjoyment of another's property. Illustrative cases defined a nuisance as: "An act by one which annoys and disturbs another in possession of his property, rendering its ordinary use or occupation physically uncomfortable to him "[*Yaffe v. City of Ft. Smith*, 10 S.W. 2d 886, 890 (1928)]; "annoyance; anything which essentially interferes with enjoyment of life or property "[*Holton v. Northwestern Oil Co.*, 161 S.E. 391 (N.C. 1931)]; "that class of wrongs that arise from the unreasonable or unlawful use by a person of his own property and conduct, working an obstruction or injury to the right of another, or of the public, and producing material annoyance, inconvenience, discomfort, or harm that the law will presume resulting damage" [*City of Phoenix v. Johnson*, 75 P.2d 30 (Ariz. 1938)] (Black, 1968, p. 1214).

Nuisances are classified as public, private and mixed. A public nuisance is one which affects an indefinite number of persons, or all the residents of a particular locality, or all people coming within the extent of its range or operation, although the extent of injury may be unequal [*Burnham v. Hotchkiss*, 14 Conn. 317 (1841)]. A private nuisance, as distinguished from public nuisance, includes any wrongful act which destroys or deteriorates the property of an individual, or few persons, or interferes with their lawful use and enjoyment thereof, *** or causes them a special injury different from that sustained by the general public [*Baltzeger v. Carolina Midland R. Co.*, 32 S.E. 358 (S.C. 1894)] (Black, 1968, p.1215). A mixed nuisance would, obviously, be one that affects both public and private rights.

3.3.2 *Trespass*

The doing of an unlawful act, or of a lawful act in an unlawful manner to the injury of another's person or property [*Waco Cotton Oil Mill of Waco v. Walker*, 103 S.W. 2d 1071, 1072 (1937)]; an unlawful act committed with violence, actual or implied, causing injury to the person, property, or relative rights of another; an injury or misfeasance to the person, property, or rights of another, done with force and violence, either actual or implied in law [*Southern Ry. Co. v. Harden*, 28 S.E. 847 (Ga. 1897)]. In its more limited and ordinary sense, it signifies an injury committed with violence, and this violence may be either actual or implied; and the law will imply violence even though none is actually used, when the injury is of a direct and immediate kind, and committed on the person or tangible and corporeal property of the plaintiff. *** of implied, a peaceable but wrongful entry upon a person's land.

Trespasses are often described as *continuing* or *permanent*. A permanent trespass is one which is in its nature a permanent invasion of the rights of another; as, where a person builds on his own land so that a part of the building overhangs his neighbour's land [*H. Hitt Lbr. Co. v. Cullman Property Co.*, 66 So. 720, 721(Ala. 1914)]. A continuing trespass is one which consists of a series of acts, done on successive days, which are of the same nature, and are renewed or continued from day to day, so that, in the aggregate, they make up one indivisible wrong (3 Blackstone's Commentaries 212, Black, 1968, p. 1674).

The term *non-trespassory invasion* is frequently applied where there has been an encroachment by gases, odours, dust, objects (as thrown rock), airborne particles, etc., which have emanated from the acts of the wrongdoer on his property and invaded the land, property or rights of another, but have involved no physical entry by the body of the wrongdoer himself.

To establish a *prima facie* claim for trespass there must be an act of physical invasion of another's property. Intent to trespass on another's land in not required, but intent to do the act that constitutes or causes the trespass is sufficient. Gross recklessness resulting in the unlawful invasion of another's property may be sufficient, even in the absence of intent. It is not necessary that the wrongdoer come on to another's land; trespasses will exist where the "trespasser" throws rocks onto the land, or floods it.

3.3.3 *Nuisance-Trespass Detail and Differences*

As may be inferred, the difference by definition between a trespass and a nuisance may be a grey area at times. A trespass is an interference with a landowner's right to exclusive possession, which includes the quiet enjoyment of his land. Still, a nuisance is also an unreasonable interference with the landowner's use and enjoyment of his land, but lacks the challenge to exclusive possession.

It is noted in the Restatement, Second of Torts, Section 826(a)(1977) that in the case of a socially useful activity, such as a factory or commercial establishment, the defendant's conduct must be "unreasonable" and cause the plaintiff "substantial" harm. The determination of reasonableness is a balancing process for the court considering the gravity of the harm and the utility of the defendant's conduct.

To consider the gravity of the harm of a nuisance, American courts consider various factors that are summarised by the Restatement (Second) Section 827 as follows:

1. the extent of the harm involved;
2. the character of the harm involved;
3. the social value which the law attaches to the type of enjoyment invaded;
4. the suitability of the particular use or enjoyment invaded to the character of the locality; and,
5. the burden on the person harmed of avoiding the harm.

The factors used to determine the utility of the defendant's conduct in an alleged nuisance are stated in Section 828 of the Restatement:

1. the social value which the law attaches to the primary purpose of the conduct;
2. the suitability of the conduct to the character of the locality; and
3. the impracticability of preventing or avoiding the invasion.

At common law, the rule was *sic utere tuo ut alienum non laedus*, or, "use your own property so as not to injure that of another", and when violated, such harm was actionable. Courts retained some discretion to determine what type of injury was actionable [*William Aldred's Case*, 9 Coke 57B, 77 Eng. Rep. 816 (K.B. 1611)]. It should be noted that "*** the common law differed from later American practice in providing a cause of action for nuisance without a showing of negligent or intentional damage". (Schoenbaum, 1985, pg. 36).

3.3.4 *Negligence*

The law of negligence is founded on reasonable conduct of reasonable care under all circumstances of a particular case [*Charbonneau v. MacRury*, 153 A. 457, 462 (N.H. 1931)]; the doctrine of negligence rests on duty of every person to exercise due care in his conduct toward others from which injury may result [*Johnson v. Grand Trunk Western R. Co.*, 224 N.W. 448, 449 (Mich. 1929)]; it is the omission to do something which a reasonable man, guided by those ordinary considerations which ordinarily regulate human affairs, would do, or the doing of something which a reasonable and prudent man would not do [*Schneider v. C.H. Little Co.*, 151 N.W. 587, 588 (Mich. 1915)].

The classification of "negligence" as "gross", "ordinary", and "slight" indicates only that under special circumstances great care, ordinary care, or slight care are required, but failure to exercise care demanded is "negligence" (39 Del. L Laws, c.26) (Black, 1968, p. 1184–1185).

3.3.5 *Injunction*

A prohibitive writ (order) issued by a court of equity, at the suit of a party complainant, directed to a party defendant in the action, forbidding the latter, or his servant or agent, to do some act which is threatened, or the continuance of an act already performed, such act being unjust, inequitable or injurious to the plaintiff, and cannot be adequately redressed by an action at law. (*City of Alma v. Loehr,* 22 P.2d 424); a judicial process operating *in personam,* and requiring a person to whom it is directed to do or refrain from doing a particular thing [*Gainsburg v. Dodge*, 101 S.W. 2d 178, 180 (Ark.1937)] (Black, 1968, p. 923).

Injunctions may be of varying durational time limits for enforcement. They may be temporary, preliminary, provisional, interlocutory, or permanent.

3.4. A Review of Litigated Claims Against Mining Damages

A review and inspection of some of the historical types of pollution claims that were made against surface mining operations and the manner in which they were treated and resolved prior to the advent of the great environmental legislation revolution reveals that there are not any new environmental wrongs in the present age; just more of them. Present day mine regulations attempt to eliminate them at their source, i.e., "by treating the disease, not just the symptoms".

3.4.1 *United Kingdom*

As will be shown in the following historical litigated claims, lawful acceptance and judicial tolerance for alleged damages from mining operations was considerably greater a century or two in the past. Mining operations were allowed much greater leeway in creating dust and gases that fouled the air and fell on surrounding nearby vegetation and residences, releasing pit and mine waters and assorted contaminants into and fouling public waterways, and in leaving the surface of abandoned or mined-out pits disturbed with accompanying spoil piles unvegetated, and the surface generally un-reclaimed or un-restored for another use. Such tolerance was generally accepted because mining was a basic and necessary industry providing essentials for the general public's welfare. Additionally, the country had a lower population density, and media coverage to dramatise and publicise nuisance grievances and events was limited in earlier days to a few newspapers. Consequently, organised activist and militant groups were lacking to provide present day "people power".

In fact, under common law an alleged act causing a private nuisance, mining or otherwise, if continued for 20 years without successful challenge, became established as a prescriptive right, therefore, legalised. Thus, any subsequent purchaser of property complaining of a mining nuisance, had "come to the nuisance" and would be required to "take subject to the nuisance" without complaint. *Wright v. Williams*, [1836] 1 M & W 77. And yet, contra, individual property rights were protected and preserved against a mining nuisance where a claim for physical damage was made which affected the enjoyment of property. In *St. Helen's Smelting Co. v. Tipping,* [1863–4], 35 L.J.Q.B. 66 (H.L.), a nuisance was found where the trees and shrubs of an adjoining property to the smelter were withered by the vapours from the smelting works. That court found that there was a distinction to be drawn between nuisances producing personal discomfort and nuisances producing physical injury to property. It was also noted that it mattered not if the complainant came to the nuisance where there was injury to property.

In the 1851 case of *Walter v. Selfe*, [1851] 4 DeG. & S. 315, where a brick kiln had been erected within 50 yards of a residence, and smoke, vapours and floating substances (dust) became mixed with the air of the house and pleasure grounds, Viscount Justice K. Bruce, said "the plaintiff, although not entitled to air as fresh, free, and pure as at the time of building his house, was yet entitled to have the air kept in such a condition as was compatible with physical comfort; and the nuisance not being fanciful, but very real and sensible, and materially interfering with the ordinary comfort of existence, he be granted the injunction". *Walter v. Selfe* established that where a nuisance is proved, an injunction generally will be granted unless damages are shown to be an adequate remedy (also, see *Luscombe v. Steer,* [1867] 15 W.R. 1193) (Bainbridge, 1900, pp. 433–434).

Similar nuisance principles were found in the cases of: iron works (*Shotts Iron Co. v. Inglis*, [1882] 7 A.C. 518) where the injunction set a distance limitation of one mile from a habitation; copper works (*Tipping v. St. Helen's Smelting Co.*, [1865] 1 Ch. 66); cement works (*Umfreville v. Johnson*, [1875] 10 Ch.580); chalk works or lime works (*Walter v. Selfe*, supra). (Mac Swinney, 1897, p. 484).

It is noted that in Mac Swinney's treatise on the Law of Mines, Quarries and Minerals of 1897 for the U.K:

> "But, as in the case of brick-burning, an actual substantial nuisance must be established. Sentimental grievances, or prospective, contingent, or remote, **cases of nuisance will not be allowed to interfere with the great industries of the country**, especially where the persons who complain reside in the seats of those industries. (*Shotts Iron Co. v. Inglis,* supra). (Emphasis added)
>
> "The provisions of the Public Health Act, 1875, do not extend to mines, **so as to interfere with or obstruct the efficient working of the same; or to the smelting of ores and minerals; or to the calcining, puddling and rolling of iron and other metals; or to the conversion of pig iron into wrought iron, so as to obstruct or interfere with any of such processes**. (emphasis added) [*Re. Dudley,* (1881) 8 Q.B.D. 86) On the face of it, this exemption applies only where there is an interference with, or obstruction of, the working or the processes. (*Patterson v. Chamber Coll. Co.* [1892] 8 T.L.R. 278). And it extends only to liability under the provisions of the Act, and is not a protection against the common law liability. (*A.-G. v. Logan*, [1891] 2 Q.B. 100). A decision under the repealed Act. (Mac Swinney, 1897, pp. 482–485)". (Emphasis added)

Early claims for pollution of water were termed as the "fouling of water". It was established under riparian rights that an owner may use the water from his stream in a reasonable degree, or divert it in reasonable quantities (*Weeks v. Heward*, [1862] 10 W.R. 557). A mine operator might use the stream water for the purpose of working his machinery, as a means of transporting his minerals, or other uses in his mining process in a reasonable degree; and by washing his minerals or by pumping water from his mines or pits into the stream which might alter its quality to a reasonable degree. However, the miner was neither allowed to substantially diminish the quantity, nor to materially alter the quality. "And he must not impregnate it with poison or foul matter". (*Hodgkinson v. Ennor*, [1863] 4 B & S.229).

A mining operation could acquire, by grant or by prescription (over 20 continuous years), the right to foul a stream (*Carlyon v. Lovering* [1857] 1 H.&N. 784). The right of prescription could be obtained at common law or under the Prescription Act. Such was the case in *Carlyon*, supra, where both parties were riparian owners on the same stream. Lovering's tin mine was located upstream from Carlyon. In the washing process, the tin miner introduced a quantity of sand, stone and trash flowing down stream contaminating, or fouling Carlyon's flowing waters. Carlyon brought a nuisance action against Lovering. The tin miner plead a right to do so under the Prescription Act, which was upheld by the court. Another of Lovering's successful pleas was **his right as a custom immemorial and of the trade** of the stanneries of Cornwall to do the act complained of. **The court found the custom to not be unreasonable as it was necessary to the working of the mines.** (emphasis added).

In a similar water pollution action, *Wright v. Williams,* [1836] 1 M. & W. 77, Wright complained of the fouling of his watercourse by Williams' operation of a copper pit upstream. The copper miner pleaded his right to "sink pits on his own land; to fill such pits with iron; to cover the same with water pumped from the copper pit for the purpose of precipitating the copper contained in such water; and then, a right under the Prescription Act to let off the water impregnated with metallic substances into the watercourse". The copper miner's plea was upheld by the court (MacSwinney, 1897, pp. 461–462).

It is noted that in 1876 the U.K. passed the Rivers Pollution Prevention Act (RPPA). This Act provided:

> "Every person, who puts the solid refuse of any quarry into a stream, so as to interfere with its due flow, or to pollute its waters, commits an offence against the Riv. Poll. Prev. Act, 1876. Every person, who puts any solid matter from any mine into any stream, in such quantities as to prejudicially interfere with its due flow; or any poisonous, noxious or polluting solid or liquid matter proceding from any mine, other than water in the same condition as that in which it has been drained or raised from such mine; commits an offence against the Act; unless, in the case of poisonous, noxious or polluting matter, he shows to the Court, that he is **using the best means** to render it harmless". (emphasis added)

It should also be noted in the emphasis above, that even in 1876, the forerunner of our terms "best practical means" and "best practical available technology" were used in the Act.

Proceedings for court action under the RPPA could not be taken against any person who put refuse from a mine into a stream except by a sanitary authority, nor without the Local Government Board. However, if the sanitary authority refused to take legal action, an individual seeking damages might apply for an action through the Local Government Board (Mac Swinney, 1897, pp. 461–462).

A point of conflicting miner's water-use rights of the period, between the British position and an American court's position, was commented on in the case of *John Young & Co. v. Bankier Distillery Co.*, (Lord Watson) [1893] A.C. 691, 439 A.E.R. (1891–94), decided by the House of Lords. The Scotland-origin case upheld the British position as decided in *Hodgkinson v. Ennor*, [1863], supra, which held the miner was neither allowed to substantially diminish the quantity, nor to materially alter the quality; "And, he must not impregnate it with poison or foul matter".

Young was a coal miner operating a pit upstream from the Bankier Distillery which used water from the same stream. Young pumped water from the "lower strata" of the pit into the nearby stream. The water so pumped in, "although pure, was hard ('acidulated') and although it did not effect the water for ordinary purposes, it rendered it much less suitable for distilling purposes". Lord Shand stated, "A lower proprietor must submit to the flow of water which comes down upon his lands by natural force, whether flowing in a defined channel or not, or above or below the surface; but he is not bound to receive water pumped from below the surface by artificial means, which would never have reached his land by the ordinary force of gravitation".

In the course of pleading Young's defence in the lower courts, the miner's counsel introduced an American case, *Pennsylvania Coal Co. v. Sanderson,* 56 Am. Rep.89 (1866), as legal authority to justify its pumping of mine waters into the stream. *Pennsylvania Coal* was decided in the Supreme Court of Pennsylvania in a 4–3 decision. It essentially held that the owners of a (coal) mine were entitled to pump up water from the lower strata of the mine, to send it into an adjoining stream with its quality so affected as to render it totally unfit for domestic use by the lower riparian users. The Pennsylvania decision reflected the enormous value placed on mining by that State at the time. Lord Shand took note of the American court's decision which stated, "The use and employment of a stream of pure water for domestic purposes by the lower riparian owners, who were settled there before the opening of the mine ... must *ex necessitate* give way to the interests of the community, in order to permit the development of the natural resources of the country, and to make possible the prosecution of the lawful business of mining coal".

Lord Shand further commented that the American case had no application to the case at bar; that the Pennsylvania court appeared to him "to be making law rather than interpreting the law so as to give effect to sound, just and well-recognised principles as to common interest and rights of upper and lower proprietors in the running water of a stream".; and the Pennsylvania decision "affords no good legal ground for allowing the proprietor of a mine to work his minerals for his own profit as to destroy or greatly injure his neighbour's estate, by subjecting it ... to the burden of receiving water destroyed in quality, without payment of compensation or damages for the injury done".

Thus, *tunc pro nunc*, (then for now) the seeds of pollution claims resulting from surface mining operations were sewn in the English common law.

3.4.2 *United States*

Until 1936 in the decision of *National Labor Relations Board (NLRB) v. Jones & Laughlin Steel Corp.*, 301 U.S.1 (1936), the U.S. Supreme Court changed its former position where it had long held by the U.S. federal government that mining in itself was a local enterprise which Congress had no power to regulate. Consequently, "environmental", or pollution and physical damage claims were pursued in state court systems, whether private or public, except in claims involving mining activities on federal lands, federally controlled "navigable waters", or where there was a question of federal law involved.

(i) **Nuisances:** Mining had been viewed as a lawful and necessary business, as well as a reasonable use of property, and surface mining was not a *nuisance per se*, so long as the mining activity was conducted in the usual manner with customary precautions recognised by the trade. In *McCaslin v. Monterey Park*, 329 P.2d 522 (Cal.App., 2 Dist., Div.3, 1958), the California court stated the business of excavating rock and gravel is a lawful and useful operation and not a nuisance *per se.*

So long as the business was conducted in the ordinary way and with the usual and customary precautions, the operator was not accountable for incidental annoyances to others that necessarily follow the mining operations. However, where excessive smoke, fumes, or dust from the mineral operations resulted in damage to crops and vegetation, or to the discomfort of neighbouring residents, there were grounds for finding the operation a nuisance (*Shannon v. Missouri Valley Limestone Co.*, infra; *Brede v. Minnesota Crushed Stone Co.*, 143 Minn. 374 (1919). Excessive noise and vibrations from surface mine blasting and the operation of mining equipment might constitute a claim for a nuisance [*Ledbetter Bros., Inc. v. Holcomb*, 108 Ga. App. 282 (1963); *Hakkila v. Old Colony Broken Stone & Concrete*, infra; *Blackford v. Herman Construction Co.*, infra; *Lademan v. Lamb Constr. Co.*, infra;] (Am Jur 2d, 1971, Vol. 54, p. 382).

In *Hartung v. Milwaukee County*, 86 N.W.2d 475, rehearing 87 N.W.2d 799 (Wis.1958), the Wisconsin court held that quarry operations, including blasting, grinding and crushing stone, which necessarily produced stone, did not constitute a private or public nuisance where there was no showing that anyone had moved from the area because of the quarry operations (Am Jur 2d, 1971, p. 382)

Claims of "nuisance" against surface mining operations were generally for an alleged "environmental" violation of noise, air or water pollution, and occasionally for causing excessive, noisy, and unsafe traffic conditions. At times, the surface mining operation had started and been long in operation before the locale surrounding it became densely settled and prior to local zoning ordinances. After zoning regulations were in place, the zoning boards required the mining operator to obtain a non-conforming use permit to continue its operation in a non-industrially zoned area. In *Hakkila v. Old Colony Broken Stone & Concrete Co.,* 162 N.E. 895 (Mass.1928), the Massachusetts court said, "The fact that the owner of a quarry has a permit issued by the proper authority to use explosives in the operation of his quarry does not provide him with a defense against an action for maintaining a nuisance, where the blasting operations threw stones onto neighbouring premises". [Also, in *Barnes v. Graham Virginia Quarries, Inc.*, 204 Va. 414 (1963)] (Am Jur 2d, 1971, p. 382).

In earlier years, some courts in the U.S. had supported a defensive doctrine known "as coming to the nuisance" which had been long and successfully employed as a defence by surface miners against alleged claims of the open-pit mine or quarry as being a nuisance. Simply stated, where a surface mine, or other industrial operation, or foul-smelling agricultural or stockyard location, had been in place prior to the complainant's coming to an adjoining area, the alleged nuisance claims were made invalid because the complainant had "come to the nuisance" knowing full well, or should have known (expected), that such conditions complained of existed before his arrival. A related doctrine, *caveat emptor*, or buyer beware, supported the "coming to the nuisance" doctrine. Those who bought properties in the area of an existing quarry, or some objectionable industry, and then complained of its operation were subject to both legal doctrines. Their claims were given little credence under those two legal doctrines.

In the case of *Barrett v. Vreeland*, 182 S.W. 605 (Ky.App.1937), where eight Kentucky residents across a river from a limestone quarry were unsuccessful in their claim of a private and public nuisance in the manner of the operation of the quarry. The quarry owners claimed their operations started in 1906, predating most of the claimants' nearby residency, and on a site where stone had been quarried since 1852. The complainants were accused of moving to the nuisance.

The property owners claimed the physical structures of their homes were injured by the shock from excessively heavy blasting, which threw stones upon the properties, filling the air with dust, and were disturbed at night by the noise of rock crushing machinery which at times was so great as to prevent ordinary conversation in their homes and prevent sleep at night. They sought an injunction against further operation of the quarry. The injunctive relief was denied on the basis that there was insufficient evidence to show injuries.

Contra to the doctrine, in 1908, the St. Louis (Missouri) Court of Appeals in *Blackford v. Herman Construction Co.*, 112 S.W. 287 (Mo.App.1908), found that the use "of explosives in a stone quarry contiguous to another's property in a large city is unreasonable, and will authorise either injunctive relief as against a nuisance, or an action for damages, though the business is entirely lawful and prosecuted with the utmost care". And, in 1927, the same court enjoined the West St. Louis Quarry Company from further operation in a dense residential area, finding it a nuisance by its blasting to cause "throw" rock to fall and damage nearby buildings, to cause cracks in walls, and causing dust, and continuous noise, all of which rendered their lives and the enjoyment of the property uncomfortable [*Lademan v. Lamb Constr. Co.*, 297 S.W. 184 (Mo.App. 1927)]. In *Fagan v. Silver,* 188 P. 900 (Mont. 1932), the Supreme Court of Montana upheld an injunctive order for a stone quarry to cease its operation in a residential area of Butte as it was found to be a nuisance to the residents.

In Kentucky, contra to *Barrett v. Vreeland*, supra, its Court of Appeals in 1937 found in *Rogers v. Gibson,* 101 S.W.2d 200 (Ky.App. 1937), that Rogers' operation (d/b/a Louisville Crushed Stone Company) of a stone quarry was a nuisance whether the acts complained were due to negligence, or not; and that a nuisance may exist with, or without, negligence. Although the quarry was outside the city limits and the neighbourhood not thickly settled, the quarry operation was enjoined from further blasting which interfered with neighbours' use and quiet enjoyment of their buildings and property.

It becomes apparent that surface mining operations were beginning to lose ground against nuisance complaints of their operations. In 1941, the New Jersey case of *Benton v. Kernan,* 21 A.2d 755 (N.J.Ct.App.1941), a compromising decision was reached by the court as indicated by its *dicta* and results. The court's syllabus stated:

> "1. Held, under the circumstances of this case, complainants are entitled to an injunction against quarry blasting that causes physical damage to their properties and that causes stones to be thrown on their premises, but not against blasting that merely jars, vibrates or shakes their buildings without physical damage thereto.
>
> 2. To justify the enjoining of a perfectly legal business on account of the noise made in conducting it, the evidence should be clear and convincing.
>
> 3. Noise sufficient to enjoin the operation of a business must be such as to affect injuriously the health or comfort of ordinary people in the vicinity of an unreasonable extent, and must pass the limits of reasonable adjustment to the conditions of the locality and of the needs of the maker to the needs of the listener". (op. cit., p. 756)

The quarry operation in *Benton* had been established 33 years prior to the complaints. Though the doctrine of "coming to the nuisance" was not put forth as a prime defence, the court gave consideration to the doctrine in its dicta:

> "While we do not hold that the fact that the quarry was in existence long before complainants moved into the locality is conclusive, it is an element to be considered in determining the reasonableness of the disturbance to them. Most of the complainants moved to the neighbourhood within four or five years prior to the filing of the bill, and the quarry had been in continuous operation for over 30 years. It was less active during the years of the depression than it has been lately, although its most active years were from 1925 to 1929. At any rate, persons moving into the vicinity of a quarry in operation had less reason to expect perfect quiet than persons in the country or a residential area remote from industrial activity would naturally expect". (ibid)

Additionally, the complainants in *Benton* sought an injunction against the noise created by the operation of trucks going to and coming from the quarry on the public streets in front of their homes. The court declined to restrain the operation of trucks.

The *Benton* court stated that "the evidence presented failed to establish that health or comfort of residents in the vicinity of the quarry were injuriously affected to such an unreasonable degree by noise incident to operation of quarry machinery as to warrant an injunction, particularly where the quarry, which was operated only from 8:00 a.m. to 4:30 p.m., had been in operation for many years before".

Although traffic, road haulage to and from the quarry, and dust therefrom was a minor issue in *Benton*, supra, it became the major issue in the 1963 Iowa case of *Shannon v. Missouri Valley Limestone Co.*, 122 N.W. 2d 278 (Iowa 1963). Neighbouring property owners sought to enjoin a nuisance arising out of the hauling of limestone from the quarry. The Court in deciding for the property owners, stated that a common law "nuisance" was created by dust raised by trucks hauling crushed rock from the quarry over a limestone and dirt surfaced three-mile road along which 40 homes were located where the dust was irritating to the skin, noise and throat, making the ordinary use of the homes and lawns impossible during dry weather, and was not merely a temporary situation. ... A limestone quarrying company was liable for and was properly required to abate the nuisance created by the trucks hauling from its quarry".

The anticipation of all the terrible injury that "will occur" when a surface mining operation is merely planned took form in the 1951 alleged nuisance action decided by the State of Washington's Supreme Court in *Turner et al v. City of Spokane*, 235 P.2d 300 (Wash.1951). "The evidence justified dismissal of action by neighbouring residents and property owners to enjoin a proposed operation of rock quarry and crushing plant and use of explosives by city on ground that danger of dust, noise and confusion, danger to wells and loss or pollution of water, and danger and annoyance to plaintiffs' comfort, health, repose and safety due to proposed operations were not of sufficient imminence at time of hearing to warrant granting injunctive relief. ... While a court of equity may enjoin a threatened or anticipated public or private nuisance, where it clearly appears that a nuisance will necessarily result from the contemplated act or thing which it is sought to enjoin, the court should not interfere where the injury apprehended is of a character to justify conflicting opinions as to whether it will, in fact, ever be realised".

(ii) **Trespass:** An action in trespass was the basis for a damaging dust claim from a nearby limestone quarry and cement manufacturing plant in the 1961

California case of *Roberts et al v. Permanente Corporation,* 10 Cal. Rptr. 519 (Cal.App.1961). The court stated that "an act which will, to a substantial certainty, result in entry of foreign matter on another's land is intentional trespass on which liability may be based. ... A statute prohibiting a private individual from enjoining the operation of manufacturing of cement permitted by a local zoning ordinance does not bar recovery of damages for trespassory invasions of another's property occasioned by conduct of such manufacturing". The lower court denied damages to the claimant because negligence was not proven. However, the appeals court reversed the decision adding that no negligence or intent was required to be proven against the trespasser.

(iii) **Negligence:** Owners or operators of quarries are liable to owners of other property when their performance becomes so careless and negligent as to cause injury to adjoining and proximate premises. However, if in their excavating for minerals, an injury occurred to the owner of adjoining land without fault or negligence on their part, some jurisdictions will find no liability. There has been a split among the states as to whether negligence was required for liability, or whether the liability was strict, i.e., without fault, negligence or intent. In *Cass Co. Contractors v. Colton*, 139 Colo.593, the court found that the defendant quarrier was guilty of negligence in causing damage by blasting, while in *Davis v. Palmetto Quarries Co.*, 212 S.C. 496 (1948), the court found no negligence but only strict liability for damages from quarry blasting.

Negligence was found by an Arkansas court in the operation of a rock quarry in *McGeorge v. Henry,* 101 S.W.2d 440 (Ark. 1937), where water wells on land near the quarry were drained five days after blasting rock in the quarry. The jury's damage awards of $750 to two plaintiffs were found to be excessive; that the well water loss was temporary; that new wells could be drilled deeper at a cost of $1 to $2.50 per foot to reach a deeper aquifer. However, in a 1966 decision favourable to a claim of loss of water in a well due to blasting, the Arkansas Supreme Court sustained a finding that negligence was not necessary for such liability for damages to the water well [*Western Geophysical Co. v. Mason,* 402 S.W.2d 657 (Ark. 1966)].

As illustrated by the two Arkansas cases, supra, interference with the flow of water in wells or streams, pollution and contamination of streams for downstream riparian owners, and flooding of adjoining properties by mining operators carry with it liabilities for damage.

A mine operator may have been liable to one whose well flow has been destroyed, or whose well has been contaminated by the proximate causes of mining. However, in *Bayer v. Nello Teer Co.*, 256 N.C. 509 (N.C.1962), a rock quarry, operating with the best practices of open-pit mining, which pumped no more percolating waters from its pit than necessary, was found not liable in damages to an adjoining landowner for contamination of his water supply from waters percolating into his well.

In *Gilmore v. Royal Salt Co.*, 84 Kan. 729 (1911), a salt mining company deposited a large quantity of refuse salt upon its land, and by action of rain upon it, the water underlying an adjacent tract was impregnated with saltwater through percolation, making it unfit for use and harmful for vegetation in irrigation, the mine operator was liable to the adjacent landowner. And, in *Sunray DX Oil Co. v. Thurman*, 384 S.W. 2d 482 (Ark. 1964), where a pit constructed to hold saltwater overflowed onto the land and killed vegetation and timber, the lessee was held liable for the damages from flooding.

Also, in *Freel v. Ozark Mahoning Co.*, 208 F.Supp. 93 (1962), an action for damages was brought against a fluorite mining company in Colorado for injury to private property and health resort from contamination and pollution of stream and flooding of such property by failure of the company to contain its mill tailings ponds. The court found that the plaintiffs were entitled to recover damages without proof that the conduct of the defendant mining company was negligent or intentional. Evidence demonstrated, however, that the defendant's conduct was negligent, willful, wanton and reckless.

The mining company had built several large tailing ponds containing harmful and noxious chemicals from the plant's flotation process, and later had failed to repair a smaller breach of the ponds containing walls. The downstream initial harm and injury to Freel was caused by leaking of the chemicals into the nearby stream. Greater injury was exacerbated on two occasions when the containing walls were breached and flooding from the mill ponds severely damaged the Freel's health resort buildings and polluted its mineral springs.

For an Arizona open-pit copper mining operation, the court found that the right under the Arizona statutes to use the water of the public streams for mining purposes did not give such user any right to send tailings and waste material from his reduction works down the stream to the destruction or substantial injury of the riparian rights of a user below for irrigation purposes [*Arizona Copper Co. v. Gillespie,* 230 U.S. 46 (1913)].

Similarly, in *Montgomery Limestone Co. v. Bearden*, 256 Ala. 269 (1951) a cause of action for a nuisance was made where the quarry operated a sump pump for the purpose of keeping the pit de-watered from flooding, run-in waters. The de-watering resulted in the deposit of debris in the river which then flowed through the complainant's premises in a polluted condition.

After World War II's great industrial expansion and with the advent of environmentalism accelerating in the 1960s, in addition to the historical type of damage claims from mining operations, as noise, vibrations and dust from blasting and mine machinery, new types of claims were entering the field of litigation, i.e., damages to public waters, stream pollution, fish-kills and wildlife. An example from 1963 was the case of *People (of California) v. New Penn Mines, Inc.*, 28 Cal. Rptr. 337 (Cal.App. 3 Dist. 1963), which was an action by the state's attorney general in the name of the state for abatement of an alleged public nuisance caused by drainage of toxic mine and mill wastes into a river, resulting in damage to fish life.

The State's complaint alleged: The Penn mine, once an extensive producer of copper and zinc, had been relatively inactive in recent years. During its operation, fluid ore tailings and mill wastes were placed in settling ponds, and mine waste rock piled in dump areas. The rock dumps and tailings ponds are rich in mineral salts. During the rainy season, surface waters flow over the dumps and ponds, picking up concentrations of minerals which drain into the Mokelumne River. The river is a seasonal spawning ground of the king salmon and steelhead trout. The mineral pollutants are extremely harmful to the fish life and have resulted in kills of salmon and trout. Injunctive relief was sought.

The Court held that while the owner of an inactive mine was not "discharging" industrial waste within the meaning of the water pollution act, yet when the surface water or some other mechanism causes drainage of accumulated mine wastes into a public stream, a condition of pollution or nuisance, actual or threatened, may occur.

New Penn Mines correctly objected on the grounds that the State in the person of the attorney-general lacked jurisdiction to bring the suit. The court upheld Penn Mine's position that the injunction must fail on the ground that such an action must be brought by the appropriate regional water pollution control board acting under the provisions of the Dickey Water Pollution Act 1949. (Also, see Ch. 7 §3.2 for Penn Mine 1993 violation of NPDES permit requirement.)

In a few years after 1936, federal jurisdiction and power was greatly expanded by the 1936 decision in (National Labor Relations Board) *NLRB v. Jones Laughlin Steel*, op.cit., thus, allowing Congress to extend its legislative enactments and control over mining which had hitherto been subject largely to regulation by the states. It had been widely recognised that the coal segment of the mining industry affects the public, while the non-coal-mining segment's affect was found to be more localised. The states, in the exercise of their "police powers" (governmental supervisory powers) for the public welfare, had been able to reasonably regulate non-coal mining.

The great growth of population and industrial production of post-World War II caused a slowly mounting proliferation of contamination and pollution claims by the 1960s, which may be attributed to the intensification of industrialisation responding to public demands for more goods. Such proliferation of damage claims made the older system of individual suits unmanageable and less effective. More extensive and new media-methods, as television, gave greater coverage and called more attention to contamination and pollution. As greater public concern grew, new and drastic regulatory and legal measures were called for, not only to deal with the large increase of claims and occurrences of nuisances and pollution from all industrial sources, but more expediently, to eliminate the causes for pollution and contamination claims at the source. Hence, the situation was ripe for environmental legislation which was based on the sovereign's right of exercising its "police powers" for the protection and benefit of the general public's welfare and health.

3.4.3 *Canada*

There appears to be a paucity of early "environmental" or nuisance claims against surface mining operations in the provinces of Canada. Much of the early mining litigation concerning trespass and nuisance occurred in British Columbia and the Yukon Territory regarding disputes between gold miners over encroachment on claims and, for example, the spilling and flooding of waters from one gold placer operation onto another gold miner's operation.

The paucity of litigation in the latter part of the 1800s until the first third of the 1900s involving nuisance claims against surface mining, particularly in the more populous eastern and central provinces where more construction materials quarries were located, suggests again, as in the U.K. and the U.S., that the courts had a much greater reluctance to hear these types of complaints. The courts and the public had a greater reluctance to litigate and a greater tolerance for putting up with potential claims of nuisances against surface mine and mill operators.

A noteworthy comment at this point concerns the possible reasoning for paucity of Canadian litigation over nuisance and "environmental" violation claims from surface mining from former decades even to the present day. A Canadian attorney and colleague in aiding the author with the search for litigated Canadian claims explained the scarcity of cases found by commenting that "Canadians are far less litigious than Americans, and where there are such contentious matters, Canadians tend to resolve them without court involvement". (Evidence for the litigiousness of Americans is offered, supra, at Ch. 5 §2.2).

Canadian judicial precedence for dealing with industrial nuisance and pollution claims had long before been set in England by earlier case decisions which generally exhibited legally established anti-claim positions. The following examples illustrate English case law of the earlier era which would have been followed by the Canadian judiciary: (1) the prescriptive right of an industrial nuisance. An alleged act causing a private nuisance, mining or otherwise, if continued and uncontested for 20 years, became established as a prescriptive right. Therefore, a long-established mine, smelter, or mill could acquire a legalised right to pollute or contaminate the local area; (2) such prescriptive right gave establishment to the doctrine of "coming to the nuisance". Thus, any subsequent purchaser of property complaining of a mining nuisance, had "come to the nuisance" and would be required to "take subject to the nuisance" without complaint [*Wright v. Williams* (1836), 1 M & W 77]; (3) as voiced in the English case, *Shotts Iron Co. v. Inglis,* (op.cit) "... cases of nuisance will not be allowed to interfere with the great industries of the country, especially where the persons who complain reside in the seats of those industries"; and (4) in the 1881 English case, *Re. Dudley,* 8 Q.B.D. 86, the Court said, "The provisions of the Public Health Act, 1875, do not extend to mines, so as to interfere with or obstruct the efficient working of the same, or to the smelting of ores and minerals; or to the calcining, puddling and rolling of iron and other metals; or to the conversion of pig iron into wrought iron, so as to obstruct or interfere with any of such processes".

Faced with such existing precedential case law, the probable futility in bringing a nuisance claim against industry, large or small, in a Canadian court would likely overwhelm and discourage potential claimants. The courts of the early era were less prone or likely to find liability for damages by the mining industry. Thus, the public tolerance of mining pollution was perhaps involuntarily placed on it. Still, for better or worse, the mining industry enjoyed a position of far greater prestige and acceptance by the nations as a whole.

A review of some of the few Canadian cases, given in chronological order, illustrate the types litigated.

(i) **Nuisance / Water Pollution:** *The Columbia River Lumber Co. v. Yuill, et al*, [1892] Vol. II B.C.R. 237: A lumber mill operator had obtained an injunction restraining upstream gold miners from fouling the waters in such a way as to prevent the proper working of the saw mill. The miners were using hydraulic mining, washing tree-roots, earth "tailings" down stream which obstructed the mill-race, blocked its flume and machinery and prevented operation of the saw mill. The miners sought to dissolve the injunction against their uncontrolled use and fouling of the stream waters. The court refused to dissolve the injunction, finding that no regulations allowed the nuisance complained of to continue injuring the rights of the downstream riparian user.

(ii) **Nuisance / Quarry blasting:** *Etobicoke v. Ontario Brick Paving Co.*, [1913], 25 O.W.R. 327, 5 O.W.N. 356: Plaintiffs were a municipality, a nearby public school and a private resident. An injunction was granted to restrain the owners of a quarry from continuing reckless blasting in such a manner as would cause a nuisance. However, the court ruled that if the quarry could be operated on a subdued basis as recommended by an explosives expert, further operation would not be considered a nuisance.

(iii) **Trespass / Water Pollution:** *Nepisiquit Real Estate & Fishing Co. Ltd. v. Canadian Iron Corp. Ltd.,* [1913] Vol. 42, 26 N.B.R. 387: The plaintiff, a sport-fishing group, was granted an injunction restraining the iron mining pit owner from discharging allegedly polluted mill-waters into a stream thereby injuring the sport-fishing rights of another riparian owner. Even though there was no evidence at trial to show that even one fish has been killed, and no evidence of poisonous chemicals, on the basis of discolouration of the stream flow (a muddy colour) the court issued a temporary injunction. Plaintiffs charged a trespass by discolouration deposited on their stream banks from the discharge waters of the iron mill. Defendants argued that "It would be an unfortunate thing if the law found that such a large industry has to give way to fishing for sport".

The *Nepisiquit* court in its holding stated that, "**... but, as the works of the defendant (i.e. the iron mine) were important**, the court orders that the injunction should not become operative for over three months, in order that the defendant might have an opportunity to prevent the pollution by alterations to its plant". (emphasis added). The Court added that if the mine did not stop the flow of polluted waters, the plaintiffs could apply for a perpetual injunction.

(iv) **Nuisance / flooding:** *Suttles v. Cantin*, [1914] Vol. 21 B.C.R. 139; In this nuisance action, one placer gold miner in the Yukon Territory allowed his stream-diverted waters to pass through his tailings piles carrying debris onto the land of another placer mine site. Plaintiff miner brought a successful action in nuisance and for damages against the defendant miner.

(v) **Nuisance / blasting:** *Fuller v. Thames Quarry Co.*, [1921] (1st Div. Ct. App.) 20 O.W.N. 374; Injunctive relief was granted to the owner of a dwelling across the street from a stone quarry in St. Mary's restraining the quarrier from working their quarry so as to cast stone on the private land by blasting. The claimant was granted he injunction and damages of one dollar with costs of court.

(vi) **Nuisance / flooding:** *Salvas v. Bell,* [1927] 4 D.L.R. 1099; A successful action by a lower riparian farmer against an upper riparian mining operation in Yale County, British Columbia, for damages done to the claimant's land by the dumping of material into the watercourse and altering the flow thereof flooding the farm lands with water and debris from the mine.

(vii) **Nuisance / blasting:** *Pilliterri v. Northern Construction Co.*, [1930] 4 D.L.R. 731; "One is liable for damage done to the premises of an adjoining occupier by reason of the escape of vibrations and the falling of stones thereon as a result of blasting operations".

(viii) **Negligence / blasting:** *Aikman v. Mills & Co.*, Ontario Supreme Ct. [1934] 4 D.L.R. 264; same results as *Pilliterri,* supra.

(ix) **Nuisance / noxious fumes:** *McNiven et al v. Crawford*, High Ct. Justice [1939] O.W.N. 414, affirmed, Ont. Ct. App. [1940] No. 27 O.W.N. 323; a nuisance action to recover damages for injuries to plaintiffs' peach orchards and crops caused by the escape of fumes from defendant's brick plant. Damages were awarded to plaintiffs.

A change in the courts' hitherto legal position and attitude, as being protective of the mining industry where nuisance claims were made, began to show evidence of change around the 1930s. Signs were showing in litigated arguments, and in the decisions handed down by the courts, that it was becoming no longer viable that complainants had to suffer and accept the inherent annoyances that accompany the operations of a great industry. The following case exemplifies the changing attitude towards mining in Canada.

(x) **Nuisance / dust:** *Kent v. Dominion Steel & Coal Corp. Ltd*, Newfoundland Supreme Ct., [1964] 49 D.L.R. (2d) 241; Where the plaintiff lived near the defendant's iron ore pit, and the company built a private, ore-haulage road passing within 100 feet of plaintiff's home and property, ore trucks "raised dust to such an extent that the house was materially damaged and the land was rendered unfit for cultivation". Plaintiff's action for damages for nuisance was dismissed at trial on the grounds, *inter alia,* that the dust in question was an unavoidable consequence of defendant's necessary operations. Also, "... that the plaintiff, as an employee of the iron mine and living in an industrial area, must accept the unpleasant consequences of his employment and residence". Thus, the exalted and protected position enjoyed by the mining industry had been preserved.

However, that hitherto protected position so long enjoyed by the mining industry crumbled on Kent's appeal. The Supreme Court of Newfoundland held that the defendant's iron ore mine "had created an actionable nuisance and the plaintiff was entitled to damages therefor". In a split decision, Justice Puddester in joining the majority opinion, and referring to the fact that defendant had built his home knowing the annoyance existed, wrote, "The plaintiff certainly did not agree to accept the risk of damage to his property and even if he built in an industrial area, and therefore may have to put up with certain conditions which would not be tolerated (*in other areas*), *** does not relieve an industry from liability for an unreasonable and unjustified interference with a person's right to the protection and enjoyment of his property ***". Thus, the older doctrine for a complainant's "coming to the nuisance" with full knowledge of its existence was no longer viable, or given weight or credence by this court. To substantiate this argument, a further statement of the presiding judge in *Kent* is offered in evidence.

Justice Puddester continued, "*** the view that certain industrial nuisances, such as noise, smoke or odour, affecting the enjoyment of property must be tolerated **in a modern society** (emphasis added) has absolutely no application to a situation where the land has been made unfit for cultivation and the dwelling made virtually uninhabitable. A nuisance occasioning such consequences, wherever it occurs, is an unreasonable use of land and an unjustifiable interference with the right of adjoining owners to the enjoyment of their property and accordingly constitutes an actionable nuisance".

3.5 Conclusions and Comments

In the search for litigated claims of groundwater water pollution from non-fuel surface mining affecting water wells and aquifers supplying individuals and communities adjacent to and near the pit, there was a noticeable lack of them, regardless of the periods of time. This was found to be so whether in the older era where a judicial and societal tolerance against mining nuisances prevailed, or in a later or more recent time of an increasingly active society against environmentally-destructive operations.

Only three litigated claims were found where an allegation was made that quarrying had affected water wells, and two of those were for loss of water due to blasting. Even there, the loss of water in the well was temporary and water was found to flow again, uncontaminated, after a brief interruption. In the third well case, and the only well-water pollution claim, *Bayer v. Nello Teer Co.*, supra, was an allegation made of well-water contamination from a stone quarry, and there, the quarrier was not found culpable or guilty. Further, the alleged contamination of subsurface water at the well was attributed to the discharge of surface water inflow from the pit affecting the downgradient landowner's well, not from groundwaters flowing from the pit itself.

Thus, the main complaints brought against surface mining, as found in this search and reported herein, were predominantly for blasting, thrown-rock, noise, and dust. Where claims were alleged in water pollution or contamination, the cause was found to be from a surface discharge of the mineral operation's processing plant, or from a settling or tailings pond leak with a breach of the pond's bank into a public stream affecting downstream riparian users. With the exception of pit blasting activity allegedly cutting off a water supply, groundwater contamination from quarrying itself appeared to be of little to no concern to the local public. All this appears to be true, not only in the past and up to the advent of universal environmental consciousness, but in the present as well. Support and corroboration is given to this argument herein by a Research Report released in 1992 by the U.K. Department of the Environment entitled "Environmental Effects of Surface Mineral Workings" wherein it referred to findings of a survey based on 41 stone quarries that "complaints were of dust, traffic noise, blasting and visual effects. *** **Water pollution is generally not reported as a problem** except where run-off (water) is contaminated with fuel, dust, etc". (emphasis added). The UK Research Report further cited, concerning sand and gravel workings, that a survey of 35 sand and gravel pits reported in descending order of significance, "traffic, noise, dust and visual effects were the main sources of complaint". It is notable again, that water contamination was not given as a leading complaint. (*Environmental Effects of Surface Mineral Working*, (U.K.Government Research Report, emphasis added) HMSO, London, 1991, p. 7).

The noticeable lack of claims for pollution and contamination of ground waters, effecting water wells and aquifers supplying persons and communities adjacent to and near non-fuel surface mines amongst the litigated cases researched and reviewed suggests that the author's theory may be viable, viz., that, in general, non-fuel surface mining has a negligible to no polluting effect on subsurface waters. This point is further corroborated by the UK Research Report (op. cit.) wherein it is stated: "The specific operational problems created by surface mineral workings based upon public complaint are, in approximate order of significance: traffic, blasting, noise, dust, visual effects. Other matters, i.e., **water**, wastes, ecology, archeology, agriculture and forestry may be equally important environmental considerations **but are less usually the subject of public protest**". (emphasis added). This appears to be true in the U.K. the U.S. and Canada.

Although, historically, only one unsuccessfully litigated claim of pollution of subsurface waters by non-fossil fuel surface mining was found, claims of potential groundwater pollution are frequently made in current hearings before the fact of actual mining. Groundwater contamination is commonly conjectured and alleged as a future possibility in zoning and mine permitting hearings, even made in the post-environmental stringent regulation period. They are commonly without basis, as is well illustrated in *Florida Rock Industries v. U.S.*, 21 Cl.Ct. 161 (1990). In *Florida Rock*, the federal government argued that quarrying in the local limestone would pose a risk of contaminating the sole aquifer supplying drinking water for the city of Miami, and Dade County, Florida. The government did not contend that limestone mining would actually contaminate the aquifer. The government's argument of speculative groundwater pollution from Florida Rock's future quarrying failed in view of the Court's noting that none of the presently operating quarries in the immediate area in the same rock formation had polluted, nor were they presently polluting, the aquifer in question.

The power of intimation and suggestion that detrimental hydrological contamination "may result" from a newly planned surface quarry if approved, as used by the federal government in *Florida Rock*, failed under the scrutiny of the examining court. Unfortunately, this same "scare" tactic without foundation in fact has been too often successful in defeating quarry permitting at many public hearings.

A very germane article by an eminent, expert engineering geologist and professional engineer substantiating this argument of the high improbability that groundwaters are affected by quarrying's excavations and activities, particularly blasting, is offered in evidence to refute the lay-public's allegations to the contrary.

BLASTING DAMAGE; DYNAMITE AND GROUNDWATER
by Allen W. Hatheway, Ph.D.

"Human irrationality runs especially high when blasting agents are mixed by bulldozer blades into a groundwater litigation cocktail. Most cases of alleged water well damage from blasting and / or excavation are filed in the name of concerned citizens who are awakened by construction activity and then begin to notice the hitherto overlooked physical and structural flaws and frailties of their own homes, places of business and water wells. ***

My estimate, however, is that a significant percentage of "visible" or media-treated construction projects employing blasting and hillside excavation are eventually confronted by such claims. By visible I mean that a considerable number of persons are able to see and take cognizance of the project and that the construction lasts perhaps more than one single season of the year.

Naturally, projects in suburban or rural areas in which residents are on residential wells or small water supply systems are most susceptible to these claims, and, more likely, the smaller and tighter-knit and intercommunicative the community, the greater the chance of claims as word of the alleged effects spreads like a common rumor.

My basic premise is that America's civil and public works contracting industry is well aware of its need to conduct blasting and hillside excavation in such a manner as to not actually incur physical and property damage to the residents bordering and surrounding their projects.

Depending on the amount of explosives employed in each detonation, along with delay conditions, the causative blast would necessarily have to be in very close proximity to the damaged well. The most susceptible of geological conditions would be those of friable and / or poorly-cemented, cohesionless weak rock. This situation can be unfavourably aggravated in older, uncased, and poorly-maintained water wells. The very same geologic conditions are more often directly responsible for deterioration of the esthetic quality of well water in the absence of blasting or road-cuts. ***

The automatic premise of blasting induced damage is a technically inferior argument to begin with. *** here are the often-heard, yet basically implausible arguments that blasting can:

1) Fracture rock masses far beyond (say, 20–40 times the shothole diameter) the pre-split limits of the actual blast; hence form competing conduits for groundwater flow around or away from the well;

2) Induce fault-like displacements along joints and bedding planes, such that flow pathways for groundwater moving toward the well are sealed or otherwise obstructed;

3) Shake wall rock into the well so as to damage or otherwise obstruct water flow into the well or to create associated pump and screen maintenance problems;

4) Release slug-like bodies of natural pollutants found at some other stratigraphic location within the flow-migration distance of the well. Commonly, natural residual petroleum compounds such as tar and oil globules are cited;

5) Undefined objectionable tastes, such as are naturally encountered in the presence of iron in the range below 10 ppm, whereby reduction bacteria produce objectionable taste, odour, greasy feel and cause staining of laundry;

6) Water wells have finite lives, especially those constructed before introduction (early 1970s) of modern non-corrosive well materials such as PVC (polyvinyl chloride. State wellhead protection laws, passed in accordance with the USEPA Groundwater Protection Strategy of 1984 and the subsequent requirement for State Wellhead Protection Plans resulting from the June 1986 amendments to the Federal Safe Drinking Water Act. State agricultural extension representatives now frequently advise those experiencing water quality or water quantity problems to close and replace older (say pre-60s or pre-70s) wells as actually or potentially unsafe sources of water.

> Of the above phenomena, old wells and iron-loving bacteria constitute the plausible and frequent cause of constructed-cited problems. Release of other natural contaminants generally can be explained as a result of changes in the groundwater regime brought about by non-construction causes. Among the suspect alternate non-construction causes is improper well maintenance, sloppy housekeeping and waste management practices in the vicinity of the well and lack of modern welled protection.
>
> *** In the author's opinion, citizens suffering damage to residential water supply are more likely to be experiencing the effects of an overlapping or coincidental array of natural and sociological conditions unrelated to construction. There is a suite of generic geological conditions that can actually be more primary in their damage-mimicking capability (table 4) (omitted). (Assoc. of Engineering Geologists News, 38/4, 1995, pp. 37–39).

In addition to supporting the argument herein that quarry excavations and blasting are unlikely causes for destroying groundwater quality and flow, the preceding article gives credence to arguments throughout this work that public hysteria of "concerned citizens" towards new surface mining sites are often unfounded and without scientific basis.

Thus, the centuries-old respect, judicial and societal tolerance for the great mining industries in the Anglo nations came to its demise around 1960. An industry, which had so long provided not only the basic raw materials, chemicals and metals for man's necessities of life, but for a multitude of conveniences for comfortable and improved living, went through successive periods from the 1920s to the 1940s of decreasing tolerance for, and increasing complaint against, surface mining During those decades few successful challenges were made against mining's accompanying annoyances. After case precedent had been established in a few court decisions of finding nuisance, negligence, trespass by surface mining operations, and issuing injunctions enjoining the surface mines from creating further nuisances, those minority position decisions gradually gave way by the 1960s to becoming the position of the majority. Public tolerance for surface mining operations had gradually and finally been reduced to near zero. The public of the late twentieth century demanded that mining no longer be allowed to disturb or soil their world.

The Anglo nations had become fully developed countries with well-established societies, and their peoples affluent. Having attained a high degree of comfortable and luxurious living for the majority for several generations after World War I until the 1960s, interrupted for short periods by the Great Depression and World War II, the majority of peoples of the Anglo countries no longer found the struggle and compulsion of their antecedent generations for wealth necessary. This observation and statement is supported by Schoenbaum in his work, *Environmental Policy Law* (op.cit.) wherein he states, "In the 1960s material goods became relatively abundant for many, and there was a new emphasis on the quality of life". In further support of the author's statements that industry met the ever-increasing demands of an affluent society for more goods and luxuries thereby creating greater volumes of wastes, Schoenbaum also notes that concurrently with society's attainment of wealth, "there was a great increase in the volume and kinds of pollutants released into the environment, including toxic chemicals and pesticides". (op.cit., p. 4). Additionally, the following supportive statement is noted in the Introduction to the U.K.'s Department of the Environment's Waste Management Paper No. 26, that "As society has become more affluent, the quantities of waste to be deposited have increased and waste types have changed and increased in number". (HMSO, 1986, p. 1).

Thus, in the century preceding 1960, mining in the Anglo nations had been widely and largely held as being a basic industry absolutely necessary to the well-being of the local and national economies, and for providing local employment and tax-based income, in addition to providing essentials and luxuries for the public at large. After 1960, in a world of accelerating population boom and general affluence, worthy causes became important to society, e.g., the environment; improving and enlarging existing parks, creating more parks, more elaborate public recreational and educational facilities, forests, scenic rivers, wildlife habitats, clean water, clean air, placing unlimited value on life for industrial safety, and cleaning up the world that industry and mining had so allegedly wasted. The cost to carry on the environmental clean-up campaign became secondary.

A *cause noblesse*, the environmental crusade, the "Green Revolution", enters the scene.

CHAPTER 4

A BRIEF HISTORY OF THE DISPOSAL OF WASTES BY EARTH BURIAL

4. 1 Introduction

Where there is consumption, there is waste. Even in nature, plants and trees consume sunlight, air and water, resulting in waste in the form of falling leaves, dead limbs, stalks and trunks; carnivorous and herbivorous wildlife consume natural matter and deposit excreta, and finally their carcasses are left on the earth's surface to be eroded as contaminants into streams and bodies of water. And so it is with man, the greatest consumer of all earth's inhabitants, who produces the greatest volume of waste. As civilisation developed, man consumed more and more raw materials, until the present day when the disposal of waste from man's insatiable consumption has become a problem that endangers water supplies, and not only his own existence, but that of other forms of life on earth.

Debris from early man's daily living has been found in caves and other early dwelling sites. As man became more prolific, more sociable, and gregarious, cumulative wastes became a communal problem. In the Minoan civilisation, which flourished on the Isle of Crete from 3,000 to 1,000 B.C., solid wastes in the capital of Knossos were placed in large pits with layers of earth at intervals (Priestley, 1968). In Egyptian biblical times, Moses (c. 1500 B.C.) proscribed burial of human body waste (Deuteronomy 23, vv.12–13). According to Dr. Allen W. Hatheway's 1993 work, a Chronological History of Industrial and Hazardous Waste Management, the Greeks were the earliest to organise the first municipal refuse disposal site in the western world in the city of Athens about 500 B.C. Two, early Roman waste dumps are also reported as being discovered at Rome, Italy; the first from the sixth century,

B.C., and another from the second century, B.C., established at Monte Testaccio (Hatheway, 1993, p. 2). However, Priestley (op. cit.) states that the Romans had no organised system of waste removal: "disposal and wastes accumulated in the streets and around towns and villages. The practice was said to have persisted until the 19th century."

Professor Rudolfo Lanciani in his 1890 sketch of the history of monuments of ancient Rome, The Destruction of Ancient Rome, describes the filling of new structural sites by placing waste to raise their base elevations. Lanciani writes, " In tracing the history of the destruction of the Rome of the Kings and of the Republic at the hands of the Emperors, three facts become prominent: (1) the complete covering over, for hygienic reasons, and consequent elevation, of large tracts of land; (2) the rebuilding, on a totally different plan, of one or more quarters of the City, after a destructive fire; and (3) the clearing of large areas to make room ..." Continuing, he states, "The first record that we have of the covering over and elevation of a large area for hygienic reasons dates from the time of Augustus (*63 B.C.–14 A.D*). A part of Esquiline Hill was occupied at that time by a 'field of death', where the bodies of slaves and beggars and of criminals who had undergone capital punishment were thrown into common pits, together with the carcasses of domestic animals and beasts of burden. *** about 75 of these pits were discovered. In some of them the animals' remains had been reduced to a black, unctuous matter; in others the bones so far retained their shape that they could be identified. The field of death served also as a dumping place for the daily refuse of the city. This hotbed of infection was suppressed by Augustus at the suggestion of his prime minister, Maecenas. The district was buried under fresh earth to the depth of 24 feet, and a public park, a fifth of mile in extent, was laid out on the newly made ground. The results proved of so great benefit to the health of the City ..." (Lanciani, 1890, pp. 12–15).

4.2 United Kingdom

There is a paucity of references and recorded information in the western world indicating whether Man had any serious concern for waste disposal for well over a 1500 years after A.D. Epidemics, as the Black Death in 1349 in England, occasionally brought about a reaction to the accumulations of refuse and filth in the cities. As early as 1297, London imposed a legal obligation on every householder to ensure that the pavement in front of his tenement was kept clear. A 1354 decree ordered weekly removal of filth deposited in front of houses, and rakers were employed to remove the weekly accumulations. The city had the power to levy a charge on householders who failed to remove their refuse. In 1387, a London committee was elected to discuss providing laystalls (local collections points) for depositing street refuse collections. Refuse from suburban laystalls was sold to farmers and market gardeners, while that from riverside laystalls was taken downstream and dumped on the Essex marshes. By 1407, Londoners were ordered to keep their rubbish indoors until it could be carried away by the rakers. King Edward III, in 1358, enlisted the aid of the Mayor and Sheriffs of London to stem the dumping of rubbish and filth into the Thames River. A 1387 ordinance forbade the throwing of refuse into the Walbrook. This local edict was closely followed in 1388 by an Act of the English Parliament prohibiting the depositing of filth and garbage "in public waterways and ditches", and that all such refuse should be carried away to appointed places. Wilson further reports that enforcement was difficult, even in 1414, requiring payments to informers to report and gather evidence against offenders casting rubbish, offal and dirt onto the streets. Typified is the 1421 example of one William atte Wood, who was arraigned for casting "horrible filth onto the highway" and making a great nuisance and discomfort to his neighbours, "the stench of which was so odious, that none of his neighbours could remain in their shops". Of note is a court record entry in 1515 at Stratford-upon-Avon which lists Shakespeare's father as being fined for depositing filth in a public street (Wilson, 1977, pp. 2–3).

It is noteworthy that about the same time, the consciousness of raw waste accumulations was spreading in Europe. Paris, France, found that garbage disposal immediately outside the walls of the city interfered with the defenses of the city (Hatheway, p. 2). It was reported that "some medieval German cities avoided the danger of being covered in their own wastes by requiring that departing wagons which had been used to bring produce into the city, return with a load of wastes to be deposited in the countryside". (Wilson, 1977, p. 1).

That a relationship existed between waste disposal, hygiene, and the periodic epidemics of Europe was not realised until early scientists introduced bacteriology, well into the seventeenth century. The disease causation relationship was determined after a Dutchman, Antony van Leeuwenhoek, the father of microscopy, described bacteria in 1693 (Hatheway, p. 122). However, the presence of wastes had become objectionable enough, and to the point that means of disposal were being searched for. In the 16th and 17th centuries, raw sewage applied to the land, both as a means of disposal and as a fertiliser, began as a practice in 1531 in Bunzlau, Germany, and in 1650 at Edinburgh, Scotland (Hatheway, 1993, p. 3).

Scientists had not yet discovered the relationship of hygienic practices to avoid or control the water- and foodborne (so-called intestinal or filth) diseases, which may be transmitted by direct contact with infected soil, compost or decaying vegetable matter, or by vectorborne agents in contact with human and animal excreta in raw sewage. It was not until 1826 that Bretonneau described typhoid fever as "dothienenteritis", or an abscess of the small intestine. The cause of the 1854 cholera epidemic in London was discovered to be a certain infected water well by physician Dr. John Snow. A nearby privy was the cause of contamination and infection. In 1856, the British physician William Budd discovered that typhoid fever is spread by carriers from human excreta (Hatheway, 1993, pp. 6, 10); but the typhoid bacillus (Eberthella typhosis) was not identified until 1884 by Eberth.

The Public Health Act of 1848, which created a General Board of Health contained provisions for improving the supply of water and for supervising sewerage and drainage. A Royal Sanitary Commission was established in 1868, whose report led ultimately to the Public Health Act of 1875.

An 1885 English case in point is *Ballard v. Tomlinson*, [1885] 29 Ch D 115, wherein plaintiff Ballard complained of an adjoining property owner placing filth (draining his water closet) and other poisonous matter into its disused water well which was carried by the subterranean aquifer's percolating waters to the plaintiff's well in use. The issue became whether anyone has the right to contaminate a common source of water, or water reservoir. The Chancery Court determined that no one has a right to use his own land in such a way as to be a nuisance to his neighbour.

Thus, the public consciousness in Europe and North America was not seriously awakened to sanitation, waste disposal, and their effects on public water supplies until the mid-to-late nineteenth century with the discovery of sanitation-related diseases. Wilson (ibid.) reports that the refuse yards of Edinburgh, Scotland, remained the same size for a hundred years in the 18–19th centuries because everything brought to the yard was sorted and eventually sold. And, in London, he reports, that "one great heap (of refuse) at the bottom of Grays Inn Lane was not moved for a century until 1815 when the dust was extracted and sold to Russia to make brick for the rebuilding of Moscow after Napoleon's invasion. As recently as 1926, 41 English brickfields were using dust from London's refuse to produce stock bricks, and some continue to do so to this day". (Wilson 1977, p. 2). The modern system of English refuse collection and disposal by local authorities derived from the Public Health Act of 1875. Appointed days of refuse collection were established and householders were to place their refuse in a movable receptacle.

In Ch. 10, §7.2, infra, a Black Country reclamation project near Birmingham, known as Bowmans Harbour, is reviewed.The project is one of cleaning up old refuse tipping in former shallow coal workings from the nineteenth century industrial revolution.

4.3 United States

In the U.S., the first known ordinance for refuse was enacted by the city of Georgetown, Virginia (now in the District of Columbia) in 1795. It prohibited extended storage of refuse on private property or the dumping of it on a public thoroughfare. The second U.S. President, John Adams, used the first rubbish hauler to remove refuse from the White House, while the third U.S. President, Thomas Jefferson, contracted for the first refuse collection from government office buildings. Public collection of refuse in the District of Columbia did not start until 1856. However, Hatheway (p. 10) reports that even in 1860, the citizens of Washington, D.C., continued to dump garbage into the streets.

Thereafter, the public began its slow awakening to the need for sanitary disposal of wastes. In 1873, " the City of Los Angeles, California, established a garbage and dead animal plot for burial three feet below ground surface". In

1883, "New York State's Governor Cornell proclaimed an energetic campaign against industrial polluters. *** Newtown Creek in New York City had become a dumping ground for stable manure, fat-boiling and bone-boiling residues, fertiliser wastes, distillery slops, and bottoms from 13 petroleum refineries, along with wastes from 30 other offensive trades; in addition to discharges of 22 miles (35.4 km) of sewers". (Hatheway, 1993, p. 13; 17). New York City's solution was to build a main sewer line to transport those wastes into the East River, in effect only carrying the wastes further away from the population centre rather than treating the sewage and reducing it to a less destructive and contaminating state. Solid wastes were loaded on barges and carried to sea for dumping. During this same era, the cities of Chicago, St. Louis, Boston, and Baltimore carted much of their refuse to open dumps.

As public concern continued to grow, "the first recorded indictment 'for industrial waste discharge nuisance' made in the United States was in 1886 in Indiana against a Mr. Mergentheim 'for discharging the water from his woolen mills into the canal at Perue …'" As an early alternative to burial of wastes, the first municipal incinerators built in 1886–1887 and used at Wheeling, West Virginia, Alleghaney, Pennsylvania and Des Moines, Iowa. Expressing the growing concern for waste disposal, even as early as 1889, the Public Health Officer for the District of Columbia (US) stated, "Appropriate places for (refuse disposal) are becoming scarcer year by year, and the question as to some other method of disposal … must soon confront us". (Hatheway, 1993, pp. 19, 21).

The U.S. Congress reacted shortly thereafter with enactment of the River and Harbour Act of 1890 which provided for a "system of permits to be issued by the Secretary of War for the discharge of all forms of insoluble substances, including rubbish, filth, and other solid wastes into navigable waters". U.S. Stat. 26, Sect. 426, 453–454. It should be noted that prior to the U.S. Act, the British Parliament created the Royal Commission on the Prevention of River Pollution in 1857 because of the gross pollution in British rivers. However, preventative legislation in the U.K. was not enacted until 1876 and again in 1890. It was in 1890 that the United States surpassed Great Britain in its volume of industrial output, which in turn, meant also greater waste (Hatheway, 1993, pp. 10, 21).

In 1899, a third Rivers and Harbours Act was passed by the U.S. Congress. Again, it made unlawful the disposal of any refuse matter onto the bank or into any navigable river, or any tributary. It included wastes from any ship, vessel, barge, or floating craft, but excluded the flow of liquid discharges from streets and sewers of municipalities of the U.S. However, just prior to the new federal act, a California court in 1898 granted an injunction against the city of Santa Rosa for emptying impure effluent from sewage irrigation into a creek (Hatheway, 1993, p. 20).

U.S. municipalities started opening "city dumps" about the turn of the 20th century, e.g., Warwick, New York in 1898; Anaheim, California, 1901 (Hatheway, 1993, pp. 26, 29). Professor Hatheway notes that the Warwick landfill accepted industrial wastes and sludges, and that key contaminants were VOCs, PAHs, phenols, other organics and metals.

In 1916, England reputedly made its first trial of sanitary landfilling. By 1920, at Bradford, England, "controlled tipping" of solid waste began with a pioneering form of sanitary landfilling, i.e., placed in layers of approximately 60 cm., with daily cover and control of windblown debris (Hatheway, 1993, pp. 48, 52). By 1932, the Ministry of Health developed a system of "recommendations" for locating suitable sites for controlled tipping of solid wastes.

With regard to mine waste disposal, 1915 marked the "beginning of alternative mining waste disposal in the Butte District of Montana. The "traditional means of disposal of tailings, mill products and smelting wastes directly into Silver Bow Creek" were abandoned in favour of pre-treatment of the wastes before introducing them to the streams.

In 1927, North Smithfield, Rhode Island, started using a former sand and gravel pit for domestic, commercial and industrial wastes. (It should be noted that this site achieved final closure in 1985.) (Hatheway, 1993, p. 64).

In all recorded history of the disposal of wastes, a most significant milestone was the U.K. Ministry of Health's 1930 regulations for controlled tipping into landfills where refuse was to be deposited in series of layers, each layer not to exceed six feet in depth, and all surfaces were to be covered with soil or material suitable to prevent exposure to air within 24 hours of its emplacement. Screening was to be placed around the area to prevent wind dispersal of paper and other light debris. The screening thereby effectively diminished problems of fire, dust, vermin, birds, insects, windblown litter and odours, and improving the hitherto unsightliness of the disposal site.

In 1932, the City of San Francisco, California, began cut-and-fill landfilling of municipal refuse. The "first US 'sanitary landfill' was introduced in Fresno County, California, with the cut-and-daily cover method of solid waste disposal". Experimentation with compaction of the piled waste in the landfills, thus saving space, became of increasing importance. One city, New York City, was faced with finding an alternative to its sea-dumping method since court actions by other coastal cities nearby were brought against it and its disposal method, which was affecting them. The litigation against New York finally culminated in a decision by the United States Supreme Court in 1933 requiring the city to cease its ocean-dumping practice (at least within the twelve-mile coast line limits (19.31 km) established at that time). [*New Jersey v. City of New York*, 290 U.S. 237 (1933)].

In the period between the World Wars, similar procedures of sanitary land filling were being tried by municipalities throughout the U.K., Germany, Canada and the U.S. During World War II, the large concentrations of men at military installations caused the military to seek and develop quick, efficient, economical and sanitary methods of waste disposal. The sanitary landfill being found successful in the U.K., the U.S. and Canada was tried. Daily compaction by heavy military equipment, as the "bullclam", increased its success. By 1945, a survey of the U.S. reported that 100 American cities had adopted sanitary land filling, and by 1960, the number of cities had grown to more than 1,400. Gradually, with its success, open-air disposal site burning decreased, e.g., banned in 1947 in Los Angeles County.

4.4 Canada

It should be noted that federal law does not regulate solid waste disposal in Canada. The individual provinces developed their own legislation and policies dealing with the management of solid waste, and this is still true. For the greater part of Canada, regulation of waste disposal within the provinces was almost non-existent until the 20th century. As in most locations of the world, the earliest regulation of waste disposal began in the larger population centers of Canada where the greatest waste accumulations were made and affected the health of large numbers of people. Consciousness of waste disposal regulation did not become a matter of great concern until the coming of the environmental revolution in the 1960–70s. Even in the present era of much environmental law, the degree of waste disposal regulation appears to be governed by the density of population. As noted in §2.3 of Ch. 2, the density of Canada's population is about 2.5 persons per square kilometre (6 persons per square mile), compared with 27 persons per square kilometre (71 persons per square mile) in the U.S. and approximately 253 per square kilometre (660 in the U.K.). Consequently, with Canada's very low density of population, the concern for contamination of the earth and groundwaters from land burial of solid waste is correspondingly small. Not to be misled by the low density figure, which is more accurate for the more sparsely settled western provinces, the areas and cities with greater densities, particularly in eastern Canada, have had some municipal waste disposal regulation from earlier dates in this century.

For the purposes of this work, and because there is a paucity of published history of waste disposal in Canada, the review is meagre and covers about half of the Canadian provinces. Communication with several of the responsible provincial waste disposal agencies yielded very scant historical information; nor was any publication found which featured information of historical reporting on the subject in Canada. Information gleaned from a sampling of provinces across Canada follows.

4.4.1 *Saskatchewan*

As reported by Saskatchewan Environment and Resources Management, "Prior to 1974 there were no provincial regulatory requirements for waste disposal grounds". It was also revealed that prior to 1972 some municipalities in the province developed waste disposal grounds in depleted gravel pits.

4.4.2 *Alberta*

As reported by the Material Management Branch of the Alberta Environmental Protection agency, municipalities within Alberta, owning and operating MSW sites are not required to report any information to the provincial authorities.

Provincial waste management regulation was apparently not instituted or standardised until 1985 when the Public Health Act, Alberta Regulation 250/85, entitled "Waste Management Regulation" was enacted. The use of open pits or surface mines for waste disposal is reportedly unknown to this agency.

4.4.3 *Manitoba*

Provincial regulation of MSW is of recent vintage, occurring about 1988. The latent concern is a reflection of the low population density for the province. Municipal landfills of two cities, Winnipeg and Brandon, currently serve more than 60% of the provincial population.

4.4.4 *Ontario*

Centrally provided regulations for this more populous province were initiated a few years earlier than in the lower population density provinces to the west. Licencing of waste disposal sites was initiated in 1970. Of interest to this work is the use of pits and quarries as landfills within Ontario. Two large open mining pits are of note, viz., the Keele Valley landfill located north of Metropolitan Toronto, and the Walker Brothers aggregate quarry-landfill operation at Thorold in the Niagara area.

4.4.5 *Newfoundland — Labrador*

The department of Environment & Lands is responsible for the regulation of MSW disposal in this province. The department draws its authority for regulation from The Waste Material (Disposal) Act, promulgated in 1973 and as amended in 1976.

With respect to waste disposal generally, and more specifically to domestic solid waste, the Minister has the authority under The Act to issue approvals for municipalities and others to operate waste disposal systems. Where there is no municipal body to take the responsibility for waste disposal in any given area, the Minister has the authority to appoint a Waste Disposal Committee or a

Franchise Holder to handle the area's waste disposal needs. The Minister also has the authority to establish, construct, take over and operate, or manage waste disposal sites that are deemed to be in the public interest to do so. In fact, the department has been active at one time in building and operating sites, but has relinquished operation to municipalities and private, licenced operators (Dominie, 1992).

4.4.6 Quebec

This province's Solid Waste Management programme operates under Division VII of the Environmental Quality Act (R.S.Q., Ch. Q-2). The province of Quebec is Canada's largest with an area of some 1.5 km^2 — approximately the size of India. The population numbers about 7.2 million, or about one quarter of Canada's total. According to its Minister of the Environment, Canadians generate almost twice as much waste as Europeans; and, in Montreal, Canada's most populous city, each Montrealer generates 1,000 lbs. of waste per year, with an annual total of one million tons per year for the city.

The Solid Waste Regulations (R.R.Q., c.Q-2, r.3.2) governing standards for design and operation of new landfills provide for use of open-pit mines and quarries as depositories for solid waste. However, of the nearly 70 landfills in the province, a large MSW site has operated since 1968 in the former Miron quarry at Montreal. To 1995, it has received 31 million tonnes of waste and is still in use. The quarry is equipped with a leachate pumping system and collects 20,000 c.f./min. of methane from 280 bio-gas wells.

4.5 Conclusions and Comments

History reveals that the most commonly used method for disposal of humankind's waste has been by depositing it in the earth, whether it be in a man-made excavation, or a natural depression in the earth's surface. Older chosen sites for depositing waste were more than likely to have been already considered waste lands, e.g., gullied, or marshland. Those living near coastal areas readily took advantage of the nearby large bodies of water, or the estuaries running to the sea, to carry off their waste where it would certainly feed the fishes of the sea, and the inorganic matter would settle to the seabed, possibly decompose in the saltwater, and finally "return to nature" on the sea floor. Early concerns for man's environment were simply based on obvious, elementary reasoning: filth and waste drew vermin, created stench and foul odors. Once removed from the local majority's living area, his breathing space was much purer and his aesthetics were improved (out of sight, out of mind), and vermin would remain at the trough (the dump).

A 20th-century encroaching problem, given little-to-no consideration until recent decades, was that the make-up of man's waste was no longer of a simple chemistry as it had been for centuries. Non-biodegradable plastics and toxic organic compounds had been discovered and played a major part in modern manufacturing. Not only had the chemistry of wastes changed drastically with the swift advances of industry and science in the mid-20th century, but the chemical changes of waste had become more deadly in infecting the earth, the seas, and groundwaters where deposits of waste were placed, thereby threatening all life on earth. Present considerations for waste disposal must take into account other factors, as greatly increased populations with the promise of even greater numbers of humans in the near future by extrapolations, rapidly diminishing space for depositories of waste, accompanied by an increasing volume of waste, et al, all of which only serve to exacerbate the solutions for the waste depositing problem. At present, the amount of concern for contamination of the earth and groundwaters by waste burial disposal still seems to be directly proportionate to the population density. Earth burial, or landfill for mankind's wastes continues to be the most acceptable and predominantly used method for disposal, particularly in the western world. Inroads to burial alternatives, as incineration, are slow in acceptance. Recycling has been more accepted and therefore, successful, but only postpones the waste disposal volume problem. Noting that the volume of mineral excavations is reportedly five times the volume of solid wastes in the U.K., utilisation of existing, ongoing, and future surface mining excavations for the deposition of wastes appears to be a most logical solution regardless of other volume reduction methods used.

CHAPTER 5

A REVIEW OF THE ENVIRONMENTAL ERA REGULATORY ACTIONS FOR SURFACE MINING WITH LITIGATED INTERPRETATIONS

5.1 Introduction

Environmental law is thought of in present times as a new area of law. An environmental lawyer or environmental barrister / solicitor, was virtually unheard of in all Anglo countries prior to 1969. In truth, however, the practice of law for environmental matters and concerns for nature and ecology did have its beginnings several hundred years ago. Those concerns had been very slowly evolving until recent decades when may finally mushroomed into a vast new area of law and regulation.

In its earliest forms, it first showed itself in various unrelated fields of law, as in public health law where periodic gross contaminations affected the public at large and, consequently, further control was deemed necessary for protection of the general public health and welfare; or, in real property law, where certain acts performed on one tract of land caused devaluating and damaging trespasses to be made on adjoining or others' lands; or in tort law, where injurious, contaminating and polluting nuisances and trespasses were made on other's persons and properties. Later, town and country planning, or zoning law, became involved with land control over industry and "bad neighbour" developments that might be undesirable sources of nuisances, contamination and pollution for a community. Thus, environmental concern, protection and law as we know it today, is a much-revised, far more organised, and particularly pervasive product compared to its earlier forms. The proponents of better health conditions, natural resources preservation and conservation of the previous century gave way to the new social force of environmentalism which emerged in super-force about 1960. During the '60s, concern rapidly grew over environmental abuse, the exhaustion of natural resources and alteration of resource lands, growing population, radiation, increasing pollution levels, waste disposal, human impact on animal population, habitats and natural landscapes. International concern was shown by the holding of the Stockholm Conference on the Environment in 1972.

As previously stated in Ch. 2, "Having attained a high degree of comfortable and luxurious living for the majority for several generations after World War I until the 1960s ... the majority of peoples of the Anglo countries no longer found the struggle and compulsion of their antecedent generations for wealth necessary. Having attained relative wealth and luxury, affluent societies *** then *** devoted their efforts to such worthy causes as the environment. "This statement is supported by Schoenbaum in his work, *Environmental Policy Law* (op. cit.) wherein he states, "In the 1960s material goods became relatively abundant for many, and there was a new emphasis on the quality of life". In further support of the author's statement that industry met the ever-increasing demands of an affluent society for more goods and luxuries, Schoenbaum also notes that concurrently with society's attainment of wealth, "there was a great increase in the volume and kinds of pollutants released into the environment, including toxic chemicals and pesticides". (op. cit., p. 4).

Although existing liability rules developed in the Anglo nations for common law, nuisance claims against private and public property damage by pollution were not made in the name of environmentalism. They were instead the earliest form of fighting damage claims made on the individual's and the public's environment and property. In the area of an alleged public nuisance claim, an individual did not have the right, or standing in court, to join with a government entity for a claim of damage or abatement, unless the individual could show an injury that was different in kind from the public's claim. With the advent of environmental protection acts in the U.S., some federal acts provided for citizen's actions. States followed suit by enacting their version of the federal statutes with like-provisions making it possible for citizens as groups or individually to initiate environmental actions against an alleged industrial violator, or against a political subdivision for an alleged failure of a government agency to protect them against violations of environmental laws, particularly where air, water and other natural resources were concerned. However, the common law tort of a nuisance was, and still is, available and used for environmental damage claims in all three countries.

Inspection and review of laws and environmentally-related litigation concerning open-pit mining and waste disposal / landfill cases in the United Kingdom, the United States, and Canada during the period from the 1950s to 1990 gives insight as to the manner and kind of environmental regulations that were formed, enforced, amended and carried out in the national attempts to clean up the environment deterioration caused by surface mining.

5.2 An Overview of Early Environmental Regulation to 1990 for Open-Pit Mining, Quarry Tipping and Backfilling with Waste Materials

5.2.1 *Public Health, Tort and Planning / Zoning Law*

The three mentioned areas of earlier law, viz., public health, torts and planning / zoning, that evolved into environmental regulation and which intensified over time to affect surface mining are reviewed. Historically, public health law had a very limited effect on mining with its chief regard for mining working conditions being safety. The only health law concern for the mining on the natural environment was for any of its discharges into either the air or waters that might seriously affect public health actually constituting a public nuisance. Early public health laws expressed greater concern for control over disposal of waste and sewage. Tort law, concerned with environmental nuisances and trespasses, continued to be pursued in law and litigation as it had in the past. Litigated case reviews following illustrate its continued presence.

5.2.2 *Additional Environmental Law Problem Areas*

In addition to the three named basic areas of law, this chapter looks at principal ensuing law-problem areas that seriously affected surface mining, particularly in the U.S., as a result of NEPA and its various offspring environmental regulations. An increased effect on open-pit mining in the U.S. is no doubt due to the greater pervasiveness of applied environmental law in the U.S., thus having more far-reaching consequences on all industry, particularly in view of its "command and control" policy. Ensuing problems for surface mining from environmental law regulations were principally found to be: (1) regulatory takings of mineral properties; (2) wetland limitations; (3) planning law or zoning regulations placing limitations on mining property and pit expansions; (4) limitations of "grandfathering" acts for active mining operations that existed before the environmental regulation; (5) intensified reclamation requirements of mining properties; and (6) legislative responses to the public's clamour for massive conservation of large and numerous land areas to be set aside and preserved in parks, forests, wildlife habitats and other areas of pristine, natural and scenic beauty which had the effect of severely limiting the land areas for mineral prospecting and exploitation of mineral deposits.

Case examples for these six problem areas are given under each country, when available, later in this chapter. However, for the U.S., an abundant number of cases in each area were available. This is thought to be due to the greater litigiousness prevailing in the U.S. Americans in the U.S. are far more litigious and prone to litigate than their American cousins in Canada who reflect the tendency for greater tolerance, in out-of-court settlements, and a lesser bickering quality of the mother country. In addition, is the fact that environmental laws of the U.S. are more subject to litigation since they are of the "command and control" type, thereby inviting more challenges to their legality.

Evidence for the truth of the statement that "Americans are far more litigious" is offered in the Findings and Purposes statement of the U.S. Common Sense Product Liability and Legal Reform Act of 1995 which stated, "our Nation is overly litigious, the civil justice system is overcrowded, sluggish, and excessively costly, and the cost of lawsuits, both direct and indirect, are inflicting serious and unnecessary injury on the national economy; ***". [104th Congress, 1st Session, H.R. 956 (1995)].

5.2.3 *Land Use — Planning and Zoning Law*

With regard to the third area of law discussed, planning and land use, urban development planning also had ancient to early beginnings. The desire for a blending of town and country scenery has long been thought and sought to be the ultimate in a planned urban environment and is evidenced by the so-called interspersed "green areas" in current suburban planning development. However, planning law made its first effectual appearance in England as an enactment in 1909 under the Housing, Town Planning, & Act. This was followed by one of the same name in 1919 in which planning schemes were made compulsory. Again, the Act was reinforced with more planning powers in 1932. In the U.S., similar concerns for planning in urban areas, more commonly called zoning, were appearing after World War I.

After World War II, when world populations were settling down and developed nations were returning to peacetime comforts and endeavours, and while the fervor of a post-war economic boom was prolonged, a general migration from the metropolitan areas surged to nearby areas of less density — the move to suburbia. As an example, between 1950 and 1969 in the U.S. the population of American suburbs had almost doubled, while the inner city populations had barely increased. By 1970, metropolitan areas had become the home for two-thirds of Americans. As Grad notes on the environmental damage that occurs with the development in suburbia in his chapter on Land Use Planning, "Building and construction practices, together with the quickened pace of development and complementary zoning, often end in severe abuse of the land and are ultimately costly to the public. The popular practice of stripping subdivisions of all cover before commencing construction destroys tree and plant cover and can trigger heavy soil runoff. Sedimentation from this runoff *** loads nearby streambeds and ultimately river channels. This can cause costly downstream dredging and upstream flood control and destruction of the esthetic quality of lakes and rivers. Public pressure for flood control projects is often spurred by suburban development along flood plains, which usually contain fertile soil supporting an abundant variety of native plant and animal life. Construction over aquifer recharge areas, where groundwater is normally replenished, accelerates rapid runoff, increasing flooding and contributes to water shortages". (Grad, 1978, §9.01). Thus, public concern for patterns of land use and zoning gathered momentum. Land use law, planning and zoning regulation began to take a larger part and concern, particularly for the future locations of industrial developments, including surface mining, that might have a hazardous effect on the environment and the general welfare of the public.

In general, extended powers for controlling and restricting land use by planning authorities were granted under the nominal "police powers" of the sovereign in controlling all land of the domain. Those powers have been passed down to the sovereign's minion local governments. Powers for planning / zoning authorities normally includes the right to refuse or allow certain development activities on private lands, and if approved, a permitting or licencing procedure follows for the permitted use. However, a problem with environmental land use regulations is that they are generally employed for preservation purposes, e.g. open space, wetlands, etc. They also often downgrade zoning classification thereby making it more difficult for a landowner to recover compensation, and conversely strengthening justification for the good of the general public.

As an example of downgrading of zoning, or "downzoning" property, see *Agins v. City of Tiburon*, 447 U.S. 255 (1980) — where a city government modified existing zoning after a landowner bought five acres of vacant land. The re-zoning placed the property in a residential planned development and open space zone. The purpose of the zoning was preservationist. The planning ordinance stated, "It is in the public interest to avoid unnecessary conversion of open space land to strictly urban uses, thereby protecting against the resultant impacts, such as *** pollution, *** destruction of scenic beauty, disturbance of the ecology and the environment *** and other demonstrated consequences of urban sprawl". The landowner had lost the value of his contemplated investment as the five acres could not be developed or built upon. Hearing the case, the U.S. Supreme Court stated "*** the zoning ordinances substantially advance legitimate government goals. ... The specific zoning regulations at issue are exercises of the city's police power to protect the residents of Tiburon from the ill-effects of urbanisation. Such government purposes have long been recognised as legitimate".

Exceptions for existing businesses, which were extant before the land use plan was adopted, may be made. One type of zoning exception is called a variance or conditional use. The general procedures in all three Anglo countries for land use consideration and pre-permitting call for public hearings where the local community is encouraged to participate and voice their opinions as to the acceptability or objections to any specifically announced land use proposal. Permitting, or licencing, of mines and quarries by local authorities, boards and commissions made the greatest early in-roads for environmental controls over mining. Cases included are intended to illustrate this point of increased control by planning and zoning of land over surface mining. (See particularly *Goldblatt v. Hempstead* and *Cioffoletti v. Planning & Zoning Commission,* infra.)

Planning for orderly land use, while beneficial to the general public, carries with it burdens to some and injuries to other individual landowners, and to certain industries, conspicuously amongst them being surface mining and waste disposal operations. Restricted uses from zoning and planning may create loss of possible value to some properties. Such injuries commonly occur in the form of depriving the owner of the intended use of his land for which it was purchased. This is particularly true for intended surface mining properties. Minerals may be mined only where they occur and are found. The need is for mining to be permitted where the minerals are located. A mining operation cannot move to just any area already zoned for mining as could some other manufacturing plant or heavy industry. Valuable minerals can be lost "forever" to an area zoned residential or for light business.

At this point, areas of law doctrines, as eminent domain, expropriation of land or property takings, are brought to the forefront to rectify wrongs done to individuals by denial of certain uses of their land. Environmental controls over wetlands and coastal lands in the U.S. have been an especially litigious area for alleged regulatory takings of property and rights from landowners where residential and resort development were the intended land use. Examples of litigated cases are given where zoning or environmental regulations have prevented surface mining from occurring, or limited the expansion of an active surface mine, thereby reducing the life of the expected operation to the detriment of the miner or mineral owner.

The exercise of eminent domain affecting mineral properties has not been one of any note in Britain, but has been a more conspicuous result of environmental regulations in the U.S, and to a lesser degree in Canada.

5.3 United Kingdom

5.3.1 *Derelict Lands*

With regard to mineral extraction in the UK in the mid-1970s, David Hughes, referring to G. Moss's 1981 work, *Britain's Wasting Acres*, states that up until the 1970s, some 150,000 acres (60,704.17 ha) of derelict land from various uses had accumulated in the UK, and that "the principal extractive industries took 12,445 to 12,595 acres (5,036.42 to 5,097.13 ha) per annum, with very little being restored after use." (Hughes 1986, p. 240). In his later edition, Hughes noted that there was a 6% rise in the total of derelict land in the U.K. between 1974 and 1982. Nevertheless, reclamation efforts have apparently been effective, for according to the Department of Environment' s 1988 Survey of Derelict Land, the total U.K. derelict acreage had been reduced to some 100,000 acres (40,469.45 ha). However, 47%, or 46,700 acres (18,899.23 ha), of that was attributed to mining and related activities (Hughes, 1992, p. 245). Examples of U.K. reclamation projects of derelict land, derived from former mineral workings and subsequent refuse tipping, is treated in detail in Ch. 10, §7.

In addition, Moss reported that in 1974, 8,016 acres in National Parks and Areas of Outstanding Natural Beauty were affected by extraction pits and ancillary plant structures and settlement ponds (Hughes, op. cit).

5.3.2 *Statutory Regulation of Mines and Quarries*

General control of mining for health, safety and welfare at mines and quarries has been regulated by the Health and Safety at Work Act 1974. The main statutes in force for regulating mines and quarries are the Mines and Quarries Act 1954 to 1971. This latter Act from 1954 to 1971 is comprised of several different statutes that have had their controls amended from time to time and consist of: the Mines and Quarries Act 1954; the Mines and Quarries (Tips) Act 1969, Part I, §§1–10, and §38(2); Mines and Management Act 1971, §4(1). Along with subordinate legislation, the group of statutory controls regulate safety, health and welfare for employees, employment of women and youths, and the security of tips associated with mines and quarries. The Mines and Quarries (Tips) Act 1969 is particularly concerned with the prevention of public danger from disused tips (Halsbury, 1980).

5.3.3 *Nuisance Claims*

Referring again to nuisance claims, both public and private, despite the presence of environmental regulations, the charge of nuisance still plays a part in the Anglo nations' legal schemes. Current U.K. law provides for statutory nuisances. The principal legislation dealing with nuisances and offensive trades is the Public Health Act 1936, Part III. Section 92(1)(f) authorised miscellaneous offensive sources to be declared statutory nuisances. In the area of mining, unfenced shafts of discontinued or abandoned mines and unfenced quarries have been declared statutory nuisances under both the Public Health Act and the Mines and Quarries Act 1954.

The usual remedy for dealing with a public nuisance is by an indictment or information to a grand jury. It is also possible for the Attorney General to bring a relator action for an injunction to order cessation of the nuisance. A local authority may bring an action for a public nuisance under the Local Government Act 1972, §222, if it is thought expedient to protect the health and interests of the local inhabitants. Although the majority of statutory nuisances are dealt with by local authorities, an individual may proceed on his own under §99 of the 1936 Health Act, which provides:

> "Complaint of the existence of a statutory nuisance under this Act may be made to a justice of the peace by any person aggrieved by the nuisance, and thereupon the like proceedings shall be had, with the like incidents and consequences as to the making of orders and otherwise, as in the case of a complaint by the local authority, but any such order made in such proceedings may, if the court after giving the local authority an opportunity of being heard thinks fit, direct the authority to abate the nuisance".

It should be noted that an individual employing §99, is able to "short-circuit" the normal procedure and is able to initiate proceedings under §94 without first issuing an abatement notice under §93. Complainants successfully using §99 are entitled to reasonable expenses, and magistrates have discretion to make a compensation up to £1,000 against a convicted person.

Nuisance claims remain the principal legal remedy for an individual against environmental pollution, hazards and damages.

Nuisance actions by local authorities are restricted in certain respects for surface mines by §18(2) of the Clean Air Act 1956 where §92 of the Public Health Act 1936 does not apply, e.g., to quarries and mines where smoke, grit and dust emanates from the combustion of refuse deposited in them. However, such offensive air nuisances may be prosecuted under §18(1) of the Clean Air Act 1956.

5.3.4 *Planning — Change of Development and Use*

There is no national or regional development plan. All counties and districts in England have planning functions which are outlined in the Local Government and Planning Land Act 1980. Under that Act, counties are primarily concerned with the preparation of structure planning to control developments for the winning of minerals and aggregates, i.e., the open pits and ancillary processing plants, for waste disposal sites, et al. However, no single legal code treats mineral extraction and infill.

A more flexible approach to "development" plans was introduced with "structure" plans and replacement of development plans with structure plans has been scheduled to take place every five years as former development plans expire. The structural plans are statutorily mandated and provide for inclusion of environmental concerns as well as for economic and social policies. §6 of the Town and Country Planning Act 1971 (TCPA) required a survey amounting to an environmental impact assessment as a prerequisite to any structure plan.

District planning authorities are involved at a lower level in formulating the local plan. These fall into three types, viz., local plans, subject plans, action plans. Subject plans are concerned with mineral extraction operations. The County authorities have certification rights for a proposed district structure plan. Further, under TCPA 1971, §13 requires that a local public hearing on the structure plan be held before a person appointed by the Secretary of State.

Under the TCPA 1971, §22(1) states: "*** development means the carrying out of building, engineering, mining, or other operations in, on, over or under land, or the making of any material change in the use of any buildings or other land ***".

Sub§(3) "*** (b) the deposit of refuse or waste materials on land involves a material change in the use thereof, notwithstanding that the land is comprised in a site already used for that purpose, if the superficial area of the deposit is thereby extended, or the height of the deposit is thereby extended and exceeds the level of the land adjoining the site ***".

Part III of the TCPA 1971 defines the meaning of "development", and requires planning permission for it to take place. "Development" may refer to "operational" development, or to an activity "making a material change of use". Courts generally define "operational" acts as those making physical alteration to the land with some degree of permanence. In *Parkes v. Sec of State for the Environment* [1979] 1 WLR 1308, the court considered the size of the operation, its permanence, and whether it was physically attached to the land.

As Professor Hughes comments, "material change of use" is "more difficult to define and is generally decided as a matter of fact and degree. It is not always necessary for a change to be one of kind for it to be "material". Changes of degree made by a marked intensification of use can be sufficient changes of character to constitute development". He gives an illustration where the annual production increase of concrete blocks rose from 300,000 to 1,200,000 as a sufficient change of character to constitute development (*Brooks and Burton Ltd v. Sec. State for the Environment* [1978] 1 All E.R. 1294). For a comparable U.S. case and change of use, see *Re. Barlow,* infra. "The practical problem for planning authorities in such cases lies in detecting changes constituted by intensification, and in determining a point at which a creeping intensification reaches such a point as to be reasonably recognisable as a change of character". (Hughes, 1986, p. 141–142).

A criticism made here, and similarly made in the American case of *Re: Barlow*, is that this characterisation of a change of use, or intensified use, by increased production serves as a deterrent to growth for business. Proprietors and private enterprise in almost all nations desire growth of their business and economies, consequently, why penalise them for accomplishing that end? That intensification of use can be carried to extremes is illustrated by Hughes' example cited in *Hilliard v. Sec. of State for the Environment* [1978] 37 P & CR 129, where intensification of use was considered for "a particular ancillary use of one farm building" as an increase in activity worthy of being labelled "intensification". Fortunately, the court found one additional farm building to be insufficient to qualify for intensified land use.

Hughes continues, "Whether change is 'material' must be assessed by reference to its effect on an appropriate area of land, known as the 'planning unit'. Generally, this will be the whole area in a landholder's ownership or occupation (for a relative U.S. case where the entire property was not permitted for quarrying, see *Wolverhampton,* infra.) *** there are three ways of dealing with the issue: (a) where an occupier has a single main purpose (with or without ancillary uses) the whole of his unit of occupation is the planning unit; (b) if there is a variety of uses, or 'composite use' where activities may vary and fluctuate, the whole unit is the planning unit; and (c) where, however, within a holding physically separate and distinct areas are occupied for different and unrelated purposes, each separate are is the planning unit. The test is whether physically and functionally there are separate uses. A unit having varying fluctuating uses may experience change without 'development', but particular marked changes, for example major intensification of individual uses, may amount to material change of use". (id. 142).

5.3.5 *Recent United Kingdom Cases*

More recent litigation in tipping cases where open pits were being used for waste disposal have been brought under the Mines and Quarries (Tips) Act 1969 and the Control of Pollution Act 1974.

(i) In *Alexandra Transport v. Sec. of State for Scotland*, [1973] 27 P & CR 352 (Property & Compensation Reports), the Court of Session-Second Division, on appeal by the quarry owner, upheld an assessment of a betterment levy affirmed by the Land Tribunal on the quarry property owner after quarrying permission had expired. Continued backfilling of quarry waste was allowed and tipping of external waste for infilling had been approved by the Glasgow Corporation. Under §10(2) of the Town and Country Planning (Scotland) Act 1947 and §99(2) of the Land Commission Act 1967, the court found that a material change of use had taken place from a quarrying operation to tipping and constituted "development" as specified in the statutes. The levy was upheld.

Criticism is made here for laws levying a tax against a quarrying operation proceeding with reclamation of the quarry site. Reclamation of the site was stipulated as a condition precedent by the initial use approval given by the Glasgow Corporation in 1950 when the quarrying operation was started, and tipping was to be considered in the reclamation process at the proper time, although requiring prior permission by the Glasgow Corporation. Reclamation of quarry and mining pit sites is desirable and to be encouraged from an environmental standpoint. Utilisation of worked-out quarries and mine pit sites for tipping as part of the reclamation process, obviously, serves two important environmental purposes, viz., (1) a most probable safe depository for accumulated refuse and wastes, and (2) serves as infill material to complete volumetric filling of the mined hole thereby allowing restoration of the ground surface to its original condition, as well as allowing for the land eventually to serve another utilitarian purpose, as building upon it.

Without refuse and waste as infill, there would be insufficient quarry waste rock and soil left over from the mining operation to back-fill the mined-out hole, and restoration of the surface to original condition and surface levels would not be possible. A body of water at the surface would be the only result, which certainly is not amenable for future building. Its only use would be for possible recreational purposes. To disallow the mined-out quarry hole for tipping would be to defeat environmentalism's goal of restoration of the land surface to its original condition and levels. Thus, while reclamation of mined-out land is betterment in itself, it is only restorative development, not a net improvement of the original land. Complete reclamation of mined-out land is desirable, thus it must be encouraged. A betterment levy serves as discouragement to fully restore the mined-out land to its original condition and lending itself to future further use. Levying a sum on the operator for "betterment" of the land, which is in reality only reclamation after mining, is harmful and a deterrent to full restoration and complete reclamation. It also discourages and defeats utilisation of man-made depositories for waste disposal. It would be less costly for the former quarry site owner to minimally comply with reclamation regulations by removing surface equipment, "tidying up the surface", and let the pit fill up with water. However, best restoration would be defeated.

(ii) In *Roberts and Another v. Vale Royal District Council and Another* [1977] 39 P & CR 514, the Roberts, owners of the Manley Quarry at Manley in Cheshire, had been given permission in 1960 under the Town and Country Planning Act 1971 for tipping of industrial and builders' waste materials into the quarry. The type of tipping material had been so limited to "prevent nuisance arising from the tipping operations" according to the controlling authority, the Vale Royal District Council.

In 1974, the Vale Council issued an enforcement notice challenging the Roberts' tipping operations and alleging that "materials have been tipped into this quarry that are of a kind other than those approved by the local planning authority" which constituted a "material change of use" in violation of the planning act. The Queen's Bench Division — Divisional Court upheld the enforcement order requiring "restoration to the position as it was before the unauthorised development".

(iii) The case of *Regina v. Derbyshire County Council, ex parte North East Derbyshire District Council* [1979], 77 L.G.R. 389 (Local Government Reports) concerned a jurisdictional squabble between the Crown and two authorised licencing entities for tipping in a worked-out surface mine, viz., the Derbyshire County Council and the Northeast Derbyshire District Council. Although the licencing procedures are of interest, greater interest in the case for the purposes herein lies in the requirements for infill of the open pit under the two granting authorities.

The plaintiff / applicant, the Northeast Derbyshire District Council, applied for an order of certiorari to quash a waste disposal licence issued under §5 of the Control of Pollution Act 1974 in 1978 by the Derbyshire County Council to Cambro Contractors Ltd. and Cambro Waste Products Ltd. (hereafter for either, Cambro) giving them permission to deposit waste materials in a worked-out mine pit at Stretton in Derbyshire. The ground of the application was that the issue of a licence was *ultra vires* the county council because no planning permission for the use in question was in force as required by §5(2) of the Act of 1974.

Section 3 of the Control of Pollution Act 1974 provides:

> "(1) Except in prescribed cases, a person shall not- (a) deposit controlled waste* on any land ... unless the land on which the waste is deposited ... is occupied by the holder of a licence issued in pursuance of §5 of this Act (in this Part of the Act referred to as a disposal licence) ... which authorises the deposit ... in question and the deposit ... is in accordance with the conditions, if any, specified in the licence".
> [author's note: * controlled waste is defined in §30 for the purposes of Part I as being "household, industrial and commercial waste or any such waste".]

In 1969 the county council granted a company planning permission to win and work fire clays, shales and associated coal measures by open cast means over a site of 130 acres (52.71 ha). Condition 4 of the planning permission required the site to be backfilled after completion of the mining with stored overburden, adding "such quantity of fill as may be necessary to make good former levels". By Condition 8, the working of the site was to be carried out in successive phases of six acres each, the restoration of each phase being completed and the land concerned being restored to agricultural use when the next phase began. Condition 8 was never observed, although some backfilling had taken place. When the mining had stopped, 21 acres (8.5 ha) remained unfilled.

In 1978, Cambro purchased the land and was granted a disposal licence under §5 of the Control of Pollution Act 1974 by the County Council, under which they could deposit waste products, including sewer sludge, on the 21 acres (8.5 ha). Condition 3 of that licence called for the construction of an access road to the site before the depositing began. Conditions 7 and 8 stipulated that walls and floor of the void (pit) shall be protected with a compact band of selected impervious material. The object of Conditions 7 and 8 was to ensure that any toxic wastes which may be deposited would not escape through the pit's walls and cause pollution to the surrounding water courses.

(iv) Another case that concerned authorised and unauthorised tipping materials deposited in a worked-out quarry occurred in *Bilboe v. Sec. of State for*

Environment [1978] 39 P.& C.R. 495, Q.B. Division-Divisional Court, July 1978; Court of Appeal decision, February 1980.

Case Facts: In 1938, an agreement was entered into between the owner of the worked-out Sandfield Quarry located at Aughton, Lancashire, and the local council, or planning authority, whereby the owner was permitted to tip "approved" materials in two parts of the quarry. No "approved" materials were specified by the Council's surveyor. Before those two areas were filled up, ownership changed hands and in June 1948, it was agreed that tipping should cease on March 1, 1950, contemplating that the two designated areas of the quarry would be filled by that date. Over subsequent years, various applications were made for further tipping of builders' rubble, i.e., inert, non-noxious brick, mortar, putty, glass, "undiseased" wood and metal. However, the local authority issued the same conditions for tipping over the subsequent years, viz., requiring the surveyor's approval for tipped materials; and,"no materials of a noxious nature, or materials likely to give rise to overheating or noxious fumes or smell shall be deposited" adding that only sand, soil, rock and clay could be tipped. Apparently, after 1950, intermittent tipping occurred until about 1975 when the Bilboes became interested in the quarry as a continued tipping site. For the record, the Bilboes sought clarification by the authority of "approved" materials for tipping. The specified materials of "sand, soil, rock and clay" were repeated. The Bilboes continued tipping builders' rubble until July 1976, about the time that stop and enforcement notices were served on them by the council.

The Bilboes sought relief from the enforcement notice by appealing to the Secretary of State, principally on the ground of §88(1) (c) of the Town and Country Planning Act 1971. The pertinent section provided that such cessation and enforcement notice "may be served only within the period of four years from the date of the breach of planning control to which the notice relates". Since that period had elapsed at the date of service, the notice could not be enforced.

Author's Comment A recurring issue in U.K. cases during the reviewed period of the 1970s to 1980s involving use of worked-out mining pits as tips appears to be in the agonising approval process as to "development" and whether there has been a change in "use" or "operations" in compliance with the Town and

Country Planning Act 1971. Regardless of the correct nomenclature to be applied under the Act, such semantic trivia for litigation only serves to fog the essential issues and bog-down and delay the critical need for permitting approval and getting on with the use of pits as a depositories for waste.

Prior to passage of the Britain's Environmental Protection Act 1990, there was no overall, well-defined national environmental policy within the U.K. No central body existed, having entire responsibility for development, oversight and operational implementation for the loose assemblage of environmental regulations in force. Various governmental bodies, authorities and agencies shared the duties and supervision of implementing environmental directives and regulations authorised under an assortment of Acts. The lack of a central environmental organisation frequently led to duplication of supervision and control, which in turn led to confusion, less effective environmental control, and certainly greater administrational costs.

5.3.6 *Mineral and Mining Land Use*

As previously stated, there is no single legal code in the U.K. for mineral extraction and reclamation. Hughes calls attention that "Centrally oversight of minerals policy is divided between the Department of Energy, concerned with fossil fuels, the Department of Trade and Industry, interested in industrial minerals, and the Department of the Environment, concerned with land use issues ***. Counties outside London and, for sites in former metropolitan areas and in London, local planning authorities, following the Local Government Act 1985, are mineral planning authorities; also, see the TCPA 1971, §1(2B), having jurisdiction over plan making and development control". However, mining of a mineral deposit is a county matter under the TCP (Minerals) A 1981.

Mining is generally classified as an act of development under the TCPA 1971, §22. Still under the same 1971 Regulations, it is stated that generally "use" in relation to development of land does not include use of land by carrying out mining operations. The hodge-podge of controls over mining, with various exceptions and assorted delegation of responsibility to several authorities for controls is confusing, even mind-boggling.

Every mining permission granted after 22 February 1982 is subject to a 60-year limitation from its date of permission. However, granting authorities may specify various time-limit periods for the mining of a mineral deposit on an *ad hoc* (for this time) basis.

Mineral Land and Mining Right Expropriations — As previously stated, expropriation of mining lands or mineral deposits by the power of eminent domain, has not been one of note in Britain. However, the Town and Country Planning Act in the United Kingdom, and as enacted by several of the British Commonwealth nations, bears a slight resemblance to a regulatory taking in the U.S. in that provision is made to recompense the injured landowner where injustice has been done by the government's act. The power of eminent domain, however, is common to all Anglo countries.

(i) Although it is not a case occurring in England, a claim for expropriation of mineral interests occurred under the TCPA in a British Commonwealth nation. It is of interest for comparison with several of the U.S. property takings cases cited further on.

In *Lopinot Limestone Ltd v. Atty-Gen of Trinidad & Tobago,* (P.C.) [1987] 3 WLR 797, Lopinot Limestone applied for a permit to develop 22 acres of its 200-acre (80.94 ha) tract as a limestone quarry under the Town and Country Planning Act 1980. The Minister refused.

Section 26(1) of the Act provides for payment of compensation if planning permission is refused. Upon refusal, Lopinot sought the compensation under §26(1) of the Act from the Minister of Finance, but was denied any. Lopinot brought an action against the Attorney-General in the High Court of Trinidad & Tobago.

On final appeal, the court found that Lopinot's application related to development of the land by carrying out mining operations; that by §2 mining was outside the statutory definition of 'use' in relation to land, and so the proposed development would cause no material change in the use of the land. Accordingly, the exemption from payment of compensation afforded by §27(1)(a) would be inapplicable to the refusal of permission for that development; that compensation would be payable to the plaintiff in respect of refusal of permission under §26(1).

The Privy Council held that compensation would be payable to Lopinot in the amount equivalent to the difference in the land's value with permission for quarrying limestone (subject to limitations) and its value without permission. Lopinot's application for quarrying permission was remanded to the Minister, and if refused again, Lopinot would be entitled to compensation subject to the provisions.

(ii) **Kirklees Metropolitan Borough Council vs. Calder Gravel Ltd.**

The Kirklees Metropolitan Borough Council tried to rescind (under the Town & Country Planning Act 1932, as amended, §2(3) of the TCP (Interim Development) Act 1943), a 40-year old permission for a local mining operation. In *Calder Gravel Ltd v. Kirklees Metropolitan Borough Council*, [1989] 60 P. & C.R. 322, the Council's challenge to stop the mining in 1984 was unsuccessful.

In 1946, Calder's predecessor had its mining application approved by the Borough's resolution. Apparently, it was done without a written record, at least, none that could be found. For 40 years, all concerned had proceeded on the basis that a valid permission existed. In 1984, the Council contended that since there was no actual document granting permission for the operation, and under the TCPA, as amended, no valid permission could have been granted. In response, Calder then submitted formal application which was denied by the Council.

The Chancery Division heard the matter and ruled in favour of Calder to continue mining. The Court held that "a presumption of regularity arose in favour of Calder from the long treatment of the case as being one in which there was permission". The Council was estopped by its 40-year conduct of permission from denying the existence of such a planning permission.

(iii) **Mining Pit Resumption Denial Reversed- unaffected by a change of use**

The Durham County Council (England) appealed an unfavourable decision to them under the Town & Country Planning Act (1971), which allowed Tarmac Roadstone Holdings Ltd. to resume excavation at its former sand and gravel pits at Ferryhill. (*Durham County Council v. Sec. State for Environment and Tarmac Roadstone Holdings Ltd*, [1989] T. & C.P. [1990] C.O.D. 209).

The Council had alleged a breach of planning control under the Act by unauthorised mining of sand and sought enforcement for reclamation of mined land to its former condition.

Tarmac had mined sand and gravel on its Ferryhill property until 1956 and then stopped mining. In 1957, upon discontinuance of mining, a change of use was granted by the Council for the disused sand pits for household refuse burial. This change was unknown to Tarmac. In 1982, Tarmac applied for renewal of its Ferryhill sand and gravel operation on an area which included the refuse disposal. Its application was denied, but Tarmac, relying on its former 1947 permission, resumed operations. The Council then served its enforcement notice of unauthorised mining.

The Secretary of State, on review, stated that the change of use granted by the council in 1957 had not extinguished Tarmac's 1947 permission to mine sand and gravel. The Council appealed that decision.

The Court of Appeal agreed finding that former rights are only extinguished when they are inconsistent with the new use. Since further **mining of sand and gravel was compatible with refuse burial in the mined-out pits**, the Council's appeal was dismissed (emphasis added).

(iv) **English Quarrier's "Existing Rights" Claim Fails For Permitting**

East Midlands Quarries Ltd and the owners of a limestone property at Banbury, Oxfordshire County, England, applied to the County Council for permission to quarry stone on an 82-acre (33.18 ha) tract. The basis for their claim to "existing rights" to mining was a grant of permission by the former Banbury Rural District Council in October 1947 to quarry the limestone. The permission had several conditions attached, one of which stipulated that: (i) access to the site would be by a designated road, and (ii) that road was to be improved to the reasonable satisfaction of the County Surveyor. It should be noted that the designated access road and the land needed for its widening were neither within the application, nor owned by the applicant. In 1949 the Oxfordshire County Council granted further permission for buildings, plant and equipment for the quarry operation. The access road was never widened, nor was any quarrying done pursuant to the Council's permission (with the exception of one day in 1979 when 20 cu.yds. (15.29 cm.) were removed).

In 1987, the owners wishing to lease to East Midlands, claimed that the permission granted in 1947 "was a valid subsisting permission (existing right) that continued to inure to the benefit of the land". The Council rejected the claim. The Court in *Mouchell Superannuated Fund Trustees and East Midlands Quarries Ltd. v. Oxfordshire County Council*, CA[1991] 1 PLR 97, upheld the Council, but on the grounds that the first part of the condition designating an access road was unenforceable, and the second part was unreasonable, *ultra vires*, and void. The court found that "since without access the quarry could not function at all, the condition, ... was not excisable from the permission was also void.

5.4 United States — In General

The U.S. Congress responded to the "green" movement of the 1960s by passage of the National Environmental Policy Act of 1969 (NEPA), effective January 1, 1970, along with the subsequent plethora of environmental laws created for industry, including mining. NEPA gave birth and impetus to a flood of federal acts in the subsequent two decades. The more important ones that affect mining are those dealing with dust, air quality, water quality, pollution, stormwater runoff, safe drinking water, noise abatement, solid and hazardous wastes and their disposal, wetlands preservation, mined land reclamation, mine and mill site and tailings clean up, protection and preservation of fish and wildlife, endangered species, wildlife habitats, increased restrictions on mineral prospecting and mining on national forests and other public lands, mine safety and occupational health, and penalties for violations.

5.4.1 *The National Environmental Policy Act 1969*

The heart of NEPA are Sections 101 and 102, called the procedural duties of NEPA, outline and govern its application. It is noted that NEPA's objective is "... to create and maintain conditions under which man and nature can exist in productive harmony".

NEPA, §101, states the environmental policies, and charges the federal government with the responsibility of the environmental concerns; and states the purposes, viz.:

(1) act as trustee of the environment and resources for future generations;
(2) assurance of safe, healthful, productive, aesthetic surroundings;
(3) attain the greatest beneficial use of Nature without degradation or risk to health or safety;
(4) preserve cultural and natural aspects;
(5) achieve a balance between people and resource use permitting a high standard of living; and
(6) maintain renewable resources and maximise recycling of depletable resources.

NEPA, §102 authorises the policy-making and promulgation of rules and regulations laws by federal agencies to carry out the purposes of the Act; and charges the executive branch's agencies with certain specified duties to carry out the purposes and objectives of the environmental act. Some of the specifics are:

(1) coordination or interdisciplinary action of agencies for all federal projects, plans etc;
(2) authorisation of the Council of Environmental Quality (CEQ) to coordinate and approve all federal development projects;
(3) establishing the procedure for all federal planning by considering and reporting:
    (i) an environmental impact statement (EIS);
    (ii) any unavoidable adverse environmental effects, if project is implemented;
    (iii) alternative proposals;
    (iv) short term use vs. long term impact on environment;
    (v) any irreversible and irretrievable detriment to the environment in case of implementation.

40 C.F.R. §1502.1 et seq. are the CEQ's regulations for the EIS and include, purpose, format, and alternatives to the proposed action (§1502.14, which has been called the heart of the EIS).

§102 further outlines utopian ideals for international and worldwide concern and coordination and cooperation; advisements and cooperation with the states for environmental quality; collection and use of ecological information in planning and development of natural resources; and to assist and coordinate actions with the CEQ.

Through NEPA, all federal agencies are required to take into consideration the adverse environmental effects of their policies, plans, programs and projects, sometimes referred to as "the four p's". Any of those 'p's' may require an EIS. The NEPA process requires environmental considerations be integrated into planning.

Agency actions can be divided into three categories, viz., (1) "categorical exclusions" which are those that clearly do not have any significant effects on the environment, and do not require an EIS; (2) planning that clearly will have a significant effect on the environment and will require an EIS; (3) planning that is unclear whether there will be any significant impact on the environment. For this category, the proper procedure is to first undertake an Environmental Assessment (EA).

The result of the EA will be whether an EIS is required for further agency action or whether it may issue a finding of no significant impact (a FONSI). Once an agency determines through an EA that a proposal will significantly affect the environment, it must prepare an EIS. The next step is called "scoping", which determines the scope of issues to be addressed. Scoping includes: (1) determination of issues; (2) notification of involved agencies, et al; (3) in depth, or significant, issues; and (4) elimination of insignificant issues . The EIS preparation is designed to be: (1) a tool for decision making; (2) analytical and concise; (3) 150 pages for "ordinary", and 200 pages for "complex" studies. The lead agency, or most involved one, prepares the EIS.

5.4.2 *Other Environmental Regulations Affecting Mining*

The NEPA subsequently produced a series of regulations for all areas of the ecology. However, all federal statutes and acts dealing with minerals and mining are effected by NEPA through rules and regulations. The more pertinent ones to mining, to name a few, were: (1) the Surface Mining Control and Reclamation Act (SMCRA); (2) amendments and rules to the General Mining Law 1872; (3) Resource Conservation and Recovery Act (1988) (RCRA); (4) Materials Act 1947; (5) Multiple Use Mining Act of 1955; (6) The Indian Mineral Development Act of 1982; (7) the Clean Air Act; (8) the Clean Water Act; (9) the Strategic and Critical Minerals Act of 1990; (10) the National Pollutant Discharge Elimination System (NPDES); and, (11) the Mine Safety and Health Act, and (12) The Endangered Species Act.

Innumerable other federal acts and agency rules are effected in lesser ways for minerals, e.g., amendments to the Stock Raising Homestead Act (1916/1988) to prohibit any person, other than the surface owner, from entering stock raising homestead lands patented under the Act to prospect for minerals or to locate a claim under the Mining Law of 1872 without first filing a notice of intent to locate a claim and providing notice to the surface owner; require payment of fees to the surface owner, posting of a bond, and filing a surface reclamation plan; protection for damage to crops and other surface features.

5.4.3 *Problem Areas for Surface Mining*

Environmental regulation particularly affecting surface mining are presented in the following order: (1) mineral land takings by environmental regulation; (2) wetland limitations; (3) planning law or zoning that placed limitations of mining property expansions; (4) limitations of "grandfathering" acts for active mining operations that existed before zoning; and (5) intensified reclamation of mining properties.

5.4.4 *Mineral Land Takings By Environmental Regulation*

After 1970, with the rapid production of environmental laws, regulations and controls, industry in the U.S. found that such laws involved a "command and control" system which encroached on previously held areas of private ownership and self-control of individual property rights. The new system proscribed environmentally harmful acts and effects requiring specific control measures to produce governmentally-desired, environmentally beneficial results for the general welfare. Where environmental regulations ran afoul of individual property rights, the U.S. courts have, at times, upheld the regulations under the "police powers" reserved to all levels of government in the name of the public welfare and best interests. In the alternative in upholding the law, courts have at times held that an injustice had been worked on the property owner by the environmental or zoning regulation. Such an injustice has been held to be "a taking" of private property, that is, an exercise of the governmental prerogative of eminent domain for which the property owner must receive just compensation.

In the past, on occasion, where a charge of government expropriation (a taking) was made by a property owner, or by the owner of a business operation, just compensation from the government was demanded for the value of the property and / or land. If the government could prove that the property or operation upon it was a nuisance, it was well-established in law that no compensation need be paid. Such public nuisance case law precedent was found in *Mugler v. Kansas,* 123 U.S. 623 (1887), for a brewery that had been outlawed by a prohibition act making the manufacture of beer a public nuisance. Public nuisances went uncompensated when taken by regulation. This doctrine was strengthened in the later case of *Hadacheck v. Sebastian,* 260 U.S. 393 (1922) where the U.S. Supreme Court upheld a city (Los Angeles) ordinance that prohibited the manufacturing of bricks and causing smoke and fumes in a residential neighbourhood. (Note: *Hadacheck established that he purchased his land for brick-making when it was well outside the city limits and not near any residential area; that he had no knowledge that his land would ever be annexed by the city.)*

The review of a few prominent and landmark 'takings' cases in the U.S. courts is made because they were caused by environmental regulations controlling and seriously encroaching on the rights for surface mining.

The power of the federal government to take private land for public use, known as the power of *eminent domain*, is derived from the Fifth Amendment of the United States Constitution, which states in pertinent part, "... nor shall private property be taken for public use without just compensation". This is not an implied power to take, provided it pays, but rather, it is a restriction which applies only if, through some other constitutional provision, the substantive power to take is independently granted. The Fifth Amendment restraint on taking private property applies to the states through the Fourteenth Amendment's Due Process clause.

As stated by Thomas J. Schoenbaum, "Environmental land use regulations are particularly subject to attack as a taking". (op. cit.) Since its inception and codification, the National Environmental Policy Act of 1969 (NEPA) has charged federal administrative agencies with the duty of taking into account any adverse environmental effects of their programs and decision-making. As a consequence of agencies' duties to comply with NEPA, their rules occasionally conflict with or prevent mineral landowners from certain formerly allowed uses that now are thought to adversely effect the environmental quality for the overall welfare and development of man. Environmental regulatory preclusion of private mineral land uses may serve as grounds for the landowner to allege a taking of his property. One of the more sensitive areas of environmental regulation affecting surface mining has been in the area of protecting wetlands. This area has produced a number of takings cases brought by mining companies against the government.

A taking has occurred when the entity clothed with the power of eminent domain substantially deprives the owner of the use and enjoyment of his property and, "Property is deemed taken ... when it is totally destroyed or rendered valueless, or when it is damaged by a public use in connection with an actual taking by the exercise of *eminent domain*, or when there is interference with the use of the property to the owner's prejudice with resulting diminution in value thereof". (Black, op. cit.). The "taking" may be for a "public use" only. A use is held to be "public" if it furthers health, welfare, safety, moral, social, economic, political or aesthetic ends, and thought to be in the public interest. The meaning of "public use" is flexible and there has been much disagreement as to its meaning by the courts. [29 Corpus Juris Secundum., Eminent Domain, Sect. 31 (1965)]. Government action that does not physically invade a property, but otherwise deprives the owner of all economically viable use may also be a taking. When a taking is alleged by the landowner through litigation, the crucial issue for the courts is whether the government action is a "taking" requiring just compensation, or merely a regulation under the police power not requiring compensation. Environmental regulatory takings do not physically invade or possess a landowner's property, but may severely deprive the owner of its intended use, thus limiting its value to the owner. As Chief Justice Holmes said in *Pennsylvania Coal v. Mahon,* 260 U.S. 393, 415–416 (1922)" ... if a regulation goes too far it will be recognised as a taking". The judicial system may review challenged agency final actions to assure that the claimed compliance with NEPA is authorised and reasonable, and also determine whether a taking has occurred (Aston, 1992).

(i) Protectable Property Interests: A threshold matter in a taking claim is whether the claimant has a protectable property interest in the allegedly taken property. The U.S. Supreme Court stated in *Keystone Bituminous Coal Association v. DeBendictis*, 107 S.Ct. 1232 (1987) at 1242, that two factors have become an integral part of their takings analysis: "We have held that land use regulation can effect a taking if it 'does not substantially advance legitimate state interests, ... or denies an owner economically viable use of his land'". The Supreme Court quoting from its decision in *Kaiser Aetna v. U.S.*, 444 U.S. 164, 175 (1979), "... this court has generally 'been unable to develop any "set formula" for determining when 'justice and fairness' require that economic injuries caused by public action be compensated by the government, rather than remain disproportionately concentrated on a few persons". Rather, it examined the "taking question by engaging in essentially *ad hoc*, factual inquiries that have identified several factors such as the economic impact of the regulation, its interference with reasonable investment backed expectation, and the character of the government action — that have particular significance". Thus, there is no established formula for determining whether a taking has occurred. More recently, precedential case guidelines have been established for the courts' fact finding in *Whitney Benefits, Inc. and Peter Kiewit Son's Co. v. U.S.,* 18 Cl. Ct. 394 (1989): (1) the character of the government action and whether it substantially advances legitimate public interests; (2) the economic impact on the landowner; and (3) the extent to which the regulation has interfered with the reasonably investment-backed expectations. It is noted that the same guidelines were used to establish values for takings of mineral property in *Florida Rock Inds. v. U.S.,* 21 Cl.Ct. 161 (1990) and for development property in *Loveladies Harbor, Inc. v. U.S.*, 15 Cl.Ct. 391 (1990).

(ii) Although the well-known *Keystone Bituminous Coal* case applied to underground coal mining, neither of which are a subject of this work, the legal principles involved in an alleged taking of minerals are important to this study. Environmental regulation for the safety and welfare of the general public lay at the heart of the *Keystone* case. The statute at issue and complained of as taking was the Pennsylvania Subsidence Act, Penn. Stat. Ann.,Tit.52, §1406.1 et seq. (Purdon Supp. 1986). The final case decision resulted in an unstated finding that mining in that instance constituted a public nuisance to the detriment of the public welfare, health and safety, and the environmental regulations were upheld. The mining company's allegation that a taking of private property, i.e., mineable coal reserves in-place was unjustified and therefore, uncompensable. The government exercise of a taking of private property was justified on the basis of the police power and not eminent domain.

A prime case example of exercise of state police power justifying environmental mining regulation is *Keystone Bituminous Coal Association v. De Benedictis,* 107 S.Ct. 1232 (1987).

During, and prior to, the era of *Pennsylvania Coal Co. v. Mahon* (1922), the legal doctrine of *caveat emptor* (let the doer beware) was much in force. Along with that caveat the related doctrine "of coming to the nuisance" served to some degree to protect mining against claims that now find ready disfavour with the public and environmental regulations. Although a pre-existing activity creating a nuisance was not a defense, prior knowledge of the protested nuisance before coming to it was mitigating if the activity or use was reasonable. A minority of courts denied recovery altogether when plaintiff came to the nuisance with knowledge; see *East St. John's Shingle Co. v. Portland,* 246 P.2d 554 (Or. 1952); and, where the use offended no one until plaintiff moved in, the use was reasonable; see *Spur Inds. v. Del E. Webb Development Co.,* 494 P.2d 700 (1972).

Those doctrines have gradually fallen in strength and use and are no longer recognised in argument against public nuisances and the growth of environmental concerns. The Kohler Act of 1921 P.L. 1198, 52 Penn. Stat. Ann. 661 at issue in *Pennsylvania Coal* and the Subsidence Act (op. cit.) of *Keystone Bituminous Coal* were essentially the same law in purpose and coverage. In 1921, no liability for the coal miners for subsidence of, or damage to, surface structures was found by the U.S. Supreme Court. The decision was based essentially on contract law where the mining companies held waivers for liability of damages from the surface property owners. The contracts were upheld by the court.

The time element of a half-century between the two cases involving essentially the same law and same liability and contract issues, coupled with recent intensive growth of environmental concerns, made the difference in reversing the decision of the earlier *Pennsylvania Coal.* Although the same waiver of liability contacts were in force for *Keystone* as in the earlier case, the court would no longer honour them as valid as they were found to be contrary to the best interests of the public's welfare and safety. Subsidence by mining was a nuisance. Nuisance claims against the mining industry are now far reaching, effective and prevailing.

In *Keystone,* the coal companies alleged a taking of coal under Pennsylvania's Subsidence Act that required the operators to leave in place some 27 million tons (24.49 M mt) of mineable coal to prevent subsidence of the surface. The coal companies argued that they were relieved of a duty to support the surface in that they had obtained waivers for damage to the surface estates' structures and land.

Analysis of noxious and public nuisance uses of coal mining land

Where the alleged taking involves mineral land use, or its intended use by the owner is to mine, that adversely effects, or would effect, the environment, the government may possibly show that it is a public nuisance or noxious use, and therefore, it is detrimental to the public health and welfare. As such, there is no taking and consequently, no compensation is due the landowner.

The government argued that the Subsidence Act was a bona fide exercise of state powers and the act was "a legitimate means of 'protecting the environment of the Commonwealth, its economic future, and its well-being' ". (*Keystone* at 1239). The *Keystone* Court noted, too, that §2 of the Subsidence Act stated it was an exercise as a police power for the protection and general welfare of the public, and for conservation of surface land areas (id. at 1242).

The noxious use argument for mining was used by the government. A "noxious use" has been defined as "that which causes, or tends to cause, injury, especially to health or morals.: [28A, Words and Phrases at 639 citing *Moubray v. G & M Improvement Co.*, 178 A.D.737; 165 N.Y.S.842, 843 (1917)]. Noxious use as applied in *Keystone* to the coal mining methods employed in the area which caused subsidence damage to the surface, was succinctly stated by Justice Harlan in *Mugler v. Kansas*, 123 U.S. 623, 668–669 (1887):

> "*** a prohibition simply upon the use of property for purposes that are declared by valid legislation, to be injurious to the health, morals, or safety of the community cannot, in any just sense, be deemed a taking or appropriation of property ... The power which the States have of prohibiting such use by individuals of their property as will be prejudicial to the health, morals, or safety of the public, is not- and, consistently with the existence and safety of organised society cannot be — burdened with the condition that the State must compensate such individual owners for pecuniary losses they may sustain, by reason of their not being permitted, by noxious use of their property, to inflict injury upon the community".

The *Keystone* Court strengthened the noxious use of land by the coal companies throughout their findings and determination, e.g.:

(1) "The court's hesitance to find a taking when the state merely restrains uses of property that are tantamount to public nuisance ..."; *Keystone* at 1245;
(2) "Under our system of government, one of the state's primary ways of preserving the public weal is restricting the uses individuals can make of their property."; id at 1245;
(3) "Long ago it was recognised that 'all property in this country is held under the implied obligation that the owner's use of it shall not be injurious to the community'", (id at 1246 quoting *Mugler v. Kansas,* at 655);
(4) "The Takings Clause did not transform that principle to one that requires compensation whenever the State asserts it power to enforce it. id. at 1245, quoting *Mugler* at 664;
(5) "As the cases above demonstrate, the public interest in preventing activities similar to public nuisances is a substantial one, which in many instances

has not required compensation". id. p. 1246; also the Court in footnote. 22, p. 1246 noted: Courts have consistently held that a State need not provide compensation when it diminishes or destroys the value of property by stopping illegal activity or abating a public nuisance;

(6) "The use to which the mine operators wish to put the support estate is forbidden. *Mugler* at 716". id. at 1240.

Though the *Keystone* Court did not label the coal mining operations as a nuisance or noxious use of land, per se, in its announced decision holdings, such a finding was evident as the actual basis for its determination that (1) there was public purpose for the Subsidence Act; (2) public interests in the legislation were adequate to justify the impact of the Act on the coal companies' contracts with the surface owners; and (3) the Act did not work an unconstitutional taking on its face.

Keystone Property Value Considerations

In *Keystone* the U.S. Supreme Court, referring to its decision in *Agins v. City of Tiburon*, 447 U.S. 255 (1980), stated "*** the determination that the government action constitutes a taking, is, in essence, a determination that the public at large, rather than the single owner, must bear the burden of an exercise of state power in the public interest" and we recognised that this question "necessarily requires a weighing of private and public interests".

If the benefit to the public of the regulation is small and the detriment to the landowner large, the regulation will not be upheld. Contra, if the regulation benefits the public, then a value must be placed on the land taken to compensate the owner.

The courts have determined that it is insufficient for the claimant to show that there has been merely a diminution in value of the property by reason of the alleged taking (*Loveladies Harbour,* id. at 392). The claimant bears the burden of proving that he has been denied of all viable use of the property and that it is virtually worthless. Property value determination tests have been developed by the courts as to whether the landowner can still derive a reasonable return on his investment.

The Court in *Keystone* did not have to fix a value on the coal left in place since no taking was found. However, the Court found the record lacked evidence that the coal companies could not continue mining the remaining coal, leaving the required coal support, so as to interfere with their investment-backed expectations.

(iii) Another prime example of a taking of mineral land by environmental regulation is *Whitney Benefits, Inc. and Peter Kiewit Sons' Co. v. U.S.,* 18 Cl.Ct. 394 (1989).

In another post-*Keystone* mineral land taking case, *Whitney Benefits,* a regulatory taking of an unmined mineral deposit was found by virtue of a subsequent Congressional enactment and denial of a surface mining permit. Environmental restrictions in the legislative act were the basis for the prohibition of mining.

In 1983 property owner and lessor, Whitney, and mine operator-lessee Kiewit, jointly filed an action seeking just compensation for a taking of their coal-bearing property in the Powder River Basin of Wyoming as a result of the Surface Mining Control and Reclamation Act's (SMCRA) prohibition of surface mining on the alluvial floor of the Basin. The Claims Court dismissed the claim because SMCRA provided for an exchange mechanism as a "method for ascertaining and paying just compensation, thus, no taking had occurred or accrued until such mechanism had failed to provide just compensation". (*Whitney Benefits,* op. cit. at 398). The Court of Appeal stated that, "... the mere existence of an exchange provision, a remedy available at plaintiffs' option, did not determine whether or not the statute had effected a taking". (id. at 398). The decision was reversed on appeal and remanded for trial. In the interim, Whitney and Kiewit (PKS) had attempted an "exchange" with the Bureau of Land Management (BLM). Before the appeal decision, Whitney had filed an action to compel an "exchange" and the federal District Court of Wyoming found that there had been unreasonable delay and issued an order compelling BLM to offer properties for an "exchange" of equal value. On further appeal, the Court stated that the exchange mechanism is one which may be negotiated, or may be rejected by the claimants and pursue a money award under the Tucker Act (id. at 399). On the remanded trial, the facts of the case concerned Whitney Benefits' land of 1327 acres (537.03 ha) which were irrigated and subirrigated by the Tongue River alluvial valley floor. The land was leased to PKS in 1974, and advanced royalties were paid to Whitney. PKS expended exploration costs of $1 million in 1976, and PKS filed a permit application with the Wyoming Department of Environmental Quality (DEQ).

A year later, SMCRA was enacted. Part of the SMCRA guidelines provided that no permit or application shall be approved if it should "interrupt, discontinue or preclude farming on alluvial valley floors that are irrigated or subirrigated ..." Thus, Whitney's right to mine the coal on its property was invalidated by the enacted legislation of SMCRA and was the basis for the alleged taking in 1983.

In 1981, PKS had requested an exchange for federal lands to the BLM. BLM offered a coal tract, Ash Creek, and PKS spent $130,000 on exploration costs on it. The BLM also offered the Hidden Water tract, which PKS refused as it had mined it in the late 40s to early 50s and was not interested in the remaining coal. PKS and Whitney proceeded with their 1983 claim under the Tucker Act for a 5th Amendment regulatory taking.

The Court's process for determining if a regulatory restriction resulted in a taking, was to consider three factors: (1) the economic impact of the restriction on the claimants' property; (2) the restriction's interference with investment-backed expectations; and (3) the character of the government's action. With regard to the first factor, the Court found that there was a market for Whitney coal; that Whitney coal was economically and technologically mineable, and that it was valuable; and that the enactment of SMCRA had a "devastating economic impact on the property". As to second factor, the Court found that investors could reasonably expect the returns on investments as projected. In place assigned reserves were valued at $1.01/ton, and residual reserves at $.20/ton. The Court found for the third factor that there were no economically viable alternative uses for the property.

The government argued that an exercise of police power, in carrying out government regulations for the health and welfare of the public can never result in a taking. The Court responded that in relying on modern Supreme Court cases, "taking law" had progressed much, and quoting from the trial court in *Florida Rock* that "it is no longer asserted that a regulation, just by its nature of being a regulation, cannot be an exercise of eminent domain". The

Court further stated that, "... the substantial public interest at stake does not outweigh the private interest so that plaintiffs must bear the full burden imposed by the government action". Upon finding that the enactment of SMCRA took Whitney's property, the Court had to establish the date of the taking, and the value for just compensation. The Court held that the effective date of the taking was the effective date of SMCRA, (August 3, 1977). The value of the property was not to be the value to the owner for his particular purposes; i.e., not for opportunities the owner loses (lost profits).

After extensive market and mining cost investigations, the Court established a final sum of $60,296,000 for the total 1977 value of recoverable Whitney Coal for an assumed annual production rate of 2.5 million tons and a cost of $2 million for backfilling. Interest was payable to Whitney from Aug. 3, 1977 to date of payment. The taking sum was established as what a willing purchaser would have paid Whitney as a willing seller, to mine the Whitney Coal after calculating all mining related costs.

On remand of *Whitney Benefits*, the U.S. Claims Court held that: (1) the enactment of SMCRA totally eliminated economic value of plaintiffs' coal and constituted a taking under the Fifth Amendment; (2) the taking occurred at the time SMCRA became effective; (3) the valuation method incorporating discounted cash flow approach offered reliable method for determining the fair market value of the coal on the date of the taking; and (4) the plaintiffs were entitled to pre-judgment interest.

The *Whitney Benefits* decision was later upheld on appeal.

5.4.5 *Wetlands Limitations*

Following are litigated cases in which wetlands and zoning regulations came into consideration for mineral related and surface mining properties.

5.4.5.1 *Florida Rock Inds. v. U.S., 21 Cl.Ct. 161 (1990) — Mining Not a Nuisance*

In *Florida Rock,* the U.S. Army Corps of Engineers (Corps) denied a Section 404 permit to Florida Rock to operate a limestone quarry for crushed stone in

Dade County, Florida. Florida Rock had purchased its 1560-acre tract in 1972 solely for mining the limestone to supply the building construction boom of the area. It began a dragline operation in 1978 during an upsurge in the market. In December 1977, just prior to the start of its operations in 1978, the Clean Water Act (CWA) had been amended to include "wetlands" giving the Corps jurisdiction in addition to its regular "navigable waters", Public Law 95–217, 91 Stat.1567 (Dec. 27, 1977) amended the Clean Water Act, 33 USC §1251(a) (1988) and expanded the jurisdiction of the Corps to regulate activity affecting navigable waters. By the same law, §404 established a mechanism for applying for permits to discharge dredged or fill material into waters covered by the CWA, 33 USC 1344 (1988).

The Corps issued a cease and desist order to Florida Rock and the operator complied, restoring the surface to its pre-existing condition. The Corps advised that a §404 permit was required before operations could resume. The Corps also advised that a permit would only cover a three-year period and the whole tract could not be permitted at once. Florida Rock applied for the maximum area allowed, 98 acres.

The §404 permit regulated discharged dredge and fill material into waters controlled under the CWA. Florida Rock applied for the permit. After two years of waiting, the Corps denied the application. Florida Rock filed a suit against the U.S. seeking compensation for a regulatory taking of their land. The 1985 Claims Court decision in favour of Florida Rock was appealed to the U.S. Federal Circuit Court, which affirmed that only the 98-acre tract could be considered as a taking, not the entire acreage, and remanded the case to the U.S. Claims Court to re-examine.

The issues for the Claims Court on re-examination were: (1) did Plaintiff / Florida Rock have a legitimate entitlement to the proposed use of its property?, and (2) if so, whether the Corps' denial of a §404 permit denied the Plaintiff the economically viable use of its land so as to constitute a taking under the Fifth Amendment; and (3) if so, the amount of compensation to which the plaintiff is entitled. (op. cit. at 165)

If a taking was found, the proper amount of compensation as damages would have to be measured by the difference in the fair market values before and after the alleged taking; whether there had been economic impact on Florida Rock, and the severity of it, would be considered in the damage determination.

Analysis of noxious and public nuisance use of limestone mining land

The government argued that even if Florida Rock successfully proved that the permit denied resulted in a loss of all economic value of its property, the denial could not be viewed as a "taking" because the Plaintiff was not legitimately entitled to use the land for quarrying. This argument was based on former court decisions that quarrying in the present case was a "noxious use" and a "public nuisance" (58: "Nuisances" Sect. 7. "Public nuisance" has been defined as the doing of or failure to do something that injuriously affects safety, health, or morals of the public ***". Whatever tends to endanger life or generate disease, and affects the health of the community, etc. *** is a public nuisance. [*State v. Turner*, 198 SC 487 (1942), 18 S.E.2d 372] and that "*** the government need not compensate 'individual owners for pecuniary losses they may sustain, by reason of their not being permitted, by a noxious use of their property, to inflict injury upon the community' ".) (*Florida Rock,* id. at 166, quoting *Keystone*, id. at 1244, relying on *Mugler*, id. 668–669).

The Claims Court stated that the government's position on the law was accurate, but added that it did not apply in this case. Florida Rock had been characterised as a moderate polluter, but certainly not to the extent of placing toxic wastes in the drinking water. The Court noted that although the government's contention that the proposed activities of Florida Rock posed a risk of contamination of the sole aquifer furnishing drinking water for Dade County / Miami, it did not contend that the limestone mining would actually contaminate the aquifer, and it had produced no evidence to quantify that risk. The Court had observed that there were many operational limestone quarries and pits in the area close to the Plaintiff's property. The Claims Court stated that, "Rock mining of the type planned for the property never has been considered a nuisance"; and such a fact "belies any claim that a nuisance is involved here". (id. at 167). In addition it noted, "If the Plaintiff had begun mining before the CWA amendments, it would have been 'grandfathered in' and no federal permit would have been required". It was clear to the Court that the "nuisance exception", which required no compensation for a taking where a noxious use or nuisance is involved, did not apply in this case. In fact, the Court observed that desirable reclamation uses had been made of the mined-out land in that area. Some former pits were turned into small lakes and overburden that had been removed had been placed to raise the ground level above the high-water level affording building sites for residential development. It was further noted that a zone of development followed depleted mining operations of former years in the same area. The Court further stated that, "here, it is clear that the nuisance exception to the Fifth Amendment's requirement of just compensation is inappropriate. Rock mining of the type at issue here has never been considered a nuisance in this area. ... If the government's use of this argument were upheld, the concept of a regulatory taking would be virtually meaningless and severely limit the protection of the Fifth Amendment". (id. at 167).

Value Determination For The Property Taking

The measure of economic impact on the claimant was determined by the "before and after permit denial" fair markets values of the property, and whether there existed any other uses for the property.

After examining the appraisals of both parties and hearing their property value arguments and evidence, the Court was convinced that knowledge of the regulatory wetlands restrictions and controls on the property severely limited its marketability among knowledgeable investors and buyers, and the Court accepted the Plaintiff's evaluation of $500/acre. The pre-denial fair market value of the land was established at $10,500/acre, and limestone mining was found to be the highest and best use for the property.

In concluding its opinion, the Court stated that for a plaintiff to prevail in a regulatory takings claim, the comparison of the before and after fair market values must indicate more than a diminution of value of the property. Otherwise, the government could not continue to pay for every property that was diminished to some extent whenever affected by a law. However, in this case, the diminution of value was not slight, but substantial, being diminished by 95% of its pre-permit denial value. Still, that was insufficient, as even "more than a substantial reduction of value is required for concluding that a 'taking' has occurred. The owner's opportunity to recoup its investment, subject to the regulation, cannot be ignored". (id. at 176). The fact that Florida Rock had to purchase other property to meet its mining needs illustrated the extent of the economic impact on the company.

In summation, the Court quoted the US Supreme Court, "When a regulation goes too far, the Fifth Amendment requires compensation". (*Pennsylvania Coal,* 260 U.S. at 413). The Court's decision in *Florida Rock* was: (1) the proposed use of the property for limestone mining would not have constituted a nuisance; and (2) the Government failed to establish that the investment market for property following a denial of permit was comprised of investors with knowledge of restrictions on land; and (3) denial of permit constituted a taking for which landowner was entitled to damages in amount of full fair market value of property at the time of taking. The amount of damages was fixed at $1,029,000 plus interest from the date of the taking. The permit denial date was the date of taking.

(i) U.S. Army Corps of Engineers' §404 Permit
In *Borough of Ridgefield v. U.S. Army C.O.E.*, No. 89-3180, slip op. (D.N.J. July 2, 1990): The Corps considered local zoning regulations in its denial. The court held that the Corps may properly do so in determining whether practicable non-wetland alternatives were available.

(ii) Wetland-Zoning Permittal Denial-Alleged Taking
In *Cioffoletti v. Planning & Zoning Commission,* 552 A.2d 796 (Ct. 1989), Cioffoletti owned and operated a 23-acre sand and gravel pit in Ridgefield, Conn. and applied for pit expansion within and adjacent to existing wetlands. The Planning Commission referred the operator's application to the local soil and water conservation district, the US Department of Agriculture's Soil Conservation Service and the State EPA, all of which approved it. However, at public hearings, an opponent expert testified to adverse impact on the wetlands and postulated the inevitable, ubiquitous clincher question that successfully defeats a mining permit, a point that the government agencies had failed to address: the possibility of damage to the groundwater supply in event of a fuel spill.

The operator was granted a special use permit with restrictions and a $100,000 bond. Cioffoletti sued Ridgefield attempting to show that the detrimental economic effect of the restrictions effected a "taking" of his property without just compensation. The trial court rejected his arguments. On appeal, the Supreme Court of Connecticut found: (1) the exclusion of the operator's economic effect evidence limited the court's ability to rule on the property-taking issue, and ordered that the taking-claim be heard; (2) the Commission's right to regulate mining and excavation in areas adjacent to wetlands under the Inland Wetlands Act was proper; (3) due process was met as claimant was given a fair hearing before the Commission; and (4) the Commission had a right to require a performance bond.

(iii) Wetlands Mining Permit Denial

In *Florida Dept. of Environmental Regulation (DER) v. Goldring*, 477 So.2d 532 (1985), Goldring appealed an order of DER denying his application for a permit to mine limestone on a site in Dade County, Florida. Goldring challenged DER's jurisdiction.

At the DER permit hearing, mining permission approval hinged on whether the presence of saw grass, a fresh water aquatic plant on Goldring's property, could establish his land as being within DER's jurisdiction within the landward or upland extent of saltwater Florida Bay. DER claimed jurisdiction on the presence of the saw grass, coupled with the flow of fresh water across his property to the state waters of Florida Bay. The Court of Appeals found that the presence of aquatic vegetation, alone, cannot establish an exchange of waters. However, the Florida Supreme Court disagreed.

The Florida Supreme Court upheld DER holding that DER's dredge and fill jurisdiction depended on the predominance of listed aquatic vegetation on the property along with an exchange of waters, either one-way or two-way. Thus, the permit for mining limestone was denied on the basis of harm to aquatic vegetation.

5.4.6 *Planning Law or Zoning Regulation Limitations*

Placing Limitations on Mining Property and Pit Expansions — Land use planning for mineral properties — zoning regulation

Mineral land use in most localities, other than some federal / public lands, is dependent, as in England too, upon local planning or zoning. Where true, mine permitting by the appropriate agency cannot be obtained until local zoning approval for a mining operation has been obtained (as in *Barrett Paving*, infra @ p. 132). While preparing for public hearings on local zoning, it is advisable to proceed with mining plans under tentative approval by the mine permitting agency. Environmental concerns are treated in the agency process which should alleviate many of the concerns of the local citizens at the public zoning hearing. The ubiquitous NIMBY syndrome (Not In My Backyard) will be encountered at the zoning hearing. Well-prepared plans to cope with environmental issues are essential in assuring the citizenry that their backyard will not deteriorate because of the presence of a new mining operation, and that, on the contrary, benefits will be reaped. (See 5.4.5.1, Analysis of noxious and public nuisance use of limestone mining land, *Florida Rock*, supra for support).

Oft-times, in spite of well-prepared information and environmentally-safe assurances for the public hearing of permitting approval of a surface mining operation, local hysteria of water pollution, et al, overcomes the public reaction leading to the denial of a mining permit. For a typical, but outstanding, case in point reviewed in *Pit & Quarry*, the 1994 Ohio appeal of *Fulton Farms, Inc.* to obtain a conditional use permit for mining sand and gravel on its privately-owned lands is given in support.

NIMBY SYNDROME CONTINUES FOR OHIO S & G OPERATOR

"Northmont Sand & Gravel Company ran out of sand and gravel reserves about 1989 at their location in the alluvial deposits along the Stillwater River at Union, Montgomery County, Ohio. Before depleting the reserves on its 90-acre tract, Northmont planned to continue its operation just across the road on an adjoining property. That's when its troubles began and continued up until the present. The Union Zoning Board denied Northmont further extension of its conditional use permit beyond its worked-out property in spite of its environmentally clean operating record for the past 23 years.

In the intervening five years, Northmont has unsuccessfully attempted to have five properties permitted for sand and gravel operation. The results of the public hearings for each zoning and permitting have been the same — **"not in my backyard"**. The area apparently has a stigma of environmental hysteria without cause or educated reason. Public hearings sound like a broken record with the same challenges — "**What if** there should be an oil spill, ... their water would be polluted", or "Mining of the sand and gravel **will** contaminate the water well field ...", etc.

The irony of Northmont's plight is that they operated at the Union site for 23 years with the town's number one water well in the pit, and a second town water well on the very edge of the their pit. During all of Northmont's mining operation, the water from #1 well was so pure that it was placed directly into the town's water system without any prior treatment. The #2 well required only treatment for excessive iron content which was a natural contaminant and not caused by Northmont.

The singular fact that Union's best and largest water producing well was right in the pit's floor, standing like a monument to good, safe environmental mining practice by Northmont, should have been sufficient evidence and reason for permit renewal by the Union zoning board. It should have been testament enough to convince the town's representatives that Northmont was an environmentally-safe mining operator worthy of staying in business in Union while keeping their water supply safe. However, it was insufficient to overcome local public environmental hysteria that their water well field might become contaminated by sand and gravel mining. Northmont has been fighting that same unfounded fear since 1989 in trying to locate a new operating site in the area.

In its latest attempt in 1992 to operate a pit in an adjoining county, Northmont had worked out an agreement with Fulton Farms, Inc., for mining of the sand and gravel on its 50-acre property. It remained only for approval of the zoning board of Elizabeth Township, Miami County, to grant a conditional use permit. The transcript of the public hearing read like an echo of the hackneyed environmental hysteria objections of those always opposed to permitting for sand and gravel mining, or any other minerals.

Reference to Northmont's environmentally clean record at its former Union site in the town's water-well field meant nothing to the objectors. The only reference to it made by the local people at the hearing was in the negative — that Union would not renew their mining permit — which of course implied that "it must have been an undesirably bad actor"; exactly the opposite from Northmont's safe record.

In spite of well-prepared information and expert testimony by the permit applicant to counter the anticipated, usual environmental arguments of objectors, the zoning board denied a conditional use permit to mine the sand and gravel. Fulton Farms filed for judicial review of the Elizabeth Township's Board of Zoning Appeal's decision in the Common Pleas Court of Miami County.

"In *The Application for Conditional Use of Fulton Farms, Inc.* (1992), the applicant challenged the Elizabeth Township's BZA with errors of: (1) its decision of denial was "arbitrary, capricious, unreasonable and unsupported by the preponderance of substantial, reliable and probative evidence on the whole record"; (2) allowing hearsay (unsupported) evidence which was arbitrarily considered; (3) the Elizabeth Township Zoning Resolution [21.08(4)(c)] was vague and unconstitutional; and (4) the BZA usurped legislative power in zoning.

The Common Pleas Court of Miami County, Ohio, (General Division), affirmed the BZA's denial of a conditional use on all charges. Fulton Farms appealed the trial court's decision. Again, the decisions of the BZA and the trial court denying the conditional use of Fulton Farms' land were upheld in September 1994 by the Court of Appeals for Miami County. The Court found the Board's decision had not been arbitrary and that the regulation was not unconstitutionally vague.

"Not in my Backyard" for unjustified fears still prevails (Aston, 1995a, Pit & Quarry, April 1995).

Until very recently, little to no thought has been given in land-use planning for future mining of known mineral reserves. Mining operations and mineral lands have very much been neglected as a planned land use category. Most often, no forethought has been given to it by planning authorities. As stated, minerals have to be mined where they are located. Valuable minerals have been lost where the surface had already been established and used for another use, as residential, or business. In the cases of minerals that can be won by a pumping process, e.g. hydrocarbons, sulphur, salt brines, subterranean leaching of uranium, et al, they may be recovered with proper caution against surface damage and subsidence. Simultaneously with pumping extraction, the surface area required is minimal and the surface may be used for other purposes.

For a supporting argument of minimising damage to the surface estate, a review of the Texas *Plainsman* case from the *Engineering & Mining Journal*, February 1995, follows. The main issue in *Plainsman* was whether the minerals found at a relatively shallow depth belonged to the surface or the mineral estate. However, prospective damage to the surface estate came to light and the mining method proposed was recovery by *in situ* solution of the uranium mineral deposit with recovery by pumping.

ASSAYS FROM THE LEGAL VEIN ™
TEXAS SURFACE-DESTRUCTION TEST RE-SURFACES
by R. Lee Aston, Attorney, Mining Engineer

"Consideration of the surface-destruction test in determining the ownership of shallow-occurring mineral deposits in Texas surfaced again in the appellate review of *Plainsman Trading Co. v. Crewes*, 875 S.W.2d 416 (Tex.App.-San Antonio 1994). (See *Assays from the Legal Vein*, June 1993, E & MJ.)

In *Plainsman Trading I* (1992) a uranium deposit at a depth between 150 and 190-feet was decided to belong to the surface estate owner (Crews) and not the mineral estate owner. The decision was based under the maverick Texas case law of the "surface destruction" test. In that court-test, where ownership of the minerals **"in and under the surface"** have been severed from the surface estate and granted by deed to the mineral estate owner, any shallow-occurring mineral deposit that is exploitable by surface mining methods which would totally consume, deplete or destroy the surface, must belong to the surface estate owner and not the mineral estate owner. Thus, by Texas case law, the mineral estate had been lowered to a depth below 200-feet, and simultaneously, the surface estate has been lowered to a depth of 200-feet.

Plainsman presented evidence to demonstrate to the court that the mining of the uranium deposit would not essentially destroy the surface estate (of some 1900-acres). Plainsman maintained that it would not use surface mining to take the uranium, instead it would use the prevailing method of *in situ* leaching or solution mining by drill holes. Consequently, the surface would not be disrupted or destroyed as assumed by the court and the surface owner. Nevertheless, the court held fast to previous Texas case law that shallow mineral deposits are part of the surface estate to a depth of 200-feet. The fact that a mining method not destructive of the surface could be employed had no effect and gives the conclusion that the surface-destruction test cannot be successfully passed by a mineral estate owner.

Plainsman charged error by the trial court in its 1992 charges to the jury for its decision. On its 1994 review, no error was found by the court and the former decision was affirmed (Aston, 1995c).

5.4.7 *Mineral Land Planning*

Virtually no forethought or foresight with regard to land use planning has been given in the past for the mining of minerals, that is, to set aside and reserve land areas that are known to contain certain valuable and unmined minerals. Great Britain gave notable early recognition to the subject when the Government published a memorandum entitled "Control of Mineral Workings in England and Wales (the "Green Book") in 1951. The publication served as a guide to the planning control of mineral working and to indicate the broad lines of policy on the planning problems raised by mineral working. A revised edition was published in 1960. Since then, several Mineral Planning Guidance (MPG) papers have been published, each treating planning in more detail, and each increasingly recognising the need for giving more emphasis to reserving mineral bearing lands. (for further discussion on MPG's see Ch. 9 §3.2). Even with the advent of more strict land regulations, mining operations, whether existing before regulation or new, are still dealt with by the majority of local planning authorities on an *ad hoc* basis as they have been in the past. Forethought to reserving land areas for the mining, particularly of construction materials, e.g., sand and gravel or rock types suitable for aggregate manufacturing, should especially be given greater consideration in more densely populated areas to avoid future conflicts with other types of zoning, e.g. residential, recreational.

As an illustration, a survey by the Chief Planning Officers' Society in November 1990 found that (on a 70% return) there exist 980 largely unworked, "ideal sites" with consents for the extraction of a whole range of minerals (Hughes, 1992, p. 256).

Mineral planning for construction minerals is necessary to a growing and expanding community and is given a proper place by the top British planning authorities. After all, it is the establishment of new businesses and industry that creates economic growth for a community to a greater degree than recreational facilitates alone. "Although the whole world be a scenic park for enjoyment of nature and sport, it would only be a habitat for our animal friends without the income from industry for humans to enjoy the environmental beauty".

So often recreational areas and parks are planned or located where construction and other minerals are found. Recreational areas can be located elsewhere, but minerals must be mined where they are found. Some resolution of the running conflict between park planning and mining interests has been made in British Columbia. (See Ch. 7 §4.3, British Columbia, infra). Further support is noted in the British Report, Environmental Effects of Surface Mineral Workings (HMSO, 1991, p. 13) that "As good planners, the Planning Authorities should consider what they could do to protect the surface workings (of mines) from encroachment by sensitive land uses during its life", i.e., to keep them apart.

The future planning of land use for mining of construction minerals in the U.S. was given an early start in 1981 in Sonoma County, California, by its Board of Supervisors. That county had the foresight in its planning to reserve some known terrace land along a local river for future mining of sand and gravel, thus enabling a future source for construction aggregates for the local area.

A case arising out of Sonoma County's planning for mining of construction materials occurred in *Sierra Club v. County of Sonoma*, 8 Cal. Rprtr. 2d 473 (Cal. App. 1 Dist. 1992).

In 1981, Sonoma County adopted an Aggregate Resources Management Plan (ARMP), a long-term plan for future gravel and hardrock mining within the county. Under the plan, managed resource lands available for future supplies of aggregate materials were specified. Lands designated for agriculture included those overlying designated mineral resource lands, and were to be preserved for their agricultural value and as groundwater recharge zones. An area of about 2,000 acres (809.9 ha) in the Russian River flood plain was designated as eligible for mining permits. The loss of agriculture lands in the area was accepted as an unavoidable impact from mining alluvial sand and gravel to depths up to 75 feet (22.86 m). Reclamation of the worked-out pits was given priority. The reclamation plan directed stream diversion be made through the pits allowing refilling of the pits by river-borne sediments. A county ordinance required each terrace mining applicant to submit his own reclamation plan, and upon which permit approval was contingent.

After ARMP was in effect, the Basalt Rock Company was permitted to mine a 50-acre (20.23 ha) tract. Syar acquired Basalt Rock's 50-acre permitted tract in 1986. In 1989, Syar applied for an amendment to trade 145-acres (58.68 ha) of their designated mineral resource land for a 145-acre agricultural tract along the river not designated for mining. Syar then applied for other amendments of ARMP to permit 50-acres of the tract for mining and to allow reclamation by refilling the pits with processing sediments and other earth materials rather than by the river diversion process.

The Board held public hearings and considered pro and con arguments. Except for Syar's proposed use of non-native earth fill material for reclamation, the Board concluded that all of the environmental impacts that might result from the proposed change had already been considered in the ARMP's EIR. Consequently, it felt that a FONSI was appropriate and the Board approved Syar's application to mine subject to an aggregate land designation decrease from 145 to 30 acres, and rejection of the non-native fill material in the reclamation plan.

When the County of Sonoma, California, approved the application of Syar Industries, Inc., (Syar) to start sand and gravel mining along the terrace of the Russian River, environmentalist organisations, the Sierra Club and the Russian River Task Force, petitioned the court for an order denying the approval, alleging that the County's Board of Supervisors (Board) had violated the State's Environmental Quality Act (CEQA) by certifying a negative declaration (i.e., a FONSI, or "finding of no significant impact") and not requiring a new environmental impact report (EIR) for a traded property made by Syar. The Sierra Club then filed its action against the Board. The Sonoma County Superior Court granted the petition ordering the Board to set aside approval of the mining project and requiring further preparation of an EIR. Syar appealed.

In *Sierra Club v. County of Sonoma*, the Appeals Court found that the mining company's proposal to mine along the river was a "separate" project, as opposed to being part of, or even a minor modification of, the single large project already studied in the ARM Plan. Therefore, the County was obligated to consider whether a new project might cause significant environmental effects that were not examined in the prior program report (Author's note: The Board claimed to have already considered this point), and if there was substantial evidence that the project might arguably have such effects, the Board should have required another EIR. In view of the fact that substantial evidence supported fair argument that the proposed project might cause significant impacts that were not previously examined, the setting aside of the permit was proper pending submission of an EIR.

It is understandable why forethought to reserving land for mining has not been given. Such planning would have required costly geological studies and maps to be made for the political subdivision in order to understand what areas might be zoned differently, or protected for future mining. Many political subdivisions do not have the funds for such a luxury in planning. However, in Canada and the U.S., all provinces and states do have geological surveyers who could provide or prepare local governments and planning boards with generalised geological maps for use in their responsible geographic area. Where such geological assistance is not available, frequently local knowledge of large outcroppings of rock, prolific boulder exposures, and sand and gravel bars on rivers, which obviously are not suitable for residential building, or surface structure development, could be zoned for future mining of construction aggregates. Such advanced planning would serve as notice to future home builders that they proceed to build at risk in an area reserved for mining. However, such advance warning notice, in view of being related to the older legal doctrine of "coming to the nuisance" and now of no legal consequence, would be of little use unless the judiciary reverted to its use, at least in this area.

In the cases of existing mining operations which were located and operating prior to restrictive zoning regulations, exception provisions with conditions were often made to allow them to continue operation. These conditional use exceptions have been an area of much problem for the mining industry. Often, too many constraints, some reasonable and others not, have been placed in the conditions for the mining to continue after restrictive zoning.

Along with this conditional use problem area, the enactment of "grandfathering" statutes were made throughout most of the U.S. to allow existing mining operations to continue in areas where zoning came into effect afterward. These grandfathering acts have also been a legal problem area for mining operations that were pre-existing to zoning regulations. Mining operations under grandfathering acts frequently come into conflict with simultaneous conditional use permits and restrictions placed upon them by zoning regulations. This separate problem area under grandfathering acts is reviewed in Ch. 3 §4.9, infra.

Following are examples of litigation of zoning problems for surface mineral lands and mining.

(i) An older example of a city ordinance (1958) regulating surface mining depth of excavation below the water table is *Goldblatt v. Hempstead*, 369 U.S. 590 (1961). Goldblatt had mined sand and gravel within the town limits for many years. The excavation has resulted in a 20-acre (8.09 ha) lake with an average depth of 25 feet (7.62 m).

The city's ordinance prohibited further sand and gravel mining. The mine owner argued that the town of Hempstead could not take the property by regulation, but could only take his property by the process of eminent domain for which he should receive just compensation.

The Supreme Court upheld the reasonableness of the ordinance stating that although government action in the form of regulation can be found in some instances to require compensation, but here, there was no evidence to show that the value of the property will be reduced by the town's prohibition of further mining. The Court found that the effect of the town ordinance was a valid police regulation.

(ii) Zoning and environmental regulations in conflict for mine permitting
In an unreported Ohio case (*Barrett)*, a sand and gravel operator was issued a conditional use permit in 1974 by the local zoning board. The owner did not attempt to operate the property until 1987. The board had never revoked the conditional use permit in the intervening years. In 1988, the board then falsified

their records in attempt to subvert their former decision by claiming a lapse of the permit and speculating that the pit might contaminate their water supply. In adding to their position, the board argued that the owner had never obtained approval of the Ohio EPA as a condition of the use permit. The issue became whether the granted conditional use was void for failure of the owner to first obtain EPA approval.

In *Barrett Paving Materials, Inc. v. Board of Zoning Appeals of Union Township*, 1991 WL 116344 (Ohio App.Dist. 2), the Miami County Court of Appeals made a significant statement regarding the relationship and "purposes of local zoning and environmental regulations (which) are inherently different *** are complementary but wholly independent of one another. The EPA is solely concerned with the environmental protection and protection of human health from pollution and improper waste disposal. A local zoning board *** is primarily interested in land usage *** affecting the development of the community".

The Court found that although EPA approval was a condition to the use permit, it was a condition subsequent, that is, to be fulfilled at a later and proper time. The EPA will not approve an application for until an actual proposed plan for the operation of a gravel processing plan has been submitted. Therefore, the approval of a zoning certificate is not made contingent on the approval of an EPA permit ...." The Court added, "that it could find nothing in the Township's Resolution which indicates an applicant seeking a zoning certificate must obtain prior approval of the Ohio EPA". The Appeals Court reversed the lower court finding that Barrett held a valid conditional use permit to mine gravel.

(iii) Denial of Variance of Non-conforming Use for Quarrier to Recover Stone
Appeal of Eureka Stone Quarry, Inc., 539 A.2d 1375 (Pa.Cmwlth. 1988):
A Pennsylvania aggregate stone quarrier applied for a variance to a non-conforming use to recover stone reserves from an area it had quarried in prior years. The area was now regulated by a county ordinance proscribing a 400-foot (121.92 m.) setback area from physical structures as public roads and a railroad. Eureka was denied the variance to a non-conforming use. Eureka then challenged the local zoning ordinance that prohibited expansion into the 400-foot setback area where the quarrier had mined in prior years.

The court found that the quarry did not have a non-conforming use in the non-conforming area since its activity there had ceased before the enactment. Because of that cessation of activity, it only had a non-conformity which could not be expanded.

The court upheld the permit denial unless the quarrier could show that: (1) there were unique physical circumstances peculiar to that property; (2) the property could not be developed in strict conformity with the provisions of the ordinance; (3) the ordinance created unnecessary hardship; and (4) that any variance would not be detrimental to the public.

Eureka had argued that a variance to reduce the setback would increase the quarry reserves by 13% and increase its life by 14 years. However, their argument for undue hardship resulting from the ordinance failed. The court said that the mere fact of making an operation more profitable was insufficient to support a variance grant.

(iv) Zoning Ordinance Limits Quarry Expansion Rights

In *Kibblehouse v. Marlborough Twnshp.*, 630 A.2d 937(Pa. Cmwlth. 1993), a crushed stone quarry, in continuous operation since 1916, was limited in 1990 for its expansion plans under a 1970 township zoning act. The property is divided into two parts, the northside, containing the quarry operation. The other, southside part, was across a road.

In 1970, only part of the northside tract containing the quarry was zoned "Limited Industrial" (LI), where quarrying was a permitted use. The remainder of the northside tract was zoned "Residential-Agricultural" (RA-1) and encircled the LI zone. The southside tract fell entirely within the RA-1 zone.

In 1990, Kibblehouse requested that the Township declare the quarry operation on the northside as a valid non-conforming use and permit its expansion over the entire property, including the southside. It was determined that as a non-conforming use, the quarry-use could not expand over 25% of its 1970 limits. Kibblehouse appealed the determination to the Zoning Appeal Board.

The Board decided that a non-conforming use had not been established in 1970 as the quarry operation had been placed within the LI zone where quarrying was a permitted use. The Board also declared that under a non-conforming use, an expansion of land activities was not permitted by the ordinance. Kibblehouse's application was denied.

Kibblehouse appealed the Board's decision to the Court of Common Pleas arguing primarily that its decision was contrary to law, and an abuse of discretion. On April 8, 1992, the trial court's order reversed the Board holding that "Kibblehouse's quarry-use was a non-conforming use, and that since the ordinance 'contains no restrictions on the extension of the non-conforming use of the land', Kibblehouse could use its entire property for a quarry".

The trial court partially rescinded its first order stating that the quarry may be expanded on the northside, but sustained the Board that no expansion could take place on the southside. Appeal by both parties followed.

In *Kibblehouse v. Marlborough Twnshp.*, the two important industry issues were: (1) "whether quarrying was an established non-conforming use on the northside; and (2) whether the owner of a non-conforming extractive-use is entitled to expand the use throughout the landowner's entire property where the zoning ordinance is silent on the right to expand non-conforming uses of land".

Kibblehouse was able to establish the existence in 1970 of a non-conforming use for that portion of the northside that had not been included in the 1970 LI zoning portion with the quarry. The review record showed that the quarrying operation had not been entirely confined to the LI zone in 1970. Therefore, quarrying activity had been occurring on land that was zoned RA-1 by the ordinance. The Appeal Court agreed that expansion should be allowed on the entire northside because of activity prior to zoning.

The Appeal Court considered Kibblehouse's argument that the established non-conforming use may be expanded throughout the entire property, specifically with regard to the southside. The Court first supported Kibblehouse's right to expand on his own land by stating that it recognised "that the right to expand a non-conforming use is a constitutional one which may not be prohibited by a local zoning ordinance". However, it then followed with another statement that withdrew strength from the above stated protection for a quarryman's right to expand its operation on its quarry-reserve lands. Citing a 1969 Pennsylvania case, the Court noted that "a zoning ordinance may not impose an absolute prohibition on a necessary right of expansion although the municipality may enact 'reasonable restrictions' on the right of expansion".

Kibblehouse further argued the legal doctrine of a "diminishing asset" for the right of expansion as a non-conforming use. The Court noted that such argument was "only accepted by a small number of states in the Midwest and West ***. Under that theory the quarrying enterprise is 'using' all of the land owned for the extraction purpose, not withstanding the fact that a particular portion of the property may not yet be under actual excavation". Kibblehouse further cited the pertinent Pennsylvania case of *Cheswick Borough (1945),* in which the Pennsylvania Supreme Court stated that "*** it was not essential that the use, as exercised at the time the zoning ordinance was enacted, should have utilised the entire tract". This Appeals Court then stated that the Pennsylvania "courts have recognised the right to expand a non-conforming use to provide for the natural expansion and accommodation of increased trade 'is a constitutional right protected by the due process clause'. [*Silver v. Zoning Hrg. Bd.*(1969)]".

But again, seemingly intent to encroach on the landowner's constitutional right to expand which it just supported with a case cited, the Appeal Court added, "However, the natural right of expansion is not unlimited. A municipality has the right to impose reasonable restrictions on the extension of a non-conforming use".

Kibblehouse argued that because the rock on the southside tract is of excellent quality, the natural growth of the business would logically expand to the southside; and maintained that the southside was always devoted to the future expansion of the quarry. In response, the Appeal Court stated that the review of the record shows that nothing beyond mere intention to use the southside was ever established. The Court found that "mere intent" to quarry in the future was "insufficient". Thus, the Pennsylvania Appeal Court upheld the Zoning Board's decision to deny expansion of the quarry on to the southside of the owner's property.

(v) Zoning Regulations Denying Quarrying Found Unconstitutional

Davidson Mineral Properties v. Monroe County (Ga.), 257 Ga. 215 (1987): An aggregate stone quarrier was denied a permit for development of a new quarrying site. At the time Davidson applied for its permit, no land use plan was in effect, and there were no restrictions on the intended quarry property. However, the County Commissioners Board had enacted a "Moratorium on

Commercial Development" until a land use plan and zoning ordinance could be adopted. In the interim period, so that commercial development would not be halted, the Board approved permit applications on an ad hoc basis. Out of 24 building / development permit applicants prior to the quarrier's, all but one were approved. The Zoning Board denied the quarrying permit following a public hearing where a number of people voiced opposition to the quarry.

Davidson challenged the Board's actions alleging that there were no objective criteria or standards provided in the Commissioners' resolution, or elsewhere in the record for issuance of a building permit. The Moratorium was found by the Georgia Supreme Court to lack sufficient standards to meet due process requirements; lacked objective criteria for permit approval; and gave Board uncontrolled discretion. Zoning was subsequently approved for a quarry.

(vi) Local planning ordinance upheld for quarrying controls

The following case illustrates the non-conformity and lack of control by state mining agencies amongst the states in the U.S. for controls over surface mining and related environmental regulations.

In a more recent state case, the New Jersey Supreme Court in *Bernardsville Quarry v. Bernardsville Borough,* 608 A.2d 1377 (N.J. 1992), gave full support to local authorities over mining control, licensing and quarry regulation. In *Bernardsville Quarry*, the quarry owner brought an action challenging a local municipal ordinance which imposed a licensing requirement for quarrying operations limiting the depth below which the property could not be mined. The quarry owner sued contending that the ordinance effectively prohibited their property's use as a quarry and, therefore, constituted a governmental regulatory taking of its property.

The property had been operated continuously as a quarry from 1931 until 1985 when it closed on filing for bankruptcy by the former operator. The Borough adopted its first land-use ordinance in 1949, and in 1963 the quarry was declared a non-conforming use. The present owner (BQI) obtained the property at foreclosure sale in 1987, at which time the owner applied to the Borough for a Continued Certificate of Occupancy (CCO) stating that it intended to continue using the property as a quarry and to process stone as had its previous owner. The Borough promptly denied the application stating that the only current permissible use of the property was as a bituminous concrete plant and not as a quarry. The Borough stated BQI had failed to submit a site plan under the Land Use Ordinance, and that there were a number of serious environmental concerns, mainly, the potential for pollution of subsurface aquifers which served as a source of local drinking water.

The Borough's fear of water pollution was based on its finding that parts of the property had been illegally used for asphalt production, as a transfer station for consolidating garbage, and as a landfill containing machine parts which were leaking contaminants. [Comment: Such reason given as "illegal use as a garbage transfer station" is not justifiable reason to prohibit a legitimate and properly operated mine from operation. Whether such allegations were true or not, investigation, enforcement and controlling of environmental issues for mining operations should be vested exclusively under a state agency where uniform regulations may be applied state-wide, not as each local community governing body feels proper.]

The Borough council reviewed BQI's application several times and finally issued a Temporary Certificate of Occupancy (TCO) in June 1987 allowing for operation of a concrete plant only. Upon additional information from BQI, the Borough issued a second TCO permitting BQI to crush "loose" stone and a limited amount of blasting. However, nullifying its seeming permission, the Borough at the same time adopted a new Quarrying Licensing Ordinance which limited the depth of quarrying, restricted the hours of operation, required buffer zones, and imposed a license requirement. The second TOC was made contingent upon BQI's filing another application in compliance with the new ordinance. In August 1987, the Borough ordered suspension of all blasting at the quarry based on the Borough's engineers report that asbestos had been disposed of on the property at some time in the past.

In February 1988, BQI filed a detailed application for a quarry license. It sought approval for quarrying below the ordinance-allowed depth and in areas of the property which had been prohibited by limitations in the 1963 nonconforming use agreement. BQI's submitted quarry plan would expand the quarry area over 45-acres (18.21 ha) over a 20-year period resulting in a 45-acre lake with an average depth of 200 feet (60.96 m). The height of the surface above the lake's surface would be 175-feet (53.34 m). The Borough denied BQI's application in April 1988. Denial was based on: violation of the

1963 non-conforming use limitations; the asbestos contamination; unacceptable surface water discharge; unacceptable number and depth of monitor wells; failure to guarantee compliance with the N.J. Pollution Discharge Elimination System; failure to provide sufficient rock mass information; failure to evaluate groundwater flow; and, failure to provide hydrogeologic characterisation for the development of a database for groundwater monitoring wells.

In May 1988, BQI filed a complaint against the Borough charging that the Quarry Licensing Ordinance was unreasonable, particularly in its depth restrictions, and constituted an unconstitutional taking of private property without just compensation. In addition, BQI charged that the Borough could not validly require licenses for quarrying, that the license denial had been improper, and its civil rights had been violated under federal law.

5.4.8 *State Control of Mining versus Local Control*

Upon trial, the lower court ruled against BQI on its challenge to the quarrying depth limitation and that no unconstitutional taking of private property had occurred as a result of the ordinance. The court held that licensing of the quarry was within the scope of a municipality's authority and the license denial was valid. Only the ordinance's limitation on the quarry's hours of operation was found invalid. However, the denial of BQI's application was found to be arbitrary and capricious, based on an inadequate record, and remanded to the Borough's council for reconsideration. The Borough appealed and the Court of Appeals affirmed the lower court's decision. Both parties filed for review by the New Jersey Supreme Court.

With regard to the licensing of a quarrying operation, BQI had argued that registration of the quarry operation under New Jersey's Mine Safety Act pre-empted licensing by the Borough. The Supreme Court found that the Mine Safety Act was mainly concerned with the safety of mine and quarry workers. The State legislature had given no clear indication in its passage of the act that it intended to prohibit municipalities from regulating and licensing quarries for health, safety and welfare of the community. There was no statutory basis from which to infer that the State had pre-empted the field of quarry regulating.

The N.J. Supreme Court therefore found that a municipality may regulate a business under it general police power and require a license to defray the costs of such control.

An issue of a Fifth Amendment (regulatory) Taking of Private Property

The Court noted that "under our constitutions, federal and state, government cannot take private property without paying just compensation". Both the U.S. and New Jersey constitutions afford coextensive protection for a taking. "Without question private property can be effectively taken through regulatory measures that do not amount to physical occupation or appropriation ... However, there is no precise formula that courts use to determine whether a compensable 'non-invasive' (*regulatory*) taking has occurred". As determined by the U.S. Supreme Court, such type of cases must be determined on an individual basis *(ad hoc)*, i.e. case by case.

To aid in the determination of whether a compensable, non-invasive or regulatory taking has occurred, the U.S. Supreme Court has formulated three factors of "particular significance": "(1) the economic impact of the regulation on the claimant; (2) the extent to which the regulations have interfered with distinct investment-backed expectations: and (3) the character of the government action". [*Connolly v. Pension Benefit Guaranty Corp.*, 475 U.S. 211, 224, 225 (1986)]. In essence, the public's interest in the regulation must be weighed against the private property interests affected by it.

In general, the N.J. Supreme Court stated that in zoning schemes, "a regulation must substantially advance legitimate state interests, and it cannot deny an owner all economically viable use of the land". [*Agins v. Tiburon*, op. cit.] Neither can it excessively interfere with property rights and interests. The N.J. court cited several examples of regulatory takings cases in which laws were upheld under the "police power" and correspondingly denying compensation for a taking of private property by disallowing its intended use as being injurious to the public good, safety and welfare. The leading case cited was *Keystone*, in which a Pennsylvania law was upheld which required subsurface coal mine operators to leave certain amounts of coal in the ground as pillars to prevent surface subsidence without which substantial damage to buildings above the mines might occur, and posed environmental hazards and physical conditions that rendered surface development impossible. [Comment: It should be noted that the coal operators had contractual waiver agreements from the surface owners relieving them of liability for damages to the surface and structures. The validity of the waiver contracts was also litigated in *Keystone.* Under the U.S. Const., Art. I, Sect. 10, "No State shall pass any *ex post facto* Law or Law impairing the Obligation of Contracts". The contracts were found invalid.] That law had been enacted pursuant to the state's police powers for the protection, safety and general welfare of the people. In *Keystone,* the U.S. Supreme Court denied compensation to the coal operators for an alleged taking of 27 million tons of coal required to be left in underground pillars for surface support.

Similarly, the Court noted that in *Bernardsville*, the Borough's ordinance was passed to meet environmental safety and public health concerns. The ordinance declared its purpose was "to protect the environment by minimising air pollution and prevent surface and subsurface water pollution". The limitations as to depth of quarrying were designed "for the protection of persons and property and for the preservation of the public health, safety and welfare of the Borough".

[Comment: The reason for depth limitation would appear "impermissably vague" and therefore, invalid.] The Supreme Court upheld the trial court's finding that there was sufficient evidence to justify a compelling need to limit quarrying in depth. The New Jersey courts accepted as the gravest threat "the potential of surface pollution through careless deposits of oil or other pollutants during quarry operations". The Borough's evidence established for the court a connection between water in the quarry and water in subterranean aquifers, and a substantial risk of exposure of those aquifers to potential pollutants that could exist in the proposed lake.

The New Jersey Supreme Court held that under the Borough's police powers it had enacted a valid ordinance to limit quarry operations in the interests of public health, safety and welfare, and that the ordinance advanced a substantial, genuine and legitimate public purpose. As such, it concluded that the ordinance did not effect an unconstitutional taking of private property without due process of law.

Author's Conclusion:

It is not advocated that the local community should have no voice in environmental protection to their locality after giving zoning approval to a mining use. However, it is advocated that after initially approving, or zoning an area for surface mining, public participation should end and the environmental protection and regulation of the surface mining area should be left to the state's mining and environmental engineering professionals. British procedure for land use development would appear to be a preferable, less contentious and a more orderly method for licencing permission in land development. As noted by Hughes, in England "Planning law has not historically been generally over-concerned with giving third-parties rights at the development control stage. Nevertheless, there are publicity and notification requirements in the TCPA 1990". (op. cit. 1992, p. 126) Public involvement in the planning and permission approval process for land development is limited. Hearings may be held at the discretion of the Secretary of State for the particular ministry, but may limit those participating in the hearings. The Secretary of State has wide discretion in the hearings, and may or may not hold them. The ultimate decision for planning permission, whether contentious or not, is placed in the discretion of the Secretary.

The British procedure for land use permission approval may be removed from the contentiousness of the general public which makes for less friction in hearing an application for a development. This procedure is advocated for more orderly and smoother processing of land development in the U.S.

The British procedure for challenging a Planning Council's decision is also recommended for use in the U.S. Under §78 of the TCPA 1990, "a disappointed developer may appeal to the Secretary of State, who has wide powers to allow or dismiss appeals, and to vary or reverse any part of the original decision, and may deal with the application as if it had been made to him in the first instance". As an alternative to an inquiry or hearing for review by the Secretary of a land development application, the appealing party may "opt for the matter to be settled by way of written representations, a procedure used in the majority (75%) of all cases". (op. cit., 1992, p. 143.) The discretion allowed the Secretary under §35A of the Act is an interesting procedure which may circumscribe public involvement. "Section 35A further empowers the Secretary of State to 'call in' proposals before adoption. He then has a wide discretion as to the fate of the proposals, though he must consider any objections made in due form together with any other matters he considers relevant, and in this connection a wide power of consultation is given. *** Section 35B further requires that before adopting proposals the authority must, unless the Secretary of State otherwise directs, cause 'an examination in public' to be held of such matters concerning the proposals as they consider should be examined. Similarly, where the Secretary of State calls in proposals, he may hold an examination in public of any specified matter. An examination is held by persons specifically appointed by the Secretary of State, but **it is not a public inquiry in the full sense**, even though §1(1)(c) of the Tribunal and Inquiries Act 1971 applies, no person has a right to be heard, ***". (emphasis added) (Hughes, 1992, p. 112).

In *Bernardsville*, the issue of quarrying depth and its consequential affect on a local aquifer is a question to be determined by state professional engineers who understand mining and hydrogeology. In trial, experts for hire are too likely to testify as to what their environmentally-fearful employers want to hear. In *Bernardsville*, the Court noted the "divergence of opinion by so many experts, *** demonstrates the inability of a governing body (the Borough) to predict with certainty the effect of the removal of so much stone". [Comment: undoubtedly "much stone" had already been quarried over 54 years from the Bernardsville quarry, and apparently with no previous dire effect or contamination of the Borough's water supplying aquifer.]

As to the issue of quarry expansion, in the New Hampshire case of *The Town of Wolfeboro v. Smith* (1989), (see Ch. 5 §4.9.1, infra) local authorities allowed only vertical expansion and prohibited lateral expansion of a grandfathered operation on quarry-owned property held since 1950. "Grandfathering" for quarries was stated to be meaningless if expansion is limited. However, contra, in *Fletcher Gravel Co. v. Jorling (DEC)*, 179 A. 2d 286 (N.Y.1992), the New York Court of Appeal stated, "The expansion of mining activity into different areas is consistent with the nature of mining activity and does not constitute a significant change in permit conditions". The California Supreme Court agreed with the New York court in its 1995 decision in *Hanson Brothers Enterprises v. Nevada Co. Bd. of Supervisors* (No. S044011) approving an extension of a grandfathered right to mine after the local board had denied the right.

In *Bernardsville*, the Borough's expert noted the "potential of surface pollution through careless deposits of oil or other pollutants during quarry operations". The public, legislators, and the judiciary have yet to realise that if such groundless propositions and assumptions as that 'expert' made in *Bernardsville*, one could argue that "What if - accidental spills of oil at quarrying operations occurred" are given universal credence, all quarrying operations in the nation could be shut down under state police powers for the general public's good, welfare, health and safety. To carry the logic of the argument further, "What if" oil and petrol spills are made at service stations? Shall the building of new stations be prohibited? And, because of the Valdez, Alaska oil spill, shall all ocean transporting of petroleum be prohibited?

If quarrying is to be denied on such fantasy, imaginary and remote supposition, there would be no crushed stone operations in the U.S. in existence. We are indeed "in serious danger", as Justice Dooley of the Vermont Supreme Court said, "of expanding 'not-in-my-backyard' into 'not anywhere' ".

(vii) Former Sand & Gravel Operator Compelled To Reclaim Illinois Pit Under County Rule: McHenry County, Illinois, local landowners near a former, unreclaimed sand and gravel pit, brought suit against the County, the landowner on which the pit was located, and the former pit operator, successfully winning an order to enforce the County's zoning ordinance requiring its reclamation.

The pit operator, FRAMS, began its operation over six years before 1979 when McHenry County adopted a zoning ordinance, (§508) requiring "all operators extracting and / or processing earth materials" to apply for a conditional use permit. The applicant was required, jointly with the landowner, to submit a reclamation plan to the county and to post a reclamation bond to guarantee reclamation according to the approved plan.

In *Lily Lake Road Defenders, et al v. The County of McHenry, et al,* 619 N.E.2d 137 (Ill. 1993), the plaintiffs charged that FRAMS had enlarged its surface mining area by 25 acres (9.71 ha) after 1979 without submitting a reclamation plan or bond to the County. The pit ceased operation in 1988 without restoring the property. Plaintiffs sought a writ of mandamus ordering the County to enforce its 1979 ordinance for surface reclamation.

The trial court found that the county ordinance, enacted after the IEPA, was void *ab initio* (from the beginning). The ordinance was unenforceable. On appeal, the issue before the Supreme Court was "whether, and if so, to what extent the ordinance was affected by the state legislature's enactment of the Illinois Environmental Protections Act (IPEA) and the Illinois Surface-Mined Land Conservation and Reclamation Act of 1971 (ISMCRA).

The Supreme Court of Illinois reversed the lower court's decision finding that the County's zoning ordinance requiring reclamation had not been repealed by implication and passage of the IEPA 1970. The Supreme Court went on to say that, prior to the enactment of IEPA 1970, counties were empowered under the 1935 County Zoning Act to restrict the use of land for surface mining. Although the County ordinance had been enacted in 1979, after the IEPA of 1970, the doctrine of repeal by implication did not apply here.

The Court said the IEPA "presupposes that counties will continue to exercise such zoning powers ***. A county which exercises its statutory authority to regulate and restrict the use of land pursuant to the Zoning Act does not necessarily violate the terms of the IEPA. Both statutes may be given effect without thwarting the legislature's intent".

The Supreme Court gave further consideration to the argument that the county's ordinance had been repealed by the doctrine of preemption by the IEPA, 1970. In 1981, ISMCRA and EPA were amended and provided: "The issuance under this Act of a permit to engage in the surface mining of any resources other than fossil fuels shall not relieve the permittee from its duty to comply with any applicable local law regulating the commencement, location, or operation of surface mining facilities". Therefore, after 1981, the McHenry County reclamation ordinance became enforceable. Since FRAMS' operations did not cease until 1988, it was subject to the county's reclamation requirements. The operator will have to reclaim the 25-acres (9.71 ha) of the worked out pit.

(viii) Permitting for Continued Mining and Expansion "Grandfathering" Act Introduced, *Atlantic Cement Co. v. N.Y.*, 516 NYS. 2d 523(AD3 DEPT. 1987): Atlantic operated a limestone pit and cement manufacturing facility on approximately 2,000 acres (809.39 ha) of land since 1961. The New York Mined Land Reclamation Law became effective in 1975. In that year, Atlantic was required to obtain a mining permit from the NY Department of Environmental Conservation (DEC). In order to continue mining, Atlantic had to submit, *inter alia,* property maps, mining and reclamation plans, and reclamation bond. Atlantic received permit renewals in 1978 and 1981 with only minimal paperwork. In 1984, DEC rejected Atlantic's normal application and required a State Environmental Quality Review (SEQR) and additional information for DEC's Life of the Mine Review Policy.

Following Atlantic's refusal to provide the information, DEC issued notice in April 1985 for a draft EIS, and if not received by May 1, 1985, their permit renewal would be denied. Atlantic began Article 78 proceedings challenging the authority of DEC for its added requirements for permit renewal. DEC denied the permit renewal. The decision of the lower court was: (1) DEC had taken longer than the required 15 days to notify Atlantic of insufficiency of application information; (2) Atlantic's mining activities were grandfathered under the SEQR Act and Atlantic was not required to file an EIS; and (3) DEC had already determined in previous years Atlantic's operation would have no significant environmental impact (FONSI). The Court of Appeals affirmed the finding that DEC's permit denial was illegal, arbitrary and capricious.

5.4.9 Limitations of "Grandfathering" Acts for Active Mining Operations

In the mid-1970s, after NEPA had been enacted and the tide of environmental regulations was beginning to filter down to state levels and flow, permitting regulations were enacted by many of the states for control of surface mining operations. Promulgating of regulations and rules for new mining sites yet to develop were easier to enact. The miners-to-be had little choice other than to conform to the new rules in force when starting up a new quarrying operation.

Regulating existing surface mines, which had often been in business decades before, was another matter. It would be more difficult to make them conform all at once to the same stringent controls by virtue of their pre-rule existence. This would be more true for their locations, which, at times, were becoming troublesome, if not nuisances, to burgeoning communities in which they might be located. This was particularly true if the quarries had been situated long before the communities began to grow and still remained operative, still serving a vital need by supplying construction materials for the local growth. Thus, to deal with them and yet extract some effective degree of compliance with the new environmental and site permitting requirements, most states either enacted Grandfathering acts, or appended them as clauses to their mining acts for new surface mining operations. In some cases, under the acts or clauses, they were not required to obtain mining permits for their present operation. However, if they grew beyond their limit at the date of enactment, they were required to obtain mining permits for the new growth or expansion. This frequently became a point of contention and litigated. The decisions in various states depended on the court's interpretation of the particular state's grandfathering act.

Local planning authorities, when finding a pre-existing or operating surface mine as undesirable or problematic in their zoning schemes, could either give the mining property the required industrial zoning, or alternatively, grant the mine a non-conforming use licence to continue. Non-conforming use permits are a way of avoiding giving a mining property the zoning it needs for long-term growth. A non-conforming use permit is properly thought of by some authorities as a "phasing out" permission to continue in business. As the term implies, the surface mine is expected to phase out, and expansion of that business is generally prohibited as it does not conform to the general land use plan. Nevertheless, mining operations are businesses that like to grow for profit as do all other businesses, and, in that era of the 1970s, growth was still booming and requiring minerals. Logically, quarries desired to grow along with the rest of industry, and mineral products are needed for construction and growth of the community and nation.

Some of the contentious problems brought about by grandfathering acts for surface mining are illustrated in the following examples.

5.4.9.1 *Grandfather act — New Hampshire's test for permitting of expansion of operations*

In *Town of Wolfeboro (Planning Bd.) v. Smith*, 556 A.2d 755 (N.H. 1989), Smith owned a sand and gravel pit on 35-acres (14.16 ha) that had been in continuous operation since 1950. By 1989 only ten acres (4.05 ha) remained to be mined. The Town of Wolfeboro filed a petition for a cease and desist order to prevent further pit expansion on Smith's land without a permit. At trial, Smith claimed he was not required to obtain a mining permit as he was "grandfathered" in when New Hampshire's law (RSA Chapter 155-E:2 Laws 1979, 481:3) took effect on August 24, 1979. The statute allows the owner of an "existing excavation to continue such existing excavation without a permit". At the time of effectiveness of the law, only eight of the 35-acres had been mined. Additional excavation had continued from 1979 to 1989 without a mine permit until only ten acres remained to be mined At trial there was no dispute between the parties that Smith was entitled to continue mining the existing pit. The sole issue was whether the statute's clause entitled Smith to excavate the remaining ten acres without a mining permit. The town argued "that by its very language, the provision exempts only the 'land area' which is being used, as of August 24, 1979, and, therefore, exempts only the land already excavated as of the effective date of the chapter". Thus, according to the Town, the operator could only continue mining in depth and not laterally because no enlargement of the "exempt" area was allowed. The trial court disagreed and ruled that Smith was entitled under the grandfather clause to mine his entire parcel without a permit. On appeal, the Supreme Court of New Hampshire also found difficulty in accepting the Town's interpretation, stating "If the phrase 'continue such existing excavation' as found in the grandfather clause were understood to allow only vertical and not lateral expansion, such an interpretation would lead many owners ... to find that they could not continue their existing operations very long, if at all. Only by allowing the continued excavation of land previously appropriated for that use would the owner truly be able to continue an excavation which he had begun". The Court further stated that the lateral expansion of an existing pit "... will be considered a continuation of a previous excavation" if the land was appropriated for excavation prior to the effective date of the law.

What seemingly was shaping to be a decision in favor of grandfathered surface mining operations was quickly reversed by the Court's subsequent condition: that the operator had to "objectively" show appropriation of the land for mining before the law's passage. "Intent alone is not enough". The Court believed that the phrase "continue such existing excavations "contained some limitation on increases in area or intensity of mining. It stated that "An increase in the intensity which serves to change the character of purpose of the non-conforming use will be considered to have changed the use". It added that a great increase in the size or scope of a mining operation could end its exemption from the law as a grandfathered use.

In its conclusion and holding, the New Hampshire Supreme Court established a three-pronged test for operators wishing to continue mining without a permit in New Hampshire:

(1) The operator must prove the activities were actively being pursued when the law became effective;
(2) He must prove the area he wishes to excavate was clearly intended to be excavated, as measured by objective actions, not by subjective intent; and
(3) He must prove continued operations do not and will not have a substantially different and adverse impact on the neighborhood.

Question and Comment on the Wolfeboro Decision

Under the rationale and final test of the N.H. Supreme Court, how was the operator able to expand mining from the 8 acres (3.24 ha) already mined at the date of the effectiveness of the law, for ten more years without a mining permit to the date that Wolfeboro demand no further mining of the last ten acres (4.05 ha), without a mining permit? It would appear that the mining of the 17-acres (6.88 ha) in between the two dates would imply that it was more than "*subjective intent*" to mine the whole 35-acres (14.16 ha); and the act of having mined 17-acres after August 24, 1979, shows "by *objective* actions" that the whole 35-acres was "clearly intended to be excavated". (For further argument on permitting of surface mining, see "A Quarryman's Reply" to the *Wolfeboro* decision, Ch. 8 §5, United States — Weaknesses of Surface Mine Permitting Procedures, infra.)

5.4.9.2 *Mining permit required for removal of 24-year old stockpiles*

Crowell Constructors v. State, 393 S.E. 2d 312 (N.C. App. 1990): In 1984, the North Carolina Department of Environmental Health and Natural Resources (DEHNR) required, with a Notice of Violation (NOV) an asphalt manufacturer to apply for a mining permit to remove 24-year old stockpiles of sand from a property purchased to acquire the old stockpiles. The former sand mining operation had ceased in 1960 and were left in place.

Crowell contended that such short-term removal was not mining. DEHNR reviewed and found that the situation lent itself better to regulation under N.C.'s Sedimentation Pollution Control Act. Crowell was required to submit a soil erosion and sedimentation control plan. With the plan approved, Crowell continued removal of the sand. In February 1986, DEHNR notified Crowell with another NOV that it was in violation of the Mining Act for mining without a mining permit, and assessed a penalty of $5,000 per day for each day of illegal operation. Crowell continued removal until March 21, 1986. On March 27, DEHNR fined Crowell $10,000 stating that off-site sedimentation had occurred and that his restoration efforts were ineffective. The Mining Commission upheld the violation and fine. The N.C. Court of Appeals held that "mining" had occurred under three different definitions of in the Mining Act and a mining permit was required.

Author's Comment: Such a decision stretches the definition of "mining" beyond reasonableness. Recovering and loading minerals previously mined and stockpiled cannot "reasonably" be called mining. By analogy, recovery of gold from an ancient sunken ship on the seafloor is not ocean floor mining, but merely salvaging. The accumulation of sea growth over the gold does not make it part of the original sea floor, and consequently, recovery of the gold is not mining of the gold. It is simply salvaging. Similarly, Crowell's recovery of the sand is mere salvage, not mining as the North Carolina court decided.

5.4.9.3 Mining permit — attempted revocation by citizens group

Concerned Citizens v. Rhodes, 394 S.E.2d 462 (N.C.App. 1990): A citizens' action group filed an action seeking to have a permit granting a mining company the right to operate a crushed stone quarry declared void (revoked). The Superior Court declared the mining permit void. Appeal was made. North Carolina's DEHNR is not among the agencies specifically exempt from provisions of the Administrative Procedure Act (APA), so that anyone challenging the decisions of DEHNR must follow APA procedures and exhaust administrative appeals before seeking judicial review. The Mining Act provides that any affected person may contest a decision of the DEHNR to deny, suspend, modify or revoke a permit or reclamation plan. The Citizens' actions group had failed to make an administrative appeal, therefore the Superior Court had no jurisdiction to void the mining permit. The Court of Appeals reversed the permit voiding and remanded the claim for proper procedure.

5.4.9.4 *Intensified reclamation of mining properties*

Mined Land Reclamation — Surface Mining Control and Reclamation Act (SMCRA)

The federal SMCRA 1977 regulates only surface coal mining. There is no comparable federal act that regulates non-coal surface mining operations. However, a majority of states have enacted their own version of state SMCRAs for non-coal surface mining and the regulations apply to all surface mining operations within a state's boundaries, whether federal or state lands. Federal pre-emption may apply to BLM lands within a state. Litigated case examples follow.

(i) Reclamation-Environmental Impact Statement (EIS) not Required: *City of Ukia v. County of Mendocino & Ford Gravel Co.*, 238 Cal. Rptr. 139 (Cal.App.1987): In a California case, Ukia brought an action to compel the county to set aside approval of a gravel operator's reclamation plan and compel an EIS. The court ruled that the operator's reclamation plan provided sufficient substantive description of natural streambed restoration and did not require an EIS.

(ii) *Grove v. Winter,* 554 N.E. 2d 722 (Ill.App. 5 Dist. 1990): An Illinois court held that a lessor of a limestone quarrying property could assert a breach of contract against the quarrier for failure to comply with Illinois SMCRA, which states in part, "All grading ... shall proceed in conjunction with surface mining and shall be carried to completion by the operator prior to the expiration of eleven months after June 30 of the fiscal year in which the mining occurred". [Ill. Rev. Stat. 1985, ch. 96-1/2, para. 4501, *et seq.*, 4507(k)].

(iii) *Morgan v. Ga. Vitrified Brick & Clay Co.*, 397 S.E.2d 49 (Ga.App. 1990): Where the surface owner sued the mineral estate owner for failure to restore the surface, the court found that mining did not require the surface to be left in such a rough state rendering the surface estate useless.

(iv) *Barton v. Gifford-Hill & Co., Inc.,* 760 F.Supp. 98 (W.D. La. 1991): Lessor of a sand and gravel property sued the operator for breach of contract for failure to restore the surface. The Louisiana court held that the miner had a duty under the Louisiana Mineral Code to partially restore the land after its operations were finished by sloping the sides of the gravel pit, even if the leasing agreement did not contain any provision for surface restoration.

(v) SMCRA-New Performance Standards: Several federal SMCRA reclamation procedures and performance standards for post-mining lands were litigated in the 1990 case of *National Wildlife Federation v. Lujan,* 733 F.Supp.419 (D.D.C. 1990), which has been appealed. Those procedural changes in federal law were largely reflected later in state regulations for non-coal mining.

Revegetation: [30 CFR 816.116 (c)(4)/817.116(c)(4)]: The court upheld the rule that rill and gully repair can be considered normal practice that will not restart extended responsibility period for operator. 31 ERC 1617, 1621–15 (June 8, 1990).

Reforestation: [30 CFR 816.116(b)(3)/817.116(b)(3)]: The court upheld the requirement that both state and federal regulatory authority approval for reforestry and wildlife programs. The success standard that 80% of trees be in place for 60% of the applicable period of responsibility (i.e.-five years in the East and ten years in the West) 31ERC at 1625–31.

Reclamation Period of Success for Post-Mining Land: The court upheld the precipitation requirements, viz., for areas (grazing, pasture and crop lands) with more than 26-inches of precipitation, two years will meet the success standard of responsibility; for areas with less than 26-inches of precipitation the success standard period shall be met in the last two successive years of responsibility. OSM shall have discretion in accounting for weather variability during the periods. ERC at 1631–33.

Water Impoundments: [30 CFR 816.79, 817.49]: The court upheld rules for class (b) and (c) dams (those in which failure would cause loss of life or serious property damage). Temporary impoundments and spillways were not ruled on while re-evaluation by the Secretary is being made. 31 ERC 2034, 2059-61 (August 1990).

Recharge Capacity of Water for Underground Mining: 30 CFR 817.41(b)(2): The court upheld deletion of the requirement that underground mines be required to recharge capacity of disturbed area and that enforcement was legally "debatable". ERC 1617, 1633-36 (June 30, 1990).

Reclamation Roads: The court upheld the rule requiring any road used to meet reclamation performance standards be retained as a primary road regardless of frequency and nature of traffic. Operators are also responsible for prevention of mitigation of lands within the boundaries of reclamation as listed in Sect. 522(e)(1) that may result from roads into the area.

(vi) Surface Mine Site Buildings Must Be Demolished In Reclamation

The Wyoming Department of Environmental Quality (WDEQ) ordered the present owner of the former U.S. Steel Corporation's Atlantic City open-pit iron mine to post an additional performance bond to ensure reclamation of its mine-site buildings and facilities.

U.S. Steel owned and operated the iron pit in Fremont County, Wyoming, for many years pursuant to a permit issued under the Open Cut Land Reclamation Act of 1969 until 1974 when permitting authority was placed under the Wyoming Environmental Quality Act (WEQA). Mine operation continued until 1983 when the pit was closed. Instead of reclaiming the mine property itself, U.S. Steel sold the mine and transferred its mining permit to Universal Equipment Company (Universal) in January 1985. Universal obtained the salvage rights to the mining equipment, materials and buildings. In accepting the transfer, Universal assumed responsibility for reclaiming the mine-site and was required to post a reclamation performance bond of $1.8 million.

Universal contracted with various companies to carry out different reclamation phases between 1985 and 1989 when a dispute arose between it and one of the contractors (ARIX) over payment for its services. The contractor sued Universal and DEQ, and won a judgement against Universal. ARIX attempted to garnish Universal's performance bond held by DEQ. In a separate suit, the court found that funds held by DEQ were not subject to garnishment. During the period, DEQ had reduced Universal's bond amount from the $1.8 million to $300,000 in recognition of partial reclamation of the site. Under the Court's supervision, Universal was able to convert its bond to cash by substituting an irrevocable Letter of Credit for its performance bond. The Court ordered payment to ARIX for its services.

During a 1990 hearing on DEQ's petition, DEQ required an additional bond of $4 million for Universal's remaining obligations, Universal challenged DEQ's authority to require an additional bond amount to insure costs of demolishing and removing the mine buildings and structures built before enactment of the WEQA. The court ordered Universal to make the $4 million bond, or, alternatively, to submit a detailed plan in 60 days which buildings were intended to be demolished and / or the proposed use for those left. Universal failed to submit a plan and challenged the bonding.

In *Universal Equipment v. Wyoming*, 839 P.2d 967 (Wyo. 1992), the Wyoming Supreme Court found WEQA's reclamation and bonding requirements applied retroactively to mine facilities and structures built prior to WEQA's enactment; and, DEQ could require a permitted reclaimant to post additional bonding for demolition of buildings on a site.

5.5 Canada

5.5.1 *Mineral Land Takings (Expropriation) By Environmental Regulation*

(i) *The Queen in Right of British Columbia v. Tener* [1985] 1 S.C.R. 533 (B.C.): Teners' predecessor in title had obtained Crown-granted mineral claims in 1937 on land which subsequently, in 1939, was included in a new provincial park, Wells Gray, in British Columbia. There were no restrictions on mining in the newly-formed park until 1965 when restrictions began with the issuance of park use permits for natural resources. Teners received a use permit in 1973 but were denied permits thereafter and informed in 1978 that no new mineral work would be permitted. The Teners sued the government claiming that the permit denial was an expropriation of an interest in land under the Park Act. The Supreme Court of Canada held that to refuse a permit to a grantee of mineral rights to exercise rights under the grant is the equivalent in law of a compulsory taking or expropriation. The Court stated, "This acquisition by the Crown constitutes a taking from which compensation must flow". (*Tener*, ibid at 563.)

(ii) A Canadian mineral company owning mining claims since 1938 in a provincial park on Vancouver Island, British Columbia, had its rights over the decades gradually limited by subsequent legislation, culminating in the requirement of a resource use permit. Finally, in 1988, an Order in Council prohibited the issuance of any park resource use mining permits amounting to an expropriation of the property. The mineral company successfully brought an action for compensation for the taking of their mineral property.

In *Casamiro Resources Corp. v. British Columbia*, [1991] B.C.L.R.(2d) 346; 80 D.L.R. (4th), Casamiro held title on 19 Crown-granted mineral claims in Strathcona Park. Strathcona Park had been established in 1911 by an Act of British Columbia. By later changes in the law, it became lawful to locate mineral claims in the park. These claims were located in 1938–39 by Casamiro's predecessor in title pursuant to the Mineral Act, R.S.B.C. 1936. In 1946–47, the claim locator received Crown grants for the claims which allowed the owner, his heirs and assigns to "have and to hold the said minerals ... forever". An added proviso stated that the "said minerals shall be subject to the laws for the time being in force respecting mineral lands held in fee simple:" Under the Mineral Acts of 1936 and 1948, full surface rights were granted to the mineral claim owner for the purpose of winning the minerals therein and for conducting the business of mining.

By 1957, the Strathcona Park Act had been repealed and the new Department of Recreation and Conservation Act, S.B.C. 1957, enacted. By this new act, Strathcona was designated a Class "A" park and the mineral claims fell within a reclassified "recreation area". Under its provisions, "All mineral claims within the park ... and all records and grants in respect of such claims, shall, in addition to the Mineral Act be subject to further terms and conditions and restrictions ... as the Lieutenant-Governor may" prescribe. In 1971, the legislature enacted the Environment and Land Use Act, and a subsequent revision (R.S.B.C. 1979, c.110) empowered an advisory committee to make environmental recommendations to the Lieutenant-Governor.

The Minister for Parks was empowered under the Park Act, 1979, to issue resource use permits in parks. Casamiro, having acquired its predecessor's title rights, applied for a permit in 1987. However, by an Order in Council on November 25, 1988, the Minister for Parks was ordered to not issue a resource use permit for the recreation area of Strathcona Park. The Court found that the Park Act further empowered the Minister responsible for parks to purchase, acquire or otherwise take possession of land, and to expropriate "the rights of a recorded holder of a mineral title in or on a recreation area".

Additionally, the Court found that Section 17 of the current Mineral Tenure Act, S.B.C. 1988, c.5 was determinative stating, "Not withstanding any Act, agreement, free miner certificate or mineral title, no person shall locate a mineral title", explore, develop or produce minerals "in a park created by an Act ... unless authorised by the Lieutenant-Governor ***".

The Court noted that a prior case, *The Queen v. Tener* (1985) 17 D.L.R. (4th), established precedent. The holding in *Tener* was "to refuse a permit to a grantee of mineral rights to exercise rights under the grant within a park established after the date of the grant is the equivalent in law of a compulsory taking or expropriation". In keeping with "Tener rights", the court concluded that the Crown rights granted under the Mineral Act were not affected by its repeal, and assuming the Order in Council to refuse a permit was made under proper authority, it had the same effect as a refusal to grant a permit. Consequently, it amounted to an expropriation. Since the legislature did not authorise the taking without compensation, Casamiro was entitled to payment for the taking of its mineral claims.

(iii) *Cream Silver Mines Ltd v. British Columbia (B.C.C.A.)* [1993], B.C.J. No. 304: In a sequel to the *Casamiro Resource Corp. v. B.C.* suit, under a very similar set of circumstances in the same provincial park, another mining company was not so fortunate as to win compensation for the taking of its mining claims. Cream Silver Mines' rights to mine its claims located in the provincial park had likewise been gradually eroded and finally deprived. Cream Silver Mines brought a claim action against the British Columbia government seeking compensation for the taking of its undeveloped mineral claims. The trial court agreed that the precedent of compensating for a regulatory taking had been followed in *Casamiro,* and as set in the original taking case of *Tener v. The Queen* in 1985. The trial court found that Cream Silver should be compensated for the regulatory taking of its mineral claims. However, on appeal by British Columbia, the Court of Appeals reversed the decision and denied compensation.

In *Cream Silver Mines,* the B.C. Court of Appeals found a distinguishing legal difference between Cream Silver's claim and Casamiro's sufficient to deny any monetary award to Cream Silver. Casamiro's mining claims were Crown grants where Cream Silver's mining claims were mere located mineral claims located under an authorising statute. The court held that Crown grants are an interest in the land, while located claims are not and are incapable of registration under the B.C. land registry. Cream Silver's claims having been made under a statutory right of location, their interests died with statutory amendments denying mining rights and no statutory provisions had been made for compensation with their loss. Cream Silver was denied compensation for the loss of its mineral claims.

5.5.2 *Canadian Wetlands*

A search of past and current litigation involving wetlands and mining in Canada offered no cases. There probably have been none, nor will there be any until Canada has some regulation in place for wetland protection. There is very little constitutional authority permitting federal government enactment of legislation that would directly affect wetlands under §91 the Constitution Act, 1867. "Only through the seldom exercised 'peace, order and good government' clause in the preamble to §91 could the federal government develop legislation for express purposes relating to wetlands. Wetlands are, however, often incidentally affected when the federal government exercises legislative jurisdiction over fisheries, navigable waters, migratory birds, transportation, or most often, over agriculture". (Tkachuk, 1993, p. 9).

5.5.3 *Planning Law or Zoning Limitations*

Each of the provinces have their own statutes and regulation governing every aspect of law. Following are various cases from the Provinces.

(i) Province of Ontario: The Pits and Quarries Control Act, 1971 (Stats. Ontario, c.96): The purpose for enactment ostensibly was to provide certain protections to nearby residents, to require rehabilitation of exhausted lands and to correct proven problems of conflicting land uses. (See *Millar v. Min. of Nat. Resources and Preston Sand and Gravel, Ltd.*, [1978], 7 CELR 156, infra; also, see *Re. Allied Chemical of Canada Ltd. and Township of Anderdon*, [1979] infra)

(ii) The Environmental Protection Act, 1971, S.O. 1971; for a case, see *Steetley Inds. Ltd. v. A-G Ontario* [1978] 7 CELR 2 (re: air pollution by quarry)

(iii) Surface Mining on lands zoned "agricultural"
In *Millar v. Minister of Natural Resources & Preston Sand and Gravel,* the Supreme Court of Ontario (Divisional Ct.), [1978] 7 CELR 156, Preston S & G operated a sand and gravel pit on lands zoned "gravel pit". The pits were licenced pursuant to the Pits and Quarries Control Act, 1971, (Ont.) c.96, by the Minister / NR. The subject lands were designated "agricultural" in the local Official Plan. The Ontario Municipal Board had twice before recommended that no license be issued and the Minister / NR once previously had refused to grant the same licence.

On application for an order that the licence granted by the Minister/NR was null and void, the court held that the application be dismissed. It found that the Minister / NR had not exceeded his authority in granting the licence as the pit's location did not contravene the local official plan and zoning by-law. Lands designated "agriculture" are not required to meet the present and future development needs. Pits and quarries are not prohibited from such designated areas.

(iv) In *Re: Jackson and East Gwillimbury Official Plan, Amendment #8,* (Ontario Municipal Board), [1978], 7 CELR 131, an application for an amendment to the Official Plan to change the designated use from "rural" to "extractive" for mining gravel was denied for insufficient evidence and the application denied.

The application "was opposed by virtually all the residents" and by a local fishing club. Detailed evidence with regard to the social, economic and physical environment impact that the gravel pit would have was presented. The Board held that the evidence was not sufficiently conclusive, particularly with regard to the geology and hydrology to support the application. A further concern by the Board was expressed concerning the traffic problem that might exist caused by gravel trucks. The Board noted, too, that the need for gravel from this site was not established. The application was denied.

5.5.4 *Mining-Environmental Regulation Problems — Water and Pollution*

(i) **Ontario:** In *Re. Allied Chemical of Canada Ltd. and Township of Anderdon* (Ontario Municipal Board) [1979], 8 CELR 48, Allied Chemical applied to Township for approval to permit an extension of its quarrying operations beyond its lands already used.

The Board noted that a licence had been issued to Allied without a hearing pursuant to the Pits and Quarries Control Act, 1971, after operations had begun and that no objections were filed prior to its issuance. The Ontario Water Resources Commission had given approval for Allied to take water for its operation. Under the Board's zoning of restricted land use, which would thus have been in conformity with the Official Plan, an extension onto restricted lands would require approval.

The Board found that Allied had been unresponsive to complaints from members of the community regarding the lowering of the water table due to its quarrying operations, though some settlements had been reached with regard to structural damage to their properties. Consequently, the Board gave approved for only a part of the land expansion applied for by Allied.

In its reasoning, the Board found that because of the lowering of the water table by the quarrying operation, that it was incompatible with the surrounding farm and residential uses. The Board also noted that although meeting the extensive regulations of quarrying by several agencies was requisite, a responsibility to neighbours and to the community must also be met.

(ii) **Nova Scotia: — A defence to a strict liability offence of stream pollution-**

In *Regina v. Aberdeen Paving Ltd.* [1981] 11 CELR 25, violation by a stone and gravel operator of discharging suspended solids into a water course that may cause pollution or impair the quality of the water for beneficial use contrary to §16 of the Water Act, R.S.N.S. 1967, c.335, amended, was the charge.

Water used in the crushing plant for washing and cleaning the crushed stone was directed to a large settling pond surrounded by an earthen berm designed to prevent the escape of muddy water from the pond. Water was found to be leaking from the pond and the company took steps to patch the leakage. A few days later, a further discharge into the local stream was noted by Fisheries officers. The plant was shut down while repairs were made to the berm to stop the flow of muddy water. The defendant was acquitted on a charge of unlawfully causing or permitting a discharge of suspended solids into a watercourse under §16. On appeal, the case was dismissed.

Section 16 of the Nova Scotia Water Act 1967, amended by §9 of c.58 of the Acts of 1972 provides:

> 16. Unless approved by the Minister, no municipality or person shall discharge or deposit or cause or permit a discharge or deposit of any material of any kind into or in any well, aquifer, lake, river, stream, creek, pond, spring, lagoon, swamp, marsh, wetland, reservoir, or other water or water course or on any shore or bank thereof or into or in any place that may cause pollution or impair the quality of the water for beneficial use.

The provision is substantially the same as its counterpart in the Ontario Water Resources Act which was considered by the Supreme Court of Canada in *Regina v. City of Saulte Ste. Marie,* [1978], 7 CELR 53 with the court concluding that the offence is one of strict liability.

On appeal to the Supreme Court of Nova Scotia, the Court found that although the offence is one of strict liability, a defence is that of the exercise of reasonable care. Consideration was given as to whether proper procedures were used by the defendant to avoid the harm prohibited by the statute, and whether the defendant exercised all reasonable care to ensure that the procedures were carried out. The court affirmed the finding of the trial court that the defendant company officers had exercised all due diligence to prevent the offence. The company's employees, however, were found derelict in carrying out their duties, and the accused (company) was found not guilty of the offence.

(v) **British Columbia:** Discharging solids into stream from an open-pit operation

The defence of reasonable care was again considered in *Regina v. Jack Crewe Ltd.* [1981], 10 CELR 120, where a sand and gravel operator allowed silt, sand and clays as a deleterious contaminant from its washing plant operation to enter a tributary, and also directly into the Coquitlam River, being harmful to a fish habitat. Samples taken at the point of discharge of the plant into the stream, above and below the discharge, pin-pointed excessive suspended solids in the effluent from the Crewe settling pond as well as from the washing plant, circumventing the pond, and directly into the stream.

Suspended solids are defined as a deleterious substance under §§33(1) and 33. (2) of the Fisheries Act, R.S.C. 1970. Reputedly, from expert witness testimony, the range of suspended solids in streams affecting fish habitat have the following effect: 25 parts per million are ideal; 25–80 ppm is an acceptable level; 80–400 ppm would impact the productivity of a stream level; and levels over 400 ppm result in very poor fish production.

Several samples of the defendant's plant discharge, submitted in evidence, were measured well above the 400 ppm, ranging from low four digits to five digits in ppm. The court found that the operator had a program of control, adequate in the past, but the settling ponds had not been properly and adequately maintained. Surface runoff from the pit was uncontrolled and without directed ditching. The defendant could have exercised more control, and consequently, the court was not satisfied that the violator had taken all reasonable care in discharging its duties.

The *Crewe* Court, referring to the defence of "due diligence" which may be shown by a defendant charged with an environmental crime, stated that it is not up to the Crown to prove *mens rea* (intent) or negligence on the part of the violator. To avoid absolute liability, the burden is on the defendant to demonstrate that reasonable care and due diligence was taken.

(vi) **Yukon Territory:** Discharging mine waste effluents into watercourses
In *Regina v. United Keno Hill Mines Ltd.* [1980], 10 CELR 43, the defendant company operated a combined open-pit and underground mine, and was charged with depositing waste into Yukon waters in excess of effluent contaminants discharge limits prescribed by an Industrial Water licence issued by the Yukon Territory Water Board contrary to §6(1) of the Northern Inland Waters Act, R.S.C. 1970, 1st Supp., c.28 §6(1).

The *United Keno Hill Mines*' decision treats and analyses the attitudes and parameters that should be considered by a court in trying and sentencing for an environmental crime. Chief Judge Stuart's opinion is a classic analysis of environmental crime and enforcement attitude. It serves to illustrate the period attitude of the Canadian judicial system in hearing an environmental offence and in deciding the appropriate degree of penalty within statutorily defined limits. Its concern for appropriate justice for environmental crimes is a model to be regarded.

Case Facts:

United Keno Hill Mines Limited pled guilty to depositing waste in Yukon waters on May 1, 1979, in excess of the waste discharge limits prescribed by a water licence and thereby contravening section 6(1) of the *Northern Inland Waters Act*. United Keno Hill Mines Limited (hereinafter referred to as the accused) operates a combined open-pit and underground mine at Elsa. During the period of the infraction the accused was operating under an Industrial Water licence issued by the Yukon Territory Water Board. The licence specifies a maximum allowable concentration of contaminants in the effluent discharged by the accused into Flat Creek. The allowable limits of specific contaminants include copper — .03 milligrams per litre; cyanide — .05 milligrams per litre; lead — .2 milligrams per litre; silver — .1 milligram per litre; and zinc — .5 milligrams per litre. Further the licence requires that no discharge be toxic to fish. The licence was issued in April 1975. In January 1978, the accused applied for an amendment to the licence. In August 1978, the Water Board denied the application.

Throughout all monitoring periods in October 1978, the discharges in almost every instance exceeded the licence requirements. The accused maintained contact with the Department of Indian and Northern Affairs, co-operating and attempting to employ recommendations offered by DINA inspectors to rectify their inability to reduce discharge contaminants to licence standards,

On May 1, 1979, samples of effluent discharge were tested. On this day the discharge contained zinc contaminants in 1.5 times licence limits, cyanide in 17 times licence limits and the bio-assay proved toxic to the extent that in 32% concentrate of the sample discharge 9 of 10 test organisms were dead in 96 hours.

Since 1978, the accused invested approximately $370,000 in attempting to resolve persisting effluent excesses, and diverted key equipment and personnel from the pit mine operation to complete construction of a third settling pond. Despite evidence of some progress in abating the excesses, as of May 1, 1979, the problem was not resolved. Throughout this period the company repeatedly asserted the licence limits were too high and not in keeping with the levels commonly established throughout North America.

Chief Judge Stuart, writing the decision for the Territorial Court of the Yukon succinctly stated that "Pollution is a crime", adding that "each offence must be sentenced in accord with its specific facts, but pollution offences must be approached as crimes, not as morally blameless technical breaches of a technical standard".

The Court's attitude was that the severity of punishment should vary with the nature of the environment affected and the extent of the damage inflicted. In sentencing corporations for environmental offences, consideration and record review should be given to the following: (1) the criminality of the conduct, i.e., whether willful, negligent, or unintentional; (2) extent of attempts to comply with regulations; (3) genuine contribution, the size and wealth of the corporation, and profits or savings realised as a consequence of the offence; and (4) any prior criminal record of environmental offences.

In treating corporate environmental offences to control corporate activities in the public interest, Judge Stuart proposed that a variety of civil, administrative, educational and criminal devices be employed. Sanctions should be used to reaching the "guiding mind" of the corporation, for otherwise, the source of direction for the illegal activity will rarely be affected. Personal liability of the corporation's directors, officers and supervisors is necessary, otherwise, the *corporate veil** may afford a measure of protection for criminal conduct by diverting corporate criminal conduct into conciliatory processes or by ignoring the criminality of responsible corporate officers. (*Author's Note: a corporation is treated as a legal entity distinct from its shareholders where the rights and obligations of the corporation are normally separate from those of the shareholders. In many cases, particularly for medium and small-sized corporations, major shareholders are also officers and directors of the corporation, and the "guiding mind" for corporate activities. Consequently, when corporate liabilities arise, the corporate entity usually shields the shareholders from the obligations and liabilities of the corporation. This shield of protection for the stockholders is known in law as the *corporate veil.* In certain cases, the corporate veil may be "pierced" by a court in order to hold the shareholders personally liable for the corporation's actions and liabilities. "Piercing the corporate veil" is possible where it may be shown that the shareholders have been using the corporate entity as their "*alter ego*". *Alter ego* is a legal theory where the corporation's entity has been utilised by the shareholders such that, in reality, their individuality has been mixed, or no separate corporate entity has been maintained by the shareholders.) Thus, the imposition of personal responsibility necessitates an ability to identify culpable corporate officials for environmental crimes, and offers the courts the right to complete access to information related to the corporate internal allocations of responsibility and liability.

With reference to United Keno Hill Mines and its alleged water pollution, there were no prior relevant violations. No evidence was offered against Keno that it had profited through its activity of securing any competitive advantages or illegal gains, and there was no evidence of environmental damage or cause for a court supervisory order. The evidence showed that Keno was a large national corporation that had demonstrated a co-operative attitude with the governmental environmental agency to reduce its discharge levels. Keno had invested time and money in attempting to conform to discharge compliance. The government had been tolerant during a prolonged period of licence violation. The court noted a responsible demeanor of corporate officials appearing before it, having shown good faith and remorse for its acts, which the court accepted as having a positive impact on sentencing.

Taking into consideration the corporation's diligent behaviour, and absent any flagrant condemnable violation activity, the court minimised sentencing of the corporation for the pollution violation, adding that "if the corporation chooses to operate, it must live within the effluent control imposed by the licence". The court set a fine of C$1,500.

Of particular thesis interest and value in constructing a model law, where sanctions and penalties for environmental crimes are an integral part, Judge Stuart listed ten additional points for consideration by the courts in constructing appropriate sentencing to prevent future corporate environmental violations. The instructive directives are summarised as follows:

Model Environmental Law Considerations for Sentencing of Corporate Officials for Environmental Crimes:

1. Statutory imposition of an affirmative duty upon senior echelon corporate officials to know and control all corporate activities;
2. A hierarchy of penalties dependent upon the degree of willfulness or recklessness attributable to actions of a corporate official;
3. Where the required degree of willfulness or recklessness cannot be established, utilise any strict liability offences as lesser included offences which pose a lesser degree of punishment;
4. Except in strict liability offences, the defence of due diligence may excuse liability;
5. Corporate officials required to report to shareholders details of convictions;
6. Courts be empowered to levy restitution orders against corporations and its officials;
7. Empower the courts to impose sanctions against professional businessmen to suspend their participation in corporate activities where they have used their position in the commission of a crime;
8. Where a corporation is a repeated violator, empower the courts to order reshaping of corporate decision-making, close down corporate activities, or impose court-sponsored supervision in order to remove the threat of further injury;
9. The use of continuing orders should be employed, where timely and appropriate, along with sentencing to avoid further infractions of regulations. The costs to the environment and of further prosecutions may be avoided by use of this coercive court power;
10. When the negotiating process is employed to correct and compensate victims of damaging environmental offences, participation by a public interest group or damaged individuals may make the corrective negotiations, along with the corporate offender, more effective.

The *United Keno Hill Mines* Court decision is unique in that the opinion written by Judge Stuart exhibited genuine concern for treating environmental offences, for finding judicially controlled sentencing methods to effectively reduce repetitive offences, and for seeking a judicially enforceable, structured sentencing method with personally applied penalties and punishment for those persons within the corporate structure who can effectively control and bring about environmental compliance.

Such environmental offence sentencing would be made more effective than by the usual, and past employed, method of substantial fines against the corporation. Such fines are often absorbed by the corporation and passed on to the consumer through the pricing structure. Its maximum effect is where the fines may be so large as to effect its pricing in competition with non-violating and non-penalised competitors. Large fines do not so effectively mould law-abiding corporate behaviour as would punishment directed and applied to the responsible corporate actors who personally determine environmental compliance for the corporation.

5.5.5 *Planning Law, Land Use, and Mineral Properties*

The "taking", or expropriation of a shale pit for a park in Quebec in the case of *Landreville v. Ville de Boucherville*, [1978] 7 CELR 12, was decided by the Supreme Court of Canada.

Landreville was the property owner. A permit for mining had been issued in December 1965 by the municipal council of Boucherville. A year later, the town adopted a resolution which resulted in a notice of expropriation. The town had decided to take the land for a park. The owner challenged the statutory right under the Cities and Towns Act, Section 605, to expropriate for municipal purposes. The owner alleged that the town was expropriating in bad faith by its alleging that the purpose was for a park, while in reality it "simply wished to acquire the land for the sole purpose of putting an end to quarrying activities". The Quebec Superior Court found for the landowner and annulled the expropriation proceedings. The town appealed and the Quebec Court of Appeals reversed on the grounds that the landowner had not convinced the court of "blatant fraud and flagrant injustice". The owner appealed. The Supreme Court of Canada held on review that "the municipal council had committed an abuse of power, a flagrant injustice equivalent to bad faith to the owner" and annulled the expropriation proceedings.

5.6 Conclusions and Comments

Environmental Regulatory Property Takings:

A balance must be found between the nation's need for mineral production and environmental concerns. Minerals are essential to the public's health, welfare and well-being. Minerals are essential for the public's daily living requirements; for concrete construction, road building, cars, housing, medicines, optics, energy, communications, and an infinite list without which, man would return to a cave-dwelling life and severe deprivation. The industry's slogan, "If it can't be grown, it must be mined". sums up the need perfectly. At present, not only from a mineral interests perspective, the pendulum has swung far enough in favour of environmental precautions, i.e., too restrictive of mining. The environmental impact of construction minerals mining, in truth and reality, should be of minimum environmental concern or consequence since the mined minerals are relatively inert (e.g., granites, limestones, sand and gravel) and contain no groundwater or surface water contaminants as found in metal mining. As stated by a Canadian consultant in Mining Land Use & Reclamation, in his 1977 report to Environment Canada, "Despite the controversies, debates, studies, and the resulting statutes and by-laws (not only in Ontario but throughout Canada), however, very little is presently known regarding the magnitude of the impact on the land resource base caused by mineral aggregate and industrial mineral production". (Blakeman, 1977, p. 2). Hence, the environmental justification for any possible over-regulation of the industrial minerals, even if true, has been one of "better safe than sorry".

The U.S *Keystone* was a fair decision. No mining entity should be allowed a noxious or destructive-use activity of their property that harms another's property (even with a waiver), or the public health and welfare in general. In addition, as to alleged values for compensation of losses for less than full use of its properties and recovery of the coal reserves required to be left in place to support the surface, the Court's analogy was right on point: that a statutory "requirement that a building occupy no more than a specified percentage of the lot on which it is located could be characterised as a taking of the vacant area as readily as the requirement that coal pillars be left in place. Similarly, *** one could always argue that a set-back ordinance requiring that no structure be built within a certain distance from the property line constitutes a taking because the footage represents a distinct segment of property for takings law purposes". (*Keystone*, id. at 1249).

Obviously, as stated by Kusler and Meyers, "Not every square foot of private property can be used for its 'highest and best use'". (Kusler, 1990).

Environmental extremism in mining regulation is to be avoided — by both environmental advocates and industry.

In *Keystone*, extreme legal protection for the coal mining industry was rejected by the U.S. Supreme Court. The coal industry's former protection against subsidence damage claims by virtue of surface land owner's consent and agreed waivers to damage claims had been a legally accepted one for many preceding decades. Contractually and constitutionally, it was tested by the U.S. Supreme Court in 1922 in *Pennsylvania Coal Co. v. Mahon* and found legally supportable. Sixty-five years later and well into the environmental revolution, the same Court found the former constitutionally supportable subsidence waiver contracts to be socially unacceptable, harmful to the public's well-being, and, therefore, void under a state's police powers. Injury with resulting damage to the well-being of the public was found beyond the rights of the contracting waiver parties, which made the contract void. Hence, extreme legal protection of the mining industry was struck by a reasonable court decision.

At times, environmental regulation has surpassed the point of reasonability, and by the same token of exercising good sense as exhibited in the *Keystone* case, needs to be reversed and struck down, too. A recent case in point where extreme environmental regulation needed to be tempered with common sense is that of *American Mining Congress* (AMC) *v. U.S. Army Corps of Engineers (and National Wildlife Federation, Defendant-Intervenors, Sierra Legal Defense Fund, National Audubon Society)*, 951 F.Supp. 267 (D.D.C., 1997).

The case involved the "Tulloch" rule under the Corps of Engineers' regulations for a §404 permit of the CWA (U.S.C. §1344) for discharging dredged or fill material into the navigable waters of the United States at specified disposal sites. The Act defines a "discharge" as "any addition of any pollutant to navigable waters from any point source".

Very briefly, the Tulloch rule evolved out of a settlement of a North Carolina case. Under the §404 permit, when a navigable waterway is being dredged for cleaning out accumulated sediments to restore shipping depth of the navigable channel, the sediments dredged must be transported to an approved and specified disposal site (ordinarily, temporarily placed on the waterway's banks and reloaded by trucks for haulage to the disposal site). Upon promulgating its own rules, in conjunction with the EPA (who also jointly administers the §404 permitting program), the Corps designated incidental fallback material from the dredging buckets (which were cleaning out the navigable channel) into the waters being dredged, as being violative of the §404 permit by placing pollutants into the waters of the United States.

Any argument for enforcement of such a rule is obviously ridiculous. Common sense dictates that digging in muddy water cannot be accomplished without some incidental spilling of watery, mud sediments as the buckets are pulled from the stream. The Tulloch rule case may be more appropriately referred to as the "dripping bucket" case.

Still, ridiculous or not, the Corps of Engineers, which is the largest remover of sediments from navigable waters in the U.S., is effectively exempt from the rule. It is only applied against private citizens who are dredging.

In *AMC v. U.S. Army Corps of Engineers*, the District court found that the Tulloch rule provided "the term does not include *de minimus* incidental soil movement occurring during normal dredging operations, defined as dredging to maintain, deepen, or extend navigational channels in the navigable waters of the United States, *** ".

The court stated, "Incidental fallback is the incidental soil removement from excavation, such as the soil that is disturbed when dirt is shovelled, or the back spill that comes off a bucket and falls back into the same place from which is was removed". In its ruling, the Court held that incidental fallback is not the "addition of a pollutant" and §404 does not cover incidental fallback. It further stated that "Congress did not intend to cover incidental fallback of soil from excavation under the CWA permitting provision, and did not intend to regulate removal of material from water such as land-clearing, ditching, and channelisation under that permitting process".

Here again, is a case expensively and frivolously litigated over a trifling matter which was obviously attempting to enforce ridiculous extremism in environmental regulation. The so-called violative "pollutants" into the waters of the United States were none other than the "drippings" from the digging bucket. Those drippings, or incidental fallback sediments, simply fell back into the hole from which they were removed; no more. The environmental supporters of the Army Corps of Engineers for enforcement of the frivolous, extremist "dripping bucket" rule only impugned their own integrity and sincerity for the environmental cause.

Temperance and Avoidance of Environmental Extremism in Regulation is Urged:

When a young American family is deprived of its property right to enlarge its own home with additional bedrooms for its growing number of children, under the Endangered Species Act, then the pendulum has swung too far to extremism for the environmental cause. Such was the case in California where human property rights were denied by the EPA and superseded by animal rights of a species of rat that had made the human's property part of their habitat. The rats were granted superior rights to the property and the human owners of the property could not enlarge their house because it would disturb the rats' habitat.

All the world cannot be a wildlife habitat or a scenic park

Minerals and mining must be appropriately included in the environmental order as essential to mankind's existence, or man will ultimately be himself placed on the "endangered species" list. This is not an extremist statement for, as previously discussed in the Introductory chapter, mining provides essential minerals and chemicals for the health and well-being of all mankind. Environmentalism cannot be carried to the point where no mining is be tolerated. Minerals must be mined where they are found, or located. Yet, there are those groups, who in the name of "environmentalism", would go so far as to prohibit all mining, or at least, minimise mining to the detrimental effect of the human races.

A case in point is the recent *Virginia Vermiculite v. W. R. Grace & Co.-Conn.*, 965 Fed.Supp.902, (W.D. Va. 1997). The environmentalist group, The Historic Green Springs, Inc. (HGSI), has pointedly made known that its goal was to prohibit mining in an entire county known to contain valuable mineral deposits of vermiculite; "Since the 1970s HGSI's stated purpose has been to halt all vermiculite mining and other land development that adversely impacts land (values) in Louisa County, Virginia".

The mineral vermiculite, in the words of the federal district court, "is a unique mineral in that it is fireproof and rapidly expands approximately ten fold in volume when heated to produce a low-density material. Vermiculite has valuable uses in fire safety, energy conservation, construction, environmental products, food processing and horticulture". It is also "one of the promising new applications for detoxification of water and soil, nuclear waste containment and removal, and industrial spill containment and clean-up".

"In the United States, vermiculite is currently mined in South Carolina and Virginia, the only two states where substantial vermiculite reserves are known to exist, other than Montana, whose deposits are believed to be contaminated". (Mining of vermiculite in Montana has been discontinued.) The Court also noted that in 1992, a significant portion of 18% of vermiculite consumed in the United States was imported from South Africa in 1992. Note here, HGSI's NIMBY-attitude; let the U.S. requirements for vermiculite be mined in someone else's backyard.

Virginia Vermiculite, Ltd. had been mining vermiculite in Louisa County, Virginia, since 1976. In 1992, its share of the market was approximately 23% of all sales. "Its only formidable domestic competitor is W.R. Grace & Company which owned 80% of mining rights in Virginia and South Carolina and was responsible for 57% of all vermiculite sold in the United States". The environmental group, HGSI, has since successfully obtained W.R. Grace's extensive vermiculite mining rights in Louisa County as a tax-deduction maneuver. Virginia Vermiculite is effectively stopped from mining future ore reserves of vermiculite in Virginia. Thus, an environmental group has partially driven mining out of Louisa County, Virginia. Virginia Vermiculite continues mining from its remaining properties in Louisa County, but has recently had to start mining vermiculite in South Carolina where the area is receptive to mining of its deposits. When the South Carolina deposits are eventually exhausted, the Virginia deposits will be the only remaining sizable vermiculite reserves in the U.S. known at this time. Again, mineral deposits must be mined where they occur. All the world cannot be a scenic park or wildlife habitat if mankind is to continue.

Moderation for environmental causes is required by all parties concerned. Extremism must be avoided.

CHAPTER 6

A REVIEW OF THE ENVIRONMENTAL ERA REGULATORY ACTIONS FOR LANDFILLING WITH LITIGATED INTERPRETATIONS

6.1 Introduction

A primary, important, related problem that accompanies landfilling is the danger of contaminating ground waters, and even surface waters, from leachate escaping from the landfill. Another prominent problem is the gas generated from landfills, primarily in the form of methane and carbon dioxide. The major areas of regulation concerning landfill and solid waste disposal are reviewed.

By reviewing landfill sites and operations that have taken place in the recent past along with ensuing allegations of environmental damage in court cases, it is hoped that insight may be obtained from former and recent practices. The past has value in serving to steer a truer course in the future. Therefore, a review of earlier areas of law which grew to form modern environmental law leads us into areas of, viz., public health law, torts of nuisance and trespass, and zoning or planning law.

6.2 United Kingdom

Intervening amendments over the decades since the Public Health Act of 1875 resulted and yielded the more comprehensive Public Health Act 1936 which in turn was supplemented by the Public Health Act of 1961 and the Act of 1969 (Recurring Nuisances). Part III of the Public Health Act 1936 still deals with nuisances and offensive trades. The 1969 Act deals with refuse disposal.

water and clean air, et al. In a country where 90% of the waste is tipped as landfill, the Public Health law reached major development with the enactment of the Control of Pollution Act 1974, in which basic environmental protection was established.

In England, waste disposal had long been a "county matter" handled by the county councils, while in the metropolitan areas the districts are the authorities. In Wales, waste disposal was generally a function of the district councils. From 1974, operation of landfills was governed by the TCPA and COPA 1974 under §§3-11 (1976). From 1976, waste disposal sites required licencing. Licencing conditions could be varied, e.g., duration, type and quantities of waste. Hughes (1992 op. cit.) reports that there was mounting criticism of the law of waste disposal throughout the 1980s under the Control of Pollution Act (COPA) 1974. New provisions were expected to correct former shortcomings of COPA with the coming into force of the Environmental Protection Act (EPA) 1990 by November 1991, or shortly thereafter.

For a period in 1991, as a result of *Leigh Land Reclamation Ltd. v. Walsall Metropolitan Borough Council* [1991] 3 JEL 281, controls and enforcement conditions for landfills were limited by that case law to situations where only a direct breach of a condition of waste deposit or treatment was involved. During the interim, from the *Leigh Land* decision and until the EPA 1990 was implemented in 1992, a number of waste disposal sites were reported as ineffectively controlled.

6.2.1 *Waste Disposal under the EPA 1990*

The new act was designed to make improvements for waste handling authorities and to tighten up on weaknesses in procedures under COPA 1974 for waste collection and disposal.

Starting with a revised definition of "controlled waste", which had been defined under COPA 1974 as household, industrial and commercial waste, it was expanded to include scrap, effluents, unwanted surpluses from processes, broken or worn out items, spoiled or contaminated material (but not anything explosive) and anything discarded. It should be noted that under both the older and newer definitions of commercial waste, waste from mines and quarries are excluded. Although mining wastes are excluded under Part II of EPA 1990, the Secretary of State has power under §§63 and 75(8) to deal with this exclusion.

A new series of authorities has been provided for, e.g., Waste Regulation Authorities (WRAs), Waste Disposal Authorities (WDAs), Waste Collection Authorities (WCAs) and Waste Disposal Contractors (WDCs). The specific responsibilities of each of the authorities are given under the new act. The WRAs are the principal control agencies for waste treatment and disposal in their respective areas. It should be noted that the WRA must confer with the National River Authority (NRA) in the preparation of plans for concern of water pollution before a disposal site can be approved.

Licencing of waste disposal sites, although not new, requires that an operator holding a Waste Management Licence (WML) and functions under a new requirement of a "Duty of Care" (DOC). Failure to comply with the waste regulations is an offence. Where an offence is proven regarding controlled waste, the person found guilty may be, on summary trial, imprisoned up to six months or fined up to £20,000, or subject to both. Where special waste is involved, imprisonment may be increased up to 5 years.

The many new changes made in the EPA 1990 were designed not only to make for greater efficiency of waste disposal, but hopefully **to instill greater public confidence in landfill management**. (emphasis added).

Mandated recycling duty

A new feature of the EPA 1990 is §49, which imposes a duty on the WCAs to plan for recycling of household and commercial wastes collected. The plan must state the kinds and quantities of such anticipated waste to be purchased and prepared for recycling; arrangements for waste disposal with contractors, plant and equipment to be used; and the estimated costs and / or savings attributable to the recycling under the plan.

Recycling of MSW is discussed further in Ch. 10 §2, Current Methods to Relieve Urgency for Landfill Space, and §3, Current Landfill Information.

6.2.2 *Evidence that Sand and Gravel Pits Are Compatible for Landfilling*

The compatibility of landfilling and sand and gravel mining in the U.K. was brought to light in *Durham County Council v. Sec. State for Environment and Tarmac Roadstone Holdings Ltd*, [1989] T. & C.P., discussed earlier in Ch. 5 §3.6 (iii), where the worked-out sand and gravel pits were being used by the County for disposal of refuse and the County opposed further mining, the Court of Appeal stated in its finding of granting former mining rights, "*** mining sand and gravel are only extinguished when they are inconsistent with the new use". Since further mining of sand and gravel was compatible with refuse burial in the mined-out pits, the Council's appeal was dismissed.

Further examples of the use of sand and gravel pits for MSW landfilling in the U.K. are entered into evidence are: Allerton Park Pit, infra, Ch. 11, §8.1.2, and Sutton Courtenay, pit, Oxfordshire.

In the U.S. and Canada, evidence of sand and gravel pits serving as landfill sites are mentioned at various parts of this thesis; in the U.S, e.g., a 125-acre sand and gravel pit in the San Gabriel Valley, California (infra, Ch. 13 §4.1), Lemons Gravel pit, Dexter, Missouri; in Canada, e.g., Hale's sand and gravel pit located in the town of Newcastle, Regional Municipality of Durham, Ontario [infra, Ch. 6 §4(i)].

6.3 United States

In 1976, Congress enacted the Resource Conservation and Recovery Act (RCRA), 42 U.S.C. §6901 et seq., to deal with municipal solid waste and hazardous waste disposal. As a result, an office of Solid Waste was established within the EPA. Under the Code, federal financial and technical assistance is made available to state and local governments for the development and implementation of solid waste plans and programs that meet federal standards. The states must define boundaries for regional solid waste management and identify state regional and local agencies responsible for implementing the plan. The term "solid waste" is broadly defined and includes not only municipal and industrial waste, but includes mining waste.

The EPA issued regulations establishing federal criteria for sanitary landfills that are environmentally acceptable. EPA also published an inventory of all facilities or sites in the U.S. that do not meet the criteria. In order for states to be eligible for financial aid, they were required to adopt a compliance schedule for bringing those sites into compliance with the criteria or closure within the first five years. This new area of regulation provided more environmental litigation to probe its parameters of control.

6.3.1 *Cases Illustrating Landfilling Conflicts Under the EPA's More Stringent Regulations*

i) A Case of Contested Landfill Expansion

The Illinois Environmental Protection Act provides for licencing of MSW sites under its §39.2. The relevant part of the regulation for the following illustrative case provides as follows:

> "(a) The county board *** shall approve or disapprove the request for local siting approval for each regional pollution control facility which is subject to such review. An applicant for local siting approval shall submit sufficient details describing the proposed facility to demonstrate compliance, and local siting approval shall be granted only if the proposed facility meets the following criteria:
>
> (i) the facility is necessary to accommodate the waste needs of the area it is intended to serve;
>
> (ii) the facility is so designed, located and proposed to be operated that the public health, safety and welfare will be protected;
>
> (iii) the facility is located so as to minimise incompatibility with the character of surrounding area and to minimise the effect on the value of the surrounding propert;
>
> (iv) the facility is located outside the boundary of the 100-year flood plain or the site is flood-proofed;
>
> (v) the plan of operation for the facility is designed to minimise the danger to the surrounding area from fire, spills, or other operational accidents;

(vi) the traffic patterns to or from the facility are so designed as to minimises the impact on existing traffic flows; ***"

The EPA further provides under its §40.1(b) that any third party may petition the Illinois Pollution Control Board within 35 days to contest the approval of any action of the county board. [Ill. Rev Stat 1989, ch.111-1/2, para. 1040.1(b).]

In Bond County, Illinois, the Board of Supervisors granted an applicant's request for expansion of a regional pollution control facility site. A citizens' group of 35 persons opposed to the expanded landfill filed a petition with the Pollution Control Board for hearing. The Pollution Control Board affirmed the County Board's decision. The citizens' group appealed.

The case is illustrative of an encounter with the NIMBY syndrome tactics brought by a local citizens' group. The Concerned Citizens Group was less concerned with potential effects of pollution and groundwater contamination than with decreased values for their properties. They naturally mustered as much resistive argument as they could to affect a defeat of a much-needed waste disposal site for their own community. In addition to strongly protesting under §39.2 (iii) above, they argued increased traffic problems, under §39.2 (vi) above. Additionally, to make the approval of the site's expansion more palatable, or more acceptable to the community, aesthetics were generously applied by the applicant in utilising only 38% of the new acreage for the actual purpose of waste burial. Such a low yield-use of land does not comport with land-use conservation. However, it appears that a better land-use percentage in keeping with conservation of land use would only incur more vehement protest by the concerned citizens thereby defeating the needed disposal space.

In *File et al v. D & L Landfill, Inc.*, 579 N.E.2d 1228 (Ill.App. 5 Dist. 1991) a civil engineer who had helped prepare the application testified for the applicant. 73 acres (29.54 ha) out of 193 (78.11 ha) would be used for trash placement. Planning included the usual landscape screening; 200-foot (60.96 m) setoff between property line and trash placement; soil borings indicated the site was underlain by a glacial till with numerous sand lenses throughout associated with groundwater. Excavation of the sand lenses was planned down to hard till and replacement with clay, as well as a bottom liner of 10-feet (3.05 m) of clay at the bottom. The applicant hoped to use existing glacial till as its clay liner. A sheet of synthetic material was to be placed on top of the glacial till, covered by a foot of sand, which would act as the leachate collection system. Trash would be deposited in layers of four feet and compacted. A daily cover of dirt on the trash would be applied; ten to fifteen monitoring wells were to be placed, with 3 or 4 upgradient for a background analysis of the groundwater. An independent firm would sample and test the ground water from wells quarter-annually for contaminants.

A report from the Illinois Geological Survey found a moderate chance of contamination from the site expansion; that sand lenses were widespread in the area and served as a source of drinking water for people near the site; that native geological materials at the site offered only a moderate to low level of protection to the shallow sand and gravel aquifers. The applicant's engineer disagreed with the State's report and the court accepted his testimony as adequate refutation.

Further extensive testimony was reviewed, including anticipated depreciation of land values. The Court agreed with the Pollution Control Board that the findings were not against the manifest weight of the evidence that the applicant had made a reasonable effort to minimise the impact of the expanded landfill on the surrounding area and properties, and affirmed the decision allowing the site permitting. The Court noted that "a determination of whether a regional pollution control facility was designed, located and proposed to be operated so that public health, safety and welfare would be protected is purely a matter of assessing the creditability of expert witnesses".

With regard to the increased traffic charge, the court found that the applicant had made a reasonable effort to minimise impact of the expanded facility on existing traffic flow and would be impacted only slightly by an insubstantial increase in truck traffic since all trucks entering or leaving the expanded landfill would be using the existing entrance.

ii) Landfill Leachate Dumping Into Navigable Waters of the U.S. Challenged

This case is offered to demonstrate that where the local, or area, environmental authorities fail to implement regulations concerning a landfill's leachate contaminating surface waters by illegal disposal, a citizen's suit may be filed to enforce regulations.

In *New York Coastal Fishermen's Association v. New York City Department of Sanitation, N.Y.C. Dept. of Environmental Protection, et al,* 772 F. Supp. 162 (S.D.N.Y. 1991), an association organised for the preservation for Long Island Sound commenced a citizens' suit under the Clean Water Act (CWA) against city and municipal agencies, challenging dumping of leachate collected from a landfill into the bay as a temporary measure in cleanup of leachate at the landfill which had been adopted pursuant to an agreement with the New York State Department of Environmental Conservation (DEC).

The Pelham Bay landfill, located in Bronx County, had been operated from 1963 to 1979 by the N.Y.C. Department of Sanitation. The landfill was then closed, but not effectively capped. Starting about 1982, complaints were lodged by local residents about leachate streams and puddling in the landfill area. After investigation, DEC declared the landfill an inactive hazardous waste site. DEC and DOS entered into an Order on Consent in 1985 requiring DOS to submit two leachate management plans to the State, one temporary and one permanent. Three years later after much delay, DOS submitted an interim proposal which was simply collection and recirculation of the leachate through the landfill. DEC rejected the plan. An alternate plan was approved and put into effect in September 1988 for leachate collection, passing it through a stormwater sewage system with discharge into Eastchester Bay, feeding into Long Island Sound. In April 1990, a second Order on Consent required completion of the permanent plan by 1995 for remediation of the landfill.

The N.Y. Coastal Fishermen's Association filed its complaint in July 1990, contending that the agencies were in violation of the CWA by dumping the leachate into the waters of the U.S., and further, that they did not have a permit for discharging the pollutants. The Association sought injunctive relief to stop further dumping of leachate and for civil penalties levied on the agencies.

The municipal agencies offered a defence against the citizen suit claiming that there had been diligent prosecution by the state to correct the pollution, thereby precluding a citizen suit. However, the court rejected their argument noting that a claim of state involvement since 1983 without correction of the problem "flies in the face" of "diligent prosecution". The court stated that immediate injunctive relief would not be granted in the citizen suit under the CWA because it would be unworkable; a permanent injunction would be entered permitting the state and municipal defendants reasonable time to develop an alternative plan for the landfill cleanup.

iii) Denial Of Permit for Landfill Expansion Based on Average Groundwater Levels

This case is offered to illustrate the point that protection of ground water is sought in landfill planning. A permit to expand an existing Type I municipal solid waste disposal site near Rosenburg, Texas, by the State Health Department was challenged by local residents for fear of ground water contamination and that the Health Department had "arbitrarily" established the ground water level for the liner design of the site's extension.

The expert engineers that testified as to the ground water level in the permit-expansion hearings varied greatly in their conclusions derived from different sets of data. One expert established the seasonal high ground water table at a depth of 60 feet (18.9 m). The other expert engineer set the level at 16 feet (4.88 m). As a result of the substantial variation of the experts' water table levels, the Health Department set the level of the seasonal high ground water table for design purposes at 33 feet (10.06 m). On the face of the testimony and evidence, it appeared that the Health Department had averaged the two experts' findings and allowed a safety margin of 5 feet (1.52 m) in arriving at the figure of 33 feet (10.06 m) for the permit's approval. The Health Department denied the residents' request for a permit re-hearing and the residents filed for judicial review. The lower (trial) court affirmed the finding of the Health Department allowing the permit. Flores appealed.

In *Flores v. Texas Dept. of Health*, 835 S.W.2d 807 (Tex.App.-Austin 1992), Flores, the local residents, the Appeals Court found that the Health Department acted arbitrarily and failed to base its finding on substantial evidence. This was substantiated by an amended answer to the court by the Health Department admitting that it had acted "arbitrarily" in establishing the ground water level. The Court, in remanding the action to the Health Department for a new determination, noted that "if the data establishing the ground water level at 16 feet (4.88 m) is accurate, … a liner designed to accommodate a seasonal high ground water table of 33 feet (10.06 m) would not be strong enough to prevent ground water contamination".

6.3.2 *Interstate Commerce Movement of MSW*

The arguments litigated in the following two cases are offered to illustrate that MSW is being transported great distances and across territorial jurisdictional lines to remote landfill sites. This point is to give support to the argument, in Ch. 10, Future Waste Disposal Trends, that many of the large, remote, non-coal, metallic mineral, open-surface mines that are not clustered near urban areas as are the construction mineral pits, should be seriously considered as landfill sites. Utilisation of these larger size, remote location pits would give space and volume relief to those smaller construction pits in the urban areas.

(i) **Out-Of-State Solid Waste Surcharge Fees Found Constitutional:** An operator of an Oregon solid waste disposal facility, the County in which the facility was located, and a Washington solid-waste disposal company challenged the constitutionality of Oregon's Department of Environmental Quality's (DEQ) rules adopted by the EQ Commission which imposed a greater surcharge for disposal of out-of-state waste at the in-state sites that on disposal of in-state waste.

In *Gilliam County, Oregon, Oregon Waste Systems, Inc., and Columbia Resource Co. v. Dept. Of Env. Quality,* 837 P.2d 965 (Or.App.1992), the petitioners argued correctly that under the federal commerce clause, that states cannot enact laws that discriminate against articles of interstate commerce unless there is some reason, apart from their origin, to treat them differently.

Petitioner Oregon Waste Systems (OWS) owns and operates a regional solid waste facility (Columbia Ridge Recycling Center and Landfill) in Gilliam County, Oregon. Both in-state and out-of-state generated solid wastes are disposed of at the site. Petitioner Columbia Resource Company (CRC), under a 20-year contract with Clark County, Washington, disposes of solid waste at the Finley Buttes Landfill in Morrow County, Oregon. Out-of-state solid waste is surcharged at $2.25 per ton, while in-state waste is capped by law at $.50/ton.

On review, the court found that the State appeared to have a reason apart from origin to treat out-of-state differently. "When state wastes reach the disposal site, it has already been subjected to state and local regulations and fees that are designed to reduce its volume and alter its character in order to limit the risks associated with dumping it on the ground". By contrast, "it first encounters out-of-state waste at the disposal site". It also found that "the legislature had made it abundantly clear that it intends only to make 'out-of-state generators pay their fair-share of the costs and no more' ". Furthermore, Oregon law had expressly forbidden EQC to impose fees on out-of-state waste that exceed the costs of management of that waste.

The higher fees for dumping of out-of-state solid waste were found to be "not excessive" and constitutional.

ii) **Constitutionality Of Waste-Flow (Interstate) Control Ordinances:** A Pennsylvania court looked at the constitutionality of a county and municipal ordinance regulating waste-flow control in Lehigh County and the included cities in *Empire Sanitary Landfill v. Comm. Dept. Environmental Resources,* 645 A.2d 413 (Pa. Cmwlth. 1994). The landfill operator, Empire, and a trash hauler, Danella Environmental Technologies, Inc. (Danella) filed an action seeking declaratory and injunctive relief against the Pennsylvania Department of Environmental Resources (DER), Lehigh County, the Lehigh County Department of Planning and Development, and the Office of Solid Waste Management.

In addition to the Act's alleged unconstitutionality restricting interstate commerce under the Commerce Clause, Empire specifically sought a declaration that its Empire-Danella contract was protected under the Act; and, that contracts for the collection, transportation and disposal of waste entered into before the county sought to implement its plan through the adoption of its waste control ordinance, were pre-existing contracts protected by the Act as well as by the Constitution's Contract Clause to not impair private contracts.

The county's ordinance provided for ten-year contracts, stipulating a fixed initial price per ton (and a fixed annual escalator), with each of the disposal facility contractors. The contracts were not "put or pay" in that the delivery of specified tonnage is not required. However, the contract provided that Lehigh County will use reasonable efforts to cause delivery to each contractor of the region's generated waste. The plan did not seek to alter the County's existing municipal waste collection and hauling system which, in large part, utilised private haulers.

Market Participant Doctrine Considered versus Market Regulator

As to its constitutionality, DER argued that the Commerce Clause should not be a constraint as long as the bidding process did not exclude out-of-state businesses. In its review of the County's ordinances under the Commerce Clause, the Court was guided by the "market participant doctrine" as developed by the U.S. Supreme Court under which states and local governmental subdivisions may engage in activities without violating the Commerce Clause. Under that doctrine, when a governmental entity is a market *participant* (to a contract) rather than a market *regulator*, the Commerce Clause is not applicable. A prime example is where governmental contracts are awarded which contain preferences for local businesses even though the successful bidder was not the low bidder.

Illustrative examples were noted in former federal case decisions. In *White v. Mass.Council Constr. Employers,* 460 U.S.204 (1983), the City of Boston was held to be a market *participant* in awarding contracts funded by the city, where the contracts were to be performed by a workforce consisting of at least 50% Boston residents.

By contrast, in *W.C.M. Window Co. v. Bernardi,* 730 F.2d 486 (7th Cir. 1984), it was held that an Illinois law requiring contractors on any public works project or improvement for a state or political subdivision employ only residents of the state, unless the contractor certified either that such workers are unavailable or incapable of performing the work, violated the Commerce Clause.

The Court distinguished *White* from *Window* because the Illinois law applied to all public construction contracts, not only to those which the state was a party.

Similar to *White,* in *Swin Resource Systems v. Lycoming County*, 883 F.2d 245 (3rd Cir. 1989), the county, as a landfill operator, was found to be a market *participant*, where its regulations gave county residents preference in the use of a county-operated landfill did not violate the Commerce Clause.

(iii) Lastly, in *Waste Recycling v. S.E. Alabama Solid Waste Disposal*, 814 F.Supp. 1566 (M.D. Ala. 1993), a governmental authority had been established by 36 local municipalities seeking to build a solid waste disposal site. Under the provisions for the Authority, the cities were required to enact flow-control ordinances requiring that all waste collected by private or public haulers must be disposed of only at the Authority's facility. The court noted that the ordinances were regulatory in nature and that the city- contracting parties could only implement the contractual provisions by their sovereign power. Private participants would not be able to implement the contractual provisions.

The Pennsylvania Court found the case at bar to be similar to *Waste Recycling*. Lehigh County did not engage in a bidding process that excluded out-of-state landfills, but it was unable to implement provisions of its contracts without adopting the county ordinances at issue. In so doing, Lehigh's ordinances subjected other non-contractual parties, e.g., haulers, other landfills and generators of solid waste to their contract. Lehigh was thus, acting as a market *regulator*, and the "market participant" doctrine did not apply.

(iv) **Consideration of Waste Regulations' "Even-handed" Treatment with Incidental Effects versus Obvious Facial or Practical Discriminatory Effects:** Waste hauling and disposal have been recognised as articles and services in interstate commerce, thus involving Commerce Clause concerns [*City of Philadelphia v. New Jersey*, 437 U.S. 617 (1978) and *Hughes v. Oklahoma,* 441 U.S. 332 (1979)]. The U.S. Supreme Court in *Hughes* relied on a 1970 decision, *Pike v. Bruce Church, Inc.*, 397 U.S. 137 (1970) where it was stated that inquiry for the Commerce Clause must consider whether the regulation being questioned regulates "even-handedly" with only "incidental" effects on interstate commerce, or if it discriminates against interstate commerce either on its face or in practical effect. Additionally, it must question whether the regulation serves a legitimate local purpose, and if so, whether alternative means could promote this local purpose as well without discriminating against interstate commerce.

Where a regulation is found to have a local purpose that outweighs its discriminatory interstate effects and there are no non-discriminatory alternatives which accomplish the local purpose intended, it may be found valid. Conversely, where it discriminates against interstate commerce and serves no superior legitimate local purpose, it will be found invalid.

In *Empire*, the Pennsylvania Court found that the contractor Empire was "precluded from challenging the discriminatory landfill designation in the plan.

The waste-flow ordinance the county adopted, by itself, does not discriminate against interstate commerce, however, the ordinance does have an incidental effect on interstate commerce by precluding, in application, the transportation of waste out-of-state".

The Pennsylvania court stated that "until recently, the U.S. Supreme Court has not squarely addressed a Commerce Clause challenge to flow-control ordinances such as "the Lehigh one at issue until it considered the case, *C & A Carbone v. Town of Clarkstown, N.Y.*, 114 S.Ct. 1677 (1994). In the *Carbone* case, Clarkstown adopted a waste flow control ordinance similar to that of Lehigh. "Clarkstown agreed to construct a solid waste transfer station to separate recyclable from non-recyclable waste and operate the station for 5 years. Under the agreement, the town would guarantee a minimum waste flow to the facility, and the contractor would charge haulers a tipping fee that exceeded the cost of disposal for non-sorted waste in the private market. The Court held that the ordinance violated the Commerce Clause.

"The (Supreme) Court stated "While the immediate effect of the ordinance is to direct local transport of solid waste to a designated site within the local jurisdiction, its economic effects are interstate in reach *** even as to waste originating in Clarkstown, the ordinance prevents everyone except the favoured local operator from performing the initial processing step. The ordinance thus deprives out-of-state businesses of access to a local market".

The Pennsylvania court concluded that although the Lehigh ordinance was not facially discriminatory, "the ordinance does have an incidental effect on interstate commerce because (1) it requires disposal of county-generated waste only at designated facilities, and (2) only facilities within the county have been designated. Thus, in application, the ordinance burdens commerce by precluding the transportation of county-generated waste to out-of-state facilities. *** Under the Pike decision, a court will strike down a regulation 'only if the incidental burden on interstate commerce is clearly excessive in relation to the putative local benefits'. The burden on interstate commerce in this case is significant — no county-generated waste may be transported to a non-designated landfill, hence, no out-of-state facility may compete for its article of commerce. Although one of the local benefits of having waste disposed at one of the designated sites is the certainty of available landfill space for the ten-year life of the county plan, that benefit does not outweigh the burden on interstate commerce". Accordingly, the Pennsylvania court found the Lehigh ordinance constitutionally invalid.

The Impairment of Contract Issue — Protected Contracts

Two types of contracts were at issue in *Empire,* viz., that made between Empire and Danella before the adoption of the county ordinance and those made between Danella and its customers. The Court gave credence to the Constitution's Contract Clause by stating that "a statute in force at the time of the making of a contract legally cannot impair the contract. *** and, statutes generally should not be applied retroactively to a contractual relationship where the application would alter existing obligations".

Empire's contract with Danella required it to keep a certain amount of landfill space available for Danella's disposal. Danella's contracts with its customers were made for set charges which were set in reliance of its agreement with Empire. Any retroactive application of the ordinance would affect and alter prior contractual obligation by those parties. Thus, the Court found that contracts entered into before the adoption of the Act are protected by the Contracts Clause. Conversely, those contracts entered into after the ordinance adoption date are unprotected. As a note of caution regarding the impairment of contracts, the Court added, "The Contracts Clause of the U.S. Constitution 'does not operate to obliterate the police power of the States', and 'The prohibition against impairing contracts is not to be read literally' ".

6.4 Canada

The earlier form of an individual nuisance claim before the existence of environmental regulations for landfills is illustrated by a Manitoba case of 1925 against a municipality for piling garbage in the town.

In *Still v. Rural Municipality of St. Vital,* [1925] 2 W.W.R. 780 (Man. K.B.), the piling by a municipality of a large quantity of garbage on the bank of the Seine River at a place within a residential district was held to constitute an actionable private nuisance. The evidence presented to the court was that "this garbage heap gives forth an odour, and is also the breeding place for rodents and flies, and has become a nuisance" and to be a health hazard in violation of the Public Health Act. The plaintiff, Still, was held to be entitled to nominal damages and to a mandatory injunction requiring the municipality to abate the nuisance "and keep it abated".

The newer practice of sanitary landfilling, requiring daily compaction of the refuse and daily placing of a soil or clay cover, had been proscribed in only a few Canadian provinces by the 1980's. Compliance was not universal, nor was it uniform throughout the provinces. For example, the dumping of waste and refuse in Manitoba Province did not come under control until 1970 with the enactment of its Clean Environment Act, R.S.M. 1970, the Municipal Act 1970 (Man.), The Public Health Act 1970 and the Planning Act, R.S.M. 1970. Waste disposal in Saskatchewan Province was unregulated prior to 1972. In 1972, the Saskatchewan Health Act was created to regulate the development and operation of landfills which numbered 800 sites in the province by January 1994. Licencing of landfill operations in Ontario was not started until 1970.

The lack of enforcement of pollution control and contamination of public waters is illustrated in Ontario by the case of *Plater v. Town of Collingwood et al,* Ontario High Court of Justice [1968], 1 O.R. 81, wherein the plaintiff Plater brought a private claim in nuisance against the municipality for burning garbage and refuse on an adjoining property from his farm where the smoke and noxious, foul and penetrating odours interfered with the reasonable use and enjoyment of his property. It should be noted that the municipality opened this new site in 1965 after closing the former city dump for reason that it was "unsanitary and threatened to contaminate Black Ash Creek and the water supply to private wells". Ontario MSW history was about to repeat itself.

The municipality hauled an average of 22 loads of garbage and refuse per day. Its volume was reduced by daily low heat burning. Covering with sand and gravel from the property was done at irregular intervals. Although evidence by an expert showed that winds brought the smoke across the plaintiff's property between 10% and 25% of the time, sufficient loss to plaintiff's tomato and cucumber crops convinced the court that damages were due plaintiff. In addition to a monetary award for diminution of crop value, the court issued a permanent injunction against the city to abate burning of refuse and an order to employ sanitary landfill methods. The court noted that the property selected for the landfill was made under the Municipal Act, 1960, (S.O.c.249, s.379(1) para. 75) and that the site was well adapted for the purpose. "It does not constitute an eyesore, being sheltered from the highway by trees, and the soil contains the necessary sand and gravel conditions adaptable to the sanitary landfill process. It is not necessary to enjoin the use of the site itself. But, if in the process of operating the dump a nuisance is committed, it is not a defence to the municipality to say that the operation of the garbage dump is beneficial to its residents, or that the chosen site is the only suitable one in the municipality; nor can it be said that open air burning is a necessary and inevitable part of the garbage disposal process. *** the modern trend is away from open dump burning and towards incineration or sanitary landfill methods, *** I believe that the sanitary landfill method is quite reasonable and feasible. The only inconvenience or expense to the defendants (the city) of the sanitary landfill method is that, there being no reduction in volume by open air burning, the life of the dumping area is necessarily shortened by as much as 50%. There is no reason, however, why the defendants should gain an economic advantage by inflicting harm on a neighbour's property".

When considering that the 1925 *Still* case in Manitoba, supra, was in the pre-environmental regulation era, it would seem unnecessary after the advent of a provincial environmental and health act for an individual in 1979 in Manitoba for a subject to have to pursue a similar claim in nuisance. Nevertheless, MSW disposal history repeated itself in *Wiebe v. Rural Municipality of De Salaberry* [1979], 11 C.C.L.T. 82 (Man. Q.B.). Wiebe purchased an 80-acre tract of land prior to the municipality's purchase of a smaller tract for an intended garbage disposal dump, adjacent to and across the road from Wiebe's. Wiebe built a home on his land with the town's knowledge and intended to start an animal feed-lot. Shortly after, the town began its refuse disposal, without notice or hearing for the local residents and Wiebe to raise objections. Dumping of refuse was open, without cover and utilised open, or low heat burning for volume reduction. The town had proceeded under its power and authorisation by the Municipal Act (Man.) which was permissive rather than mandatory. Under that Act, municipalities were directed to make provision for waste disposal, but without specific direction as to location or method used in operating such a disposal facility. Consequently, Wiebe suffered noxious odours, smoke and fumes, swarms of insects, and a proliferation of rodents, all of which diminished the quiet use and enjoyment of his home and property. In bringing his complaint against the municipality, Wiebe relied solely on the tort of nuisance, seeking only injunctive relief and damages. He had made no claims of violation of statutory duties by the town.

De Salaberry 's defence was (1) that the plaintiff "came to the nuisance"; (2) it had acted under statutory authority, viz., the Municipal Act (Man.), the Clean Environment Act and the Public Health Act (Man.), all of which directed muncipalities to make provision for waste disposal; and (3) there was an overriding benefit to the public of such necessary facilities.

The Manitoba Court (QB) found that the old doctrine of "coming to the nuisance" was not supported by the facts and that Wiebe had come first; furthermore, that doctrine is now given little credence in law and does not prevent a plaintiff from obtaining a remedy." Such coming will only be significant to the extent that the character of a neighbourhood is relevant in determining liability".

The Court found the town's defence of an overriding public benefit to be equally ineffectual, stating that "Private rights should not be subordinated to such considerations in the law of nuisance".

The most plausible defence, according to the Court, was that of acting under statutory authority of the Municipal Act to pass by-laws with a view to acquiring and operating lands for waste disposal; and under regulations of the Clean Environment Act and the Public Health Act. As noted, the Acts were permissive rather than mandatory, allowing the municipalities to act without direction as to location or method of operating the waste disposal sites. Where statutory authorisation is permissive, there is an accompanying obligation to have consideration for private rights; "and, only if the nuisance could be shown to be an inevitable consequence of such facilities (wherever situated and however conducted) would the statute provide a defence".

The Court allowed an injunction, and ordered the municipality to operate the dump through better management and supervision; to cease burning; to use earth cover frequently, and all actions to be in compliance with the Clean Environment Act. The Court allowed an award of $2,500 damages to Wiebe for a nuisance along with litigation costs.

With the passage of another six years, to 1985, in spite of the enactments of many more environmental regulations and controls over waste materials disposal, the claim of 'nuisance' for a landfill, or even as an anticipated nuisance while in the planning stage, still lingered in the courts. For example, in *Manicom et al v. County of Oxford et al* [1985] (the High Court of Justice, Ontario, Divisional Court), 52 O.R.(2d) 137, Oxford County planned to build a landfill site. A joint board appointed under the Consolidated Hearings Act, 1981 (Ont.). c.20, rejected the county's application for approval of the landfill site, but that decision was reversed by the Lieutenant-General in Council. The certificate was granted authorising the site under the Environmental Protections Act, R.S.O. 1980, c. 141.

Local residents and property owners in the vicinity of the proposed site then brought an action for an injunction to restrain the county from building the landfill based on claims of nuisance, negligence, breach of riparian rights, and for a declaration that the approval by the Lieutenant-General in Council was void under Section 7 of the Canadian Charter of Rights and Freedoms in that the proposed site threatened the life, liberty and security of the local residents. The local citizens' claim under the Charter was dismissed and on appeal, the High Court of Justice allowed the dismissal. However, the High Court found the common law claims, particularly for an anticipated nuisance, to be meritorious. The Court only allowed the nuisance claim to be stayed, i.e., in the event conditions are not complied with by the landfill project during its operation, plaintiffs may re-instate the claim for nuisance.

Judge Potts, writing a dissenting opinion in *Manicom* stated that "It is important that the county consider the welfare of the community at large, but it is the duty of the court to protect individuals from unauthorised and serious encroachments on their rights. *** Not only does the county not have the statutory authority to cause the alleged nuisance, it is prohibited by statute from doing so, Section 16 (1) of the Ontario Water Resources Act, R.S.O. 1980, c.361, makes it an offence for a municipality or other person to: "*** deposit *** any material of any kind that may impair the quality of the water of any well, lake, river, pond, spring, spring, reservoir, or other water or water course *** ".

The *Manicom* opinion referred to an Ontario case of three decades earlier, *Stephens v. Village of Richmond Hill*, [1950] O.R. 806, 4 D.L.R. 572, in which the municipality's sewage disposal plant emptied its effluent and water from storm sewers into a local stream. The plaintiff's allegations, corroborated by witness testimony, of destruction of the fresh water, fish life and watercress growth, replacing it with foul odours, trash (toilet paper, etc.) caught on the vegetation, slimy, murky waters, that no longer allowed swimming, were awarded damages by the court for the nuisance claim. An injunction was allowed against further pollution, but the Village was allowed a period of time for abatement of the problem.

Significant in the case was that the municipality had put forth an argument sounding in governmental immunity, claiming that when governmental acts are authorised, as the installation of a waste disposal site, there can be no claim of nuisance and damages made against it. The Ontario courts declined to support the immunity arguments made in both 1950 and 1985.

A series of similar claims in Ontario cases occurred in the period of the 1950s to 1960s where down stream water rights were invaded by public-owned treatment works (POTW's) discharging polluted waters or effluent, e.g., *Burgess v. City of Woodstock* [1955] O.R. 1955, 814 (where the sewage disposal plant of a 1922 vintage, built for a town of 9,000 had not been properly maintained or operated for a town that had grown to 16,000 persons by 1950); *B.C. Peagrowers Ltd. v. City of Portage La Prarie* [1963], 45 WWR 513, Man. 1963, [1965] affrmd. Supreme Court of Canada, 54 WWR 477 (where the city failed to properly maintain a sewage lagoon and seepage invaded private property and plaintiff succeeded in a claim of nuisance for damages).

Lest individual rights be not sacrificed and lost for the greater concern of environmentalism and a cleaner world, it is worthy to note Justice Stewart's admonition in his opinion in *Village of Richmond Hill,* concerning the importance of the welfare of the public at large, stated that: "I conceive that it is not for the judiciary to permit the doctrine of utilitarianism to be used as a make-weight in the scales of justice".

Ontario individuals, as well as the general public, should have found statutory protection and relief of bringing private nuisance claims against governmental agencies under Section 13(1) of the Ontario Environmental Protection Act, R.S.O., 1980, c.141, amended, which provides:

> "13(1) Not withstanding any other provision of this Act or the regulations, no person shall deposit, add, emit or discharge a contaminant or cause or permit the deposit, addition, emission or discharge of a contaminant into the natural environment that,
>
> (a) causes or is likely to cause an impairment of the quality of the natural environment for any use that can be made of it;
>
> (b) causes or is likely to cause injury or damage to property or to plant or animal life;
>
> (c) causes or is likely to cause harm or material discomfort to any person;
>
> (d) adversely affects or is likely to affect the health of any person;
>
> (e) impairs or is likely to impair the safety of any person;
>
> (f) renders or is likely to render any property or plant or animal life unfit for use by man;
>
> (g) causes or is likely to cause loss of enjoyment of normal use of property; or
>
> (h) interferes or is likely to interfere with the normal conduct of business".

The regulations affecting landfills are found primarily in Ontario Regulation 347. According to a personal communication with a Professional Engineer and Senior Policy Advisor with the Policy and Programme Development Section, Ontario Ministry of Environment and Energy, "This regulation contains definitions of waste management terms, waste classification and general standards for the operation of landfills. The regulations do not specify minimum technology standards for the design of landfills nor do they specifically prohibit landfills from being located in pits or quarries. Landfills are designed and approved based on site specific hydrogeological studies and engineering designs submitted by the applicant. Attenuation landfills are permitted provided these studies demonstrate the natural attenuation or protection capabilities of the subsurface are adequate to protect the environment. To determine if the potential impacts are acceptable, the applicant must demonstrate compliance with the Ministry's Reasonable Use Policy. This policy specifies stringent limits for impact on ground water quality which a landfill must meet to obtain approval. *** Presently, pits and quarries are being used as landfills in the province".

i) Evidence of an Ontario Sand and Gravel Pit as a Landfill with Natural Attenuation

The following case is offered as evidence to prove that construction materials pits may provide convenient and large capacity disposal space for MSW with natural attenuation to protect groundwaters from landfill contamination. Sand and gravel deposits are commonly underlain by clay beds affording natural attenuation for protection of groundwater, thereby acting as a natural clay liner.

In *Ontario v. Hale/Hale Sand and Gravel,* [1983] 13 C.E.L.R.19, the Attorney General for Ontario sought an injunction restraining Hale Sand and Gravel from continuing "any use, operation, establishment, alteration, enlargement or extension of a waste disposal site ***". in its sand and gravel pit located in the town of Newcastle, Regional Municipality of Durham.

Hale had operated the surface pit for sand and gravel for some years prior to the early 1950s when he commenced using it as a landfill site. Until 1970, there was no licencing requirement for its operation. However, in 1970 Hale applied for and received a provisional certificate of approval for a 20-acre site with a disposal capacity of 457,000 tons. In subsequent years, Hale's licence was repeatedly renewed and the operation continued. In the interim years, the Ministry kept continuous sampling of a nearby creek's flow of water and the monitoring of 7 wells dug to detect the escape of any harmful contaminants from the dump. To the date of the action herein, no harmful contaminants were detected.

No lining of the pit was ever undertaken or leachate collection system installed. The pit was obviously located in a place of natural attenuation. The reasons for the injunctive petition were not based on any claims of groundwater pollution by the landfill. The reasons given by the Ministry to shut the operation down were: (1) Phase 1 had exceeded the approved final elevation contours by approximately 28 to 35 feet in height; (2) Phase 2 had exceeded the approved final elevation contours by approximately 5 to 10 feet in height and by approximately 38,000 cubic metres in volume; and (3) Phase 2 has been "unlawfully extended" in a westerly direction from the approved western boundary or footprint for Phase 2.

In reviewing the Ministry's permission to operate the landfill, the Court found that the permitted area had been wrongfully decreased at a later date by the Ministry, therefore, the landfill had not been over-extended. It also found that the exceeded elevation in Phase 1 was done at the direct instigation of, and under the supervision of, the Ministry. That in so directing, the defendant reasonably believed the same course of contouring was to be followed in Phase 2.

The court found that the landfill appeared to be meeting a public need in a satisfactory, safe manner and that an injunction to prevent further operation and extension was unnecessary. It also found that evidence submitted showed that the pit was not overfilled and substantial capacity remained.

ii) **A case of excavating three pits to cover one landfill**

The following case is offered to show that landfilling is an earth-disturbing operation as is mining. It further demonstrates that in opening pits for the recovery of clay to be used as impermeable material in a landfill, it has in fact created another potential waste disposal site in the mined clay pits which have the ability of natural attenuation to prevent contamination of groundwaters.

In the 1989 Ontario Supreme Court case of *Joint Board v. Municipality of Metropolitan Toronto et al,* [1989] 4 C.E.L.R. 4 (N.S.), O.S.C. (Divisional Court), or simply, *Ontario (Joint Bd.) v. Metropolitan Toronto,* two other individual claims were combined with it under the Consolidated Hearings Act, 1981, S.O. 1981, C.2, §1, viz., *Rizmi Holdings Ltd. et al v. Metropolitan Toronto,* and *Corporation of Town of Vaughan v. Metropolitan Toronto.*

The point of interest in reviewing this landfill case is to demonstrate the unnecessary added expense and time of lengthy involvement of a municipality to obtain rights to excavate clay on a proximate property, along with transportation easements across intervening properties, in order to obtain liner and cover clay for an existing landfill; also, to refute the expressed New Brunswick's Department of the Environment's position that "mineral pits do not meet the stringent requirements for a MSW depository".

Case facts: The municipality of Toronto (hereafter, Metro) operated the Keele Valley landfill located in an adjoining municipality, the Town of Vaughan. Under Metro's provisional operating Certificate of Approval (Waste Disposal Site), a clay liner was required to minimise groundwater contamination from any leachate produced by the landfill pursuant to the Environmental Protection Act, R.S.O. 1980, c.14. Hitherto litigation, Metro excavated clay from two borrow pits, known as the Southeast Borrow Area (liner and cover clay) and the Avondale Borrow Area (cover clay), both close to the landfill. The clay was hauled from the pit across the former Landfill of the Town of Vaughan, pursuant to an easement agreement. In 1989, Metro found that the remaining reserves of clay were insufficient to complete construction for the 1989–90 year of landfilling.

Metro sought to expropriate more clay-bearing land, the Lilford-Pronesti lands, owned by a private owner and lying beyond the nearby existing Avondale borrow area, and to obtain further easements across the Pronesti and the Town of Vaughan's properties to reach the Keele Valley landfill. The Vaughan property was adjacent to Metro's landfill and was a completed and closed former landfill of the Town of Vaughan.

Metro applied to the Joint Board, composed of a member of the Ontario Municipal Board and a member of the Environmental Assessment Board, to consider their applications and proposals to acquire the properties and easements by expropriation. The Joint Board stated a case to the Divisional Court and the two property owners brought applications for judicial review on the issues.

The Ontario Court held that Metro had the power to expropriate land for the extraction of clay for use in its landfill, and was unaffected by the existing easement agreement. It was questioned whether Metro could expropriate land for landfill but only use it for the extraction of clay for use at the landfill. Under the Municipal Act, R.S.O. 1980, c. 302, the court found implied power to expropriate land for clay liner and cover material as part of the granted powers and process for receiving, dumping and disposing of waste, subject only to approval of the municipality or the Ontario Municipal Board.

There was also the issue of whether Part V of the Environmental Protection Act, R.S.O. 1980, applied to Metro's proposal. The Court supported the Ministry of Environment's position that the Environment Act did not apply because the clay lands were not to be used as a "waste disposal site" or "waste management system". The evidence showed that the sole purpose of expropriation was to obtain clay deposits, and as such was only an "alteration" of part of the waste management system, but the land was neither a waste disposal site, nor part of the waste disposal operation of the landfill.

It should be noted in reading *Metro Toronto* that low permeability clay is not only desirable for lining and cover material for their landfill, but required by regulation. Metro Toronto actually mines the clay for landfill use. They had already mined it from two borrow pits (Avondale and Southeast) and were preparing to open a third clay pit for more material.

iii) An unusual decision, where enforcement by the government failed, occurred in *Regina v. Enso Forest Products Ltd.* [1992], 70 B.C.L.R. (2d) 145 (S.C.). The Supreme Court of British Columbia found, under the Waste Management Act (WMA) of B.C., there is a difference between discharging pollutants into the environment and discharging in the "works", or plant site.

In *Enso*, 41,000 gallons (155,201 litres)of oil escaped from the defendant's pulp mill due to a pump failure. The oil flowed into a runoff ditch and pooled in an adjacent landfill site. None of it left the mill site. The company was convicted of violating section 3(1.1) of the WMA, allowing "waste to be discharged into the environment".

On appeal, the court found that the ditch constituted "works" within the meaning of the Act. The term "works" included "all things used by people which either created pollution or helped to control pollution. By contrast, 'environment' suggested something that was considered helpful to life and was worth preserving". The two terms were distinct and mutually exclusive. The court held that the purpose of the ditch was to collect runoff, which was covered by "works" and not considered part of the environment. The conviction of violation of the WMA was reversed.

6.5 Conclusions and Comments

Generally speaking, there has been a tightening of regulations for controlling waste management, e.g., in the U.K., the addition and imposition of a "duty of care" on those holding a waste management licence; and, the duty imposed on WCA's to recycle household and commercial wastes collected.

Deposition of MSW is far more safe and reliable in 1996 than a decade ago with the increased regulation of waste control and with the advances made in landfill technology by waste management, e.g., preventing ground water pollution by leachates, utilisation of the concept of total containment rather than the former attenuation, landfill HDPE liners of tougher and better qualities, self-repairing of ruptured liners, treatment of contaminating waters and leachates by biotechnics, collecting and greater utilisation of landfill gas for energy production, et al. Educating and making the public aware of the vast technical advances for prevention of water contamination, along with site aesthetics, control traffic, noise and odours, should lessen their opposition to landfill sites. With decreased public opposition to siting of landfills, the urgency of future MSW depository space should also lessen. Though opposition to "bad neighbour" landfills will likely die hard, the public must be educated to place trust and confidence in the public authorities and agencies appointed and trained to protect the environment while administering the public welfare.

Inter-regional commercial movement by long distance haulage of MSW from generating points to waste site depositories has made decided in-roads against waste disposal as a local matter restricted to local government management and decisions. Similarly and simultaneously, in-roads are being made against the former position that aggregate quarries cannot tolerate long, uneconomic hauls to construction centres. The combination of more distant aggregate quarries and landfill sites makes the dual purpose of surface mining and reclamation of the land by MSW backfiling immensely attractive.

CHAPTER 7

LEGISLATIVE ENVIRONMENTAL RESPONSES — A REVIEW OF SUBSEQUENT LEGISLATION TO UPDATE THE INITIAL REGULATIONS

7.1 Introduction

The environmental awareness that had been growing primarily in the developed nations after World War II increased at varying rates within the developed nations and the continents. The degree and rapidity of responsive enactments by their individual legislative bodies varied within a period of some 20 years. As stated in the introduction, the Anglo nations of the world, particularly the United Kingdom, the United States and Canada, and to a large degree the northern European countries and the EC, were advanced in having established some measures of environmental controls for industry through legislation of laws and regulations to conserve and preserve natural resources, to minimise or eliminate pollution of water and air, and to protect mankind and the health, safety, and welfare of the environment.

Perhaps only of passing interest, and without the intention of individual accolades, it appears that of the Anglo nations, the United States took an earlier, and certainly more aggressive lead in totally regulating and establishing controls for improvement of the environment. It was the first to establish an all-inclusive, nationwide Environmental Protection Act (EPA) in 1969–70. EPA's in the Canadian provinces started appearing about 1970. England did not establish a comparable, all-inclusive EPA as such, until 1990. That is not to say that neither the U.K. nor Canada had established some sort of statutory controls for the environment before those dates, but that the ultimate, the most far-reaching, all-encompassing, and most effective national environmental controls did not materialise until after that of the U.S.

Ordinarily, in the courts of the U.K.'s offspring nations, references to and citations of older English law cases are not uncommonly given in Anglo-American litigation to prove a precedential and established point of common law. This has been true in the past and an accepted practice in Anglo-law courts, whether in Canada, the U.S., Australia, New Zealand, et al. However, with regard to the new area of law, known as environmental law, there has been a noticeable citing in Canadian environmental litigation of American environmental precedential decisions of U.S. cases and references to American environmental regulations. A prime case example is *Regina v. United Keno Hill Mines Ltd.* [1980], 10 CELR 43, Canada. The Court's arguments are replete with references to U.S. environmental case law and law review articles. This may be attributable to the fact that this new area of law has not been established long enough to have built a background of case law history. With the U.S. having been at least a decade ahead of other developed nations in being more extensively regulated in pervasive and mandatory environmental law, it is logical that other Anglo nations' systems would look at one with longer and more experience, regardless of its degree of success.

Environmental control in the U.K. has not been the sole bailiwick of any single governmental state department or agency. The English legal system in the area of environmental regulation has relied more on the agencies of public health and planning control, e.g., the Public Health Acts, the Town and Country Planning Acts of various dates, as amended; and more recently, the Control of Pollution Act 1974. In the decades since World War II, these acts have played a vigorous part in environmental aspects and pollution control through land-use control by restricted and conditioned permitting and licencing. Canada and the other British Commonwealth nations have mirrored the U.K.'s Town and Country Planning Act with their own enactments of the same name. Various other acts, periodically updated with environmentally-concerned features, have augmented planning acts in the British Commonwealth nations, e.g. Public Health acts of different dates as amended. There has not been a sweeping environmental protection act in England until 1990 bringing major aspects of environmental control that compared with the all-inclusiveness of the U.S.'s EPA of 1969. Admittedly, as in England, the same pattern of various segmented environmental controls occurred in the U.S. over years preceding enactment of the U.S. EPA. It is only that the U.S. arrived at the point sooner, giving rise to an earlier experience with all-encompassing national environmental regulation. Thus, it had established a short history of environmental regulation and enforcement experience by 1980 which its sister nations had not yet developed.

Another aspect in the development of effective Anglo-American environmental law and regulation is the differential psychology of applied law in general, and environmental law in particular, between the British viewpoint and that of the United States. British law has a more laissez-faire or discretionary aspect to it with less "command and control", i.e., British people enact laws, but are less reluctant to flex their statutory muscles when persuasion can be used instead. This British attitude is discussed later in Ch. 8 under weaknesses and strengths of legislative acts. The Canadian attitude is reflective of the British. Generally speaking, a minority of people in the U.S., seemingly, and somewhat conversely, react successfully to "friendly persuasion", with a larger number reacting only to court-administered orders, punishment, penalties and sanctions. Thus, the U.S. federal government, having gone through a reasonable period of hands-off policy prior to 1969, felt the necessity of taking control of the nation's environmental protection with federal action. It had allowed the states the self-autonomy measure of voluntary, state-regulated controls to meet federally recommended pollution standards. State actions had failed to make a marked improvement on the environment. The federal government, recognising the need for strong, centralised environmental controls, found that the only way to deal with correction for a cleaner environment was to enact strict environmental federal laws and to create a schedule of fines for violations and penalties for failure to meet scheduled standards. Initially, sanctions for environmental law violators were limited to payments of fines, frequently large and severe, unless death or physical injury was a direct result of the violation. It was later determined that even large fines were not enough to deter all environmental violations, particularly by large corporations where payment of large fines might be more economically expedient than to change the manufacturing process to conform with environmental requirements. Consequently, jail sentences for environmental "crimes" have been created in law and are increasingly being carried out for violations. (See infra, Ch. 8 §8.1).

7.2 United Kingdom

After World War II, Prime Minister Clement Atlee's government brought about certain principals included in the Town and Country Act 1947 which are still held as basic tenets in the present regulatory scheme for environmental planning control. These still valid principals may be summed up as: (1) all land is subject to the jurisdiction of planning authorities; (2) land development generally may not take place without permission of a grant by planning authorities; and, (3) permission should not be granted unless the proposed development accords with a publicly prepared plan for the land in question (Hughes, 1986, p. 17).

The British cabinet created several "super-departments" in the early 1970s, each headed by a Secretary of State, with one of them being the Department of the Environment. However, by 1976, the re-organisation of departments had degraded to a degree that made the Department of Environment's function partially illusory. This was attributed to considerable fragmentation of environmental responsibilities.

Nevertheless, the Department of the Environment had responsibilities for the areas concerned in this study for:

(1) the winning and working of minerals, which came under planning control;
(2) commercial and industrial wastes, which came under pollution control; and,
(3) water and waterborne pollution, which came under its own Water Directorate as a part of the department.

7.2.1 *The European Community Environmental Laws Effect on the UK*

The U.K., as a member of the European Community (EC), is affected by the subsequent regulations adopted by the EC.

At its conception in 1957, the Treaty of Rome set out the aims for a European economic community in its Article 2: "The Community shall have as its task, by establishing a common market and progressively approximating the economic policies of member states, to promote throughout the Community a harmonious development of economic activities, a continuous and balanced expansion, an increase in stability, an accelerated raising of the standard of living, and closer relations between the states belonging to it". The only inference that can be made which implied an environmental concern would be with regard to the phrase "raising of the standard of living".

An environmental policy for the EC was proposed at its Paris meeting in October 1972. In November 1973, the First Action Programme for the Environment was adopted which reflected the region's awareness of developed nations' need for urgent action to treat environmental degradation. The Commission stated, "To remain balanced, economic growth must henceforth be guided and controlled to a greater degree by quality requirements. Conversely, the protection of the environment is both a guarantee of, and a prerequisite for, a harmonious development of economic activities".

The first of three consecutive EC Action Programmes for the Environment, 1973, 1977 and 1983, developed aims for the environment, e.g.:

(1) the prevention and reduction of pollution;
(2) the maintenance of a satisfactory ecological balance;
(3) the rational management of natural resources;
(4) the recognition of the importance of environmental aspects of development and in planning decisions; and
(5) participation in appropriate action at an international level (Hughes-Minor, p. 100).

With a decade of Community experience behind it, the Commission published in March 1984 its statement, *Ten Years of Community Environment Policy:*

> "The prevention of environmental problems arising at all is now the crucial principle; and a vital key point for the further development of the rational environmental policy envisaged under the (third) programme. That this has major implications for industry, for agriculture, for energy production and for transport is clear. The Third Programme commits the Community to the progressive (and preventative) integration of environmental aspects into the planning execution of all actions within these and other economic sectors that can have a significant effect on the environment". (id., p. 101).

The statement evidences a shift in policy from the former reduction and :urative treatment of environmental degradation to a new commitment of ›reventative measures for a cleaner environment.

'.2.2 *The EC's Environmental Administration*

The Commission is the driving force of the EC's three major bodies, viz., the Commission, the Council and the Parliament. In promulgating environmental egulation, the Commission drafts the proposals, but only after consultative neetings with civil servants of the member national groups and other interested ntities in environmental controls and concerns. The draft measure is then ublished, first as a Commission memorandum, known as a COM document, nd secondly, in the Information and Notices section of the Official Journal, nown as the "C" volume. Opinions are then sought and taken on the proposal y a committee from representatives of trade unionists, industrialists, onsumers, environmentalists and professionals.

Although this solicitation of opinions from various groups is discretionary y the committee, the Commission takes them under advisement and may mend the proposal accordingly. The opinion of the draft proposal by the EC Parliament, however, is mandatory and has been upheld by the EC Court. In *Roquetre Frers SA v. EC Council* [1980] ECR 3333, the Court held "that failure o obtain the Parliament's opinion invalidated a regulation subsequently adopted y the Council". (id., p. 102).

Adoption is the final stage in the decision-making process. Under the EC's Articles 100 and 235, the Community is empowered to make directives, make aws, regulations and administer actions in member states. Article 235 provides hat "if action by the Community should prove necessary to attain, in the course f the operation of the common market, one of the objectives of the Community, nd this Treaty has not provided the necessary powers ***", then the Community nay take the "appropriate action".

Three types of legally binding measures are empowered by EC Article 189, iz., regulations, directives and decisions. Regulations become law in the nember states without any necessary action of corroboration by the national arliaments. The UK's European Communities Act 1972, §2(1) confirms the precedence of EC law over UK law. That such EC regulations must be given precedence and superiority over member state's national law was upheld in the *Amministrazione delle Finanzedello Stato v. Simmenthal* SpA (No.2), [1978] ECR 629, 643. Further, in the notable European Court of Justice case, *Algemene Transport — en Expedite Onderneming Van Gend en Loos v. Nederlandse Tarief Commissie* [1963] ECR 1, the Court held:

> "*** the Community constitutes a new legal order of international law the benefit of which states have limited their sovereign rights, albeit within limited fields, and the subjects of which comprise not only member states but also their nationals. Independently of the legislation of member states, Community law therefore not only imposes obligations on individuals but is also intended to confer upon them rights which become part of their legal heritage". (id., p. 104).

Directives are binding "as to the result to be achieved ... but shall leave to the national authorities the choice of form and methods". Implementation of the EC directive is left to the national authorities, but must take place within the usual two-year period, unless otherwise specified. In the UK, the European Communities Act 1972, §2(3) provides for the implementation of Community law.

Directives were for a period prior to 1974, thought and/or argued to have no direct effect on the member state. It was argued that the effect of EC directives would be undermined if individuals were not allowed to rely upon them in national courts. That point was finally ruled upon and made clear in *Van Duyn* [1974] ECR 1337, 1348, in which the Court held "it would be incompatible with the binding effect attributed to a directive by Article 189 to exclude in principle the possibility that the obligation which it imposes may be invoked by those concerned". Member state compliance and recognition of EC directives as "law of the land", called "effet utile", was thus established giving it the same weight in law and courts as other national laws of member states. The point was further secured in *Becker v. Finanzamt Muster-Innenstadt* [1982] ECR 53 when the Court pointed out that normally where a directive is correctly implemented "its effects extend to individuals throughout the medium of the implementing measures of the member states concerned".

It has been established that an individual can rely on the EC-based law against the state, i.e., "vertical direct effect". However, a remaining point that has yet to be decided is whether a law based on an EC directive in a member state can be applied by one individual against another (called "horizontal direct effect").

EC decisions are binding on member states to which they are addressed. It has been established as a fundamental principle of community law in the case of *Costa v. ENEL* [1964] ECR 58 that any direct-effective Community provision prevails over conflicting national legislation.

7.2.3 *UK Water Standards and Pollution under EC Directives*

As Jaqueline Minor points out in writing Ch. 4 of Hughes' work *Environmental Law,* water pollution control has been an object of major attention of the EC's environmental actions since its earliest days because it plays such a wide part in the lives of individuals and industry. (Hughes/Minor, 1986).

Under the 1976 EC directive number 464 (OJ L129), certain toxic and hazardous substances were listed and became the subject of attack for reduction in the inland, territorial and coastal waters. Mercury, cadmium, and other carcinogenic toxic chemicals that are persistent and bio-accumulatable were listed. Less toxic, but still hazardous chemicals were also targeted for reduction, e.g., zinc, lead compounds, cyanide and ammonia. However, as Minor points out, powers in the UK already existed to control the disposal of waste generally under the Control of Pollution Act 1974, and conferred to water authorities under its §31, enacted 31 January 1985.

A separate EC directive in 1980, number 80/68 (OJ L 20), augmented the above directive 464 by focusing on and controlling the discharge of dangerous substances into groundwater sources. However, the UK's Water Act 1945, §18, had previously empowered water authorities to prevent the pollution of groundwaters; the Water Resources Act 1963 §72 prohibited waste discharges into groundwater by wells and boreholes; and, the Town and County Planning Act 1971, §§264–265, dealt with waste produced from mines and quarries.

Additionally, under the UK's Water Act 1973, §1, the Secretary of State is empowered to restore and maintain the "wholesomeness of rivers and other inland waters". §2 of the Act imposes "a duty on regional water authorities to prepare plans for restoring and maintaining the wholesomeness of rivers and other inland and coastal waters in their area. *** These plans can then be enforced so far as rivers are concerned by use of the power to control discharges under Part II of the Control of Pollution Act 1974 relating to streams, specified underground waters, coastal waters". (id., pp. 108–111).

7.2.4 *UK Waste Disposal Standards under EC Directives*

The EC's waste management programme has established three objectives:

(1) to reduce the quantity of non-recoverable waste;
(2) to recycle and re-use waste to the maximum extent; and
(3) to dispose safely of any non-recoverable waste.

These objectives have been augmented with various EC directives, e.g. # 75/442 (OJ L194) which "encourages the prevention, recycling and processing of waste and ensure that waste disposal takes place without endangering public health or harming the environment ***. The "polluter pays" principle is to apply i.e., the net cost of disposal (after recovery or reclamation) is to be met by the producer".

Various other directives have been issued, e.g., 78/319, OJ L84 (Toxic and Dangerous Waste); 76/403, OJ L108 (Disposal of polychlorinated biphenyls, etc.); 75/439, OJ L194 (Waste Oils). (Minor, id., pp. 122–123).

7.2.5 *U.K. Environmental Impact Assessment*

As noted, supra, §7.2.1, the EC's Environmental Action Programmes emphasised the control, reduction and prevention of pollution, giving particular attention to its prevention in the planning, utilisation and management of natural resources. The planning stage would necessitate the ecological consideration for all major developments by preparation of an environmental impact assessment (EIA) To accomplish this goal, EC directives promulgated the use of an EIA in 1980, OJ C169 and by amendment in 1982, OJ C110. (id., 125)

The purposes and goals for the EIA are the same as for those described in the EIA and EIS as directed by NEPA of the U.S. (see Ch. 5 §4.1, supra). Major developments, as mining and waste disposal sites, would require an EIA upon development application in compliance with the EC directives. An EC directive instituting the use of an EIA for development was initiated in March 1985. It should be noted that whether the need is mandatory or significant for an EIA for new development is determined under the directives of Annex I and II.

The EC's EIA must include: (1) the developer's individually prepared EIA, wherein any adverse environmental effects are stated, steps for mitigation or reduction thereof, reasonable alternative sites, and reasons for their rejection; (2) a consultation with interested and concerned public agencies; and (3) the decision must be reasoned and may contain conditions for approval.

7.2.6 *UK's Environmental Protection Act 1990 (UKEPA)*

In November 1990, the British Parliament enacted an EPA which attempted to consolidate, amend and strengthen many of its other in-place environmentally protective acts. Its preamble, *res ipsa loquitur*, speaks for itself.

> "An Act to make provision for the improved control of pollution arising from certain industrial and other processes; to re-enact the provisions of the Control of Pollution Act 1974 relating to waste on land with modifications as respects the functions of the regulatory and other authorities concerned in the collection and disposal of waste and to make further provision in relation to such waste; to restate the law defining statutory nuisances and improve the summary procedures for dealing with them, to provide for the termination of the existing controls over offensive trades or businesses and to provide for the extension of the Clean Air Acts to prescribed gases; to amend the law relating to litter and make further provision imposing or conferring powers to impose duties to keep public places clear of litter and clean; to make provision conferring powers in relation to trolleys abandoned on land in the open air; to amend the Radioactive Substances Act 1960; to make provision for the control of genetically-modified organisms; to make provision

> for the abolition of the Nature Conservancy Council and for the creation of councils to replace it and discharge the functions of that Council and, as respects Wales, of the Countryside Commission; to make further provision for the control of the importation, exportation, use, supply or storage of prescribed substances and articles and the importation or exportation of prescribed descriptions of waste; to confer powers to obtain information about potentially hazardous substances; to amend the law relating to the control of hazardous substances on, over, or under land; to amend section 107(6) of the Water Act 1989 and sections 31(7)(a), 31A(2)(c)(i) and 32(7)(a) of the Control of Pollution Act 1974; to amend the provisions of the Food and Environment Protection Act 1985 as regards the dumping of waste at sea; to make further provision as respects the prevention of oil pollution from ships; to make provision for and in connection with the identification and control of dogs; to confer powers to control the burning of crop residues; to make provision in relation to financial or other assistance for purposes connected with the environment; to make provision as respects superannuation of employees of the Groundwork Foundation and for remunerating the chairman of the Inland Waterways Amenity Advisory Council; and for purposes connected with those purposes".

7.2.7 *The U.K.'s Environmental Planning Concerns for Surface Mining*

As recited in Ch. 3 §5, Conclusions, supra, surveys of the public in the U.K. have indicated a general accord that the main complaints from neighbouring residents near surface mines are for noise, dust, blasting vibration, and traffic. In one Mineral Planning Authority (MPA) survey in an area with predominantly sand and gravel pits, only 4% of the people considered "themselves or their homes to be affected". Traffic and noise were the main sources for complaint. Of the 4%, lesser concern for water, landscaping and ecology/habitats was expressed (Environmental Effects of Surface Mineral Workings, 1992, p. 6).

An innovation of the U. K.'s TCPA 1990 is found in its §106, as substituted by the Planning and Compensation Act 1991. This section provides for a developer, e.g., mine operator / landfill operator, to make unilateral undertakings, that is, to privately enter into an obligation to do certain things related to the development or the land in question by agreement, or otherwise, where the objective cannot satisfactorily be accomplished by a planning condition.

As an example, it has been stated that planning authorities may consider during the formulation of environmental criteria for planning conditions for an open cast mine site applicant, there may be instances "where the operator considers it economically unreasonable to carry out further amelioration at 'source' but where the effect on neighbours is still unacceptable. In such cases, if planning permission is to be obtained, the operator will need to consider other methods of reducing nuisance". Such voluntary compensatory measures are acceptable to the planning authorities, and simultaneously assist the applicant to lessen or eliminate resistance to permission approval by irate prospective neighbours at public hearings. As an example, the Commission suggests that "providing compensatory measures off-site, such as sound-proofing of houses, or compensation by house purchase, payment or some other compensatory action can be useful ways of meeting performance criteria". (op. cit., p. 13).

7.2.8 *U. K. Wildlife and Countryside Protection*

Although rural land regulation may not play a large part in presenting obstacles to the proposed siting of an excavating operation, there are three environmentally protected special interests that cannot be overlooked. Agricultural land in the U.K. has historically had few environmental controls with the force of law. Typically British, the voluntary approach has been favoured by both farmers and the central government. A few, limited changes have been made in policy by the central government as control of stubble burning and implementation of water pollution in the Water Resources Act 1991. Even there, the legislation favours the voluntary approach. The Countryside Act of 1968, amended in 1981, and augmented by the Wildlife and Countryside Act 1981 (WC) imposes certain limited duties on ministers when considering diversification of enterprises of benefit to the rural economy. Part II of the WC Act is concerned with the conservation of nature and protection of the countryside. §28, as amended in 1985, lays down new provisions for the protection of sites of special scientific interest (SSSI). SSSIs are those containing any outstanding flora, fauna, geological or physiographical features. Notification of the location of any of the SSSIs must be given by the Nature Conservancy Councils for England and Scotland and the Countryside Council for Wales to the landowners, the Secretary of State and the local planning authorities so that damage to the sites by any planned operations may be avoided. By 1990, there were some 5,435 SSSIs listed comprising 1,713,840 ha. of land.

In a second area, the prospector searching for a deposit of rock for aggregate mining, typically for a site with little or no overburden to remove, should be aware of a related area of natural resources protection, the Limestone Pavement Orders, which come under §34 of the 1981 WC Act. Where a large area of exposed, or outcropping limestone, lying wholly (or partly) on the surface, usually as an extensive flat exposure, it is referred to as "pavement limestone". A protective order may be obtained by filing with the local planning authority if it can be shown that it is an area of special interest by virtue of its flora, geological or physiographical features.

Where a Limestone Pavement Order has been filed and granted, the rock exposure is protected and removal or disturbance of the limestone is prohibited and deemed an offence. However, an Order may be revoked for a "reasonable" excuse.

A third area protected by environmental regulation is the coastline which may be included in the National Parks, AONBs or designated as "Heritage Coasts". By 1990, there were 1,455 km of designated coastline to be protected. The National Trust has protected over 450 miles of coastline in its "Operation Neptune". Mineral development plans along unprotected shorelines would be highly scrutinised by the authorities (Hughes, 1992, pp. 178–179).

7.2.9 *Derelict Lands and Reclamation Regulation*

The Survey of Derelict Land in England 1988 found some 40,500 ha. (100,075 acres) classified as derelict, with 78% requiring reclamation. It was found that 29% of the waste land contained spoil heaps or piles from mining operations, mainly from coal mining. Though legislation dealing with derelict lands has been on the statutory books for many years, viz., in the National Parks and Access to the Countryside Act 1949, the Local Authorities (Land) Act 1963, the Local Government Act 1966, the Local Employment Act 1972 and the Local Government, Planning and Land Act 1980, they have been surplaced by the Derelict Land Act 1982. Restoration powers are supplemented by the Mineral Workings Act 1985, §§7 and 8. Local authorities are given the power to enter derelict lands, to search, bore, and where found to be hazardous to the public, to proceed with reclamation work without consent of those interested in the land. Notice of intent must be given to those with land interests. A new policy of financial aid may be found in the Derelict Land Grant Policy issued by the DOE in May 1991 (Hughes, 1992, pp. 191–193).

Hughes' conclusions on derelict land are noteworthy: "The laws regulating land use in rural areas continue to be a microcosm of the confused state of environmental law and policy. They illustrate the fragmented nature of environmental control, with responsibility being divided, unnecessarily in many cases, between central government, local authorities and a range of statutory ad hoc agencies. The number of involved bodies could be reduced; the answerability of those remaining could be increased". (Hughes, 1992, op. cit.).

7.3 United States

Miscellaneous Environmental Statutes Affecting Surface Mining are:

(i) Coastal Zone Management Act; (ii) National Historical Preservation Act; (iii) Wildlife And Endangered Species Acts; (iv) Criminal Penalties For Environmental Violations.

(i) **The Coastal Zone Management Act (CZMA)** [(16 U.S.C. 1451–1464 (1988)] requires all coastal states to develop a management plan. There are approximately 30 states under the Act, viz., states bordering both the Pacific and Atlantic oceans; states bordering the Great Lakes; Nevada and California bordering Lake Tahoe have also joined.

A state CZMA plan, approved by the US Secretary of Commerce, gives a state great power over activities, including permitting. A federal agency cannot issue a permit if the proposed mining activity is inconsistent with the state's plan. However, the states must operate under regulations of the U.S. Department of Commerce (15 CFR 930). Those regulations require a public-interest review process. The applicant for mining must certify to the state that the project is consistent with the state's CZM plan. [See, supra *Florida Dept. of Environmental Regulation v. Goldring* in re. limestone quarry development. Under Florida CMZA guidelines, mining operations are presumed to be development of regional impact. For another CMZA case, read *California Coastal Commission v. Granite Rock Co.,* 107 S.Ct. 1419 (1987)].

As part of the Omnibus Budget Reconciliation Act of 1990, Congress enacted provisions which re-authorise and fund the CZMA until 1995.

(ii) **The Endangered Species Act** (ESA) (16 U.S.C. 1531): Federal permitting agencies are required to ensure that any proposed mining action will not jeopardise the existence of listed species, or adversely affect their habitat. Permits will be denied for mining projects that will harm any species or habitat. The applicant must provide project data and information for a "jeopardy opinion" from the U.S. Fish & Wildlife Service (USFWLS).

For example, in April 1990, the USFWLS designated the desert tortoise a *threatened* species. Mining activities on public lands in the southwest desert areas of California, Nevada, Arizona and Utah are subject to the consultation requirements of Section 7 of the ESA.

Other wildlife acts that could affect a mining project are: The Fish & Wildlife Coordination Acts; the Migratory Birds Treaty Act; The Bald Eagle Protection Act; the Wild Free-Roaming Horses and Burros Act; the Marine Mammal Protection Act; the Fisheries Conservation & Management Act, et al.

(iii) **Historic and Archeological Sites:** Amongst the recent environmental land-use controls are those protecting and conserving buildings and places of historical interest. Under the Archaeological and Historic Preservation Act of 1979 (16 U.S.C.§469a-1) and the National Historic Preservation Act Amendments of 1980 (16 U.S.C. §470), the permitting agency is required to give the Advisory Council on Historic Preservation an opportunity to comment on any proposal that may affect properties listed, or eligible for listing, in the National Register of Historic Places. It should be noted here that the forerunner of similar laws were enacted in the U.K. with the Ancient Monument Acts of 1913 and 1931.

Even more pertinent and affecting public mineral lands is The Native American Grave Protection and Repatriation Act [(Pub. L. No. 1011–106, 104 Stat. 3048 (1990)]: A provision of this new act concerns any development activity on public lands, that any person who discovers, or has reason to know, Native American cultural items in the course of activities such as mining, construction, et al, must stop all activity in the area of discovery and make a reasonable effort to protect the items, and notify the secretary of the agency with primary management authority over the land. Mining and construction activities may resume following notification and after a reasonable period of time.

National Indian Youth Council v. Andrus, 501 F.Supp. 649 (D.N.M. 1980), aff'd. 664 F.2d 220 (10th Cir. 1981), is a leading minerals case to date in this area of environmental protection. In *National Indian Youth,* a federal mining lease was involved in an area of New Mexico where the Indian Council felt that native American archaeological and paleontological culture was in jeopardy from mining. The Indians argued that studies and surveys should be completed before a lease was given. The court determined that the award of a mining lease did not constitute approval of a mining plan. The court further noted that to complete a cultural study on a 40,000-acre tract before mine planning for a mere 8-acre tract bordered on the "absurd".

(iv) **Criminal Penalties For Environmental Violations:** This subject is discussed supra at Ch. 8 §8.3, U.S. Environmental Crime Punishment.

7.3.1 *Regulated Protection for Water Quality from Mining Effects*

As preservation of water quality is an important and integral part of this work where surface mining occurs, and particularly for the subsequent recycling of open pits as solid waste landfills, a few examples of litigated cases resulting from environmental regulations concerning mining effluents and water purity controls are included.

In an exemplary regulatory response to mining, the USEPA tightened water pollution controls for mining's stream discharges. Illustrative cases follow:

7.3.1.1 *Navigable Waters Pollution — Mining Discharge Violation*

In *Rybachek v. U.S. EPA,* 904 F.2d 1276 (9th Cir. 1990), an Alaskan mine operator and the Alaska Miner's Association challenged EPA's regulations which dealt with treatment required for discharges of untreated dredged soil and rock directly into navigable streams. The EPA interprets dredged soil and rock as pollutants and requires settling pond treatment before discharge.

The Court found that EPA's classification of settleable soils as a non-conventional pollutant and subject to best available technology (BAT) standards was both a reasonable and permissible construction of the CWA. Congress had not designated settleable solids as either a conventional or toxic pollutant. As to its determination of economic achievability of technology, EPA must consider the cost of meeting BAT limitations, but need not compare such costs with benefits of effluent reduction in promulgating regulations under the CWA. EPA was held to have considerable discretion in weighing technology's costs which are less important factors than in setting best practical technology (BPT) limitations.

From *Rybachek* of 1990, tightening of water controls for mining have increased from existing operations to those anticipated mining operations where even designated mining lands have been prohibited by being found "unsuitable" in preserving aquifers.

In *Village of Pleasant City v. Div. of Reclamation, Ohio, Dept. of Natural Resources*, 617 N.E.2d 1103 (Ohio 1993), the Ohio Supreme Court held that surface mine permitting required consideration of the long-range impact of mining on aquifers and the aquifer recharging area.

In *Pleasant City*, in September 1987, the U.S. EPA had designated 1,000-acres as a sole-source aquifer for Pleasant City. Following mine permitting of a company to surface mine 100 acres located in the City's designated aquifer-source area, over the opposition of Pleasant City, four monitoring wells were installed to determine whether the mining operations were affecting the village's drinking source wells. The mining company owned mining rights on additional acreage in the designated aquifer-source area.

In September 1988, Pleasant City filed a "Lands Unsuitable Petition" with the Ohio Division of Reclamation requesting that 833-acres of land surrounding the village be designated as unsuitable for mining. By October 1989, the

Division issued a decision precluding mining in about 275-acres of the aquifer source area. That area approximated the present area of the cone of depression from which the town drew its water from the aquifer. On appeal, the Board of Review enlarged the area precluding mining, reasoning that future surface mining could negatively affect the town's cone of depression in the aquifer. Moreover, replacement of removed aquifer material with mine spoil could interrupt or alter the aquifer's recharge zone. Pleasant City appealed the Board's decision.

The appeals court found that the Board's order was contrary to law because it "merely protected the area of present usage of the aquifer and recharge areas". The court granted Pleasant City's request that all 833-acres be designated unsuitable for mining.

Appeal to the Ohio Supreme Court followed in which it held that "consideration must given to the impact of mining and reclamation could have on the long-range productivity of an aquifer and the recharge zone, not solely the impact on their current use as a water supply". However, the Supreme Court disagreed with the lower court's ruling that all the designated aquifer area may be unsuitable for mining. It stated that deference should be given to the Division's expertise in determining what additional area may be unsuitable for mining. The case was remanded to reconsider the necessary area.

7.3.2 *National Pollutant Discharge Eliminations Systems (NPDES)*

In August 1990, the USEPA developed new rules for discharges into streams for industrial (mine and landfill) waste waters under §402, the National Pollutant Discharge Eliminations Systems (NPDES), more commonly known as "stormwater runoff". NPDES permits are required for all mining and landfilling operations which may discharge waters into streams or onto the surface.

§402 regulates the pertinent category of discharges from industrial activities. The administrative burden of application preparation for an NPDES permit has been reduced considerably by general permits for certain classes of industrial dischargers. Permits are either individual, i.e., for one site, or general, for an entire group of similarly situated but separately located facilities. Aggregate operations owned by one company in a geographic area can usually qualify for the general permit because of (1) similar operations; (2) discharging the same type of waste waters; (3) have the same effluent limitations and operating conditions; (4) require the same or similar monitoring; and (5) are more appropriately controlled under a general rather than an individual permit. (Evans, 1994, p. 59)

Storm Water Discharges Associated with Industrial Activity:The term "discharges associated with industrial activity" has been defined by EPA to mean "a discharge from any conveyance that is used for collecting and conveying stormwater, and that is directly related to manufacturing, processing, or raw material storage areas at an industrial plant in any of the following eleven industrial categories: ***

3. Facilities in SIC (Standard Industrial Classification) codes 10 through 14 (mineral industry) including active or inactive mining operations *** that discharge stormwater contaminated through contact with overburden, raw materials, finished products, *** on the site of such operations; ***

5. Landfills, land application sites, and open dumps that receive or have received any industrial wastes, including facilities subject to RCRA Subtitle D;

6. Recycling facilities, including metal scrap yards, battery reclaimers, salvage yards, and automobile junkyards, including SIC codes 5015 and 5093;

7. Steam electric power generating facilities, including coal handling sites; ***

10. Construction activity including clearing, grading and excavation activities; ***

As an example of a general NPDES permit, on May 31, 1994, the EPA issued a general NPDES permit for discharges from placer mining facilities in Alaska. The permit included effluent limitations and other conditions [59 Fed. Register 28,079 (1994)].

As an example of a violation of discharging without a required NPDES permit, the east Bay Municipal Utility District and members of the California Regional Quality Control Board, as joint owners and operators of the Penn Mine were found liable for unpermitted discharges from a facility designed to capture, contain and evaporate contaminated mine runoff. In *Commission to Save the Mokelumne River v. East Bay Municipal Utility District*, Env't.Rep.

Cas.(BNA) 1159 (E.D. Calif. 1993), reaching its decision, the court found no immunity from liability and deferred to an EPA determination that a leachate collection and treatment area which impounded water prior to discharge constituted a "point source" requiring a permit. (For further information on the (New) Penn Mine, see supra, Ch. 3 §3.2

(iii) Negligence — the case of *People (of California) v. New Penn Mines, Inc.*, which was an action for abatement of an alleged public nuisance caused by drainage of toxic mine and mill wastes into a river, resulting in damage to fish life.

As a related example for permitting of discharges under the CWA, though not an NPDES permit required by §402, the following case was brought under §405 regarding sludge disposal in landfills. In *Sierra Club v. EPA,* 992 F.2d 337 (D.C. Cir. 1993), the environmental groups petitioned for review of EPA's regulation controlling the disposal of toxic substances and solid wastes at landfills. The Court of Appeals held that the EPA had acted reasonably and adequately explained its decision to dispense with numeric limits for toxic substances co-disposed with sewage sludge in municipal solid waste landfills. The court found that the EPA could not measure effects of chemical interactions between pollutants and sludge or between pollutants and solid waste, and that the EPA had insufficient data about the chemical composition of debris in typical municipal landfills and, thus, could not establish scientifically defensible numeric limits (Id. at 340) (Garrett, 1994, p. 256).

7.3.3 *Acid Mine Drainage*

The subject of acid mine drainage (AMD) is thought to be of little relative import to the primary subject use and recommendation of non-metallic surface mines as landfills. Drainage from aggregate quarry spoil piles is essentially inert and thought to be ineffectual on ground waters. Common rock (granites and limestones), as used in aggregates, contain ineffective and negligible amounts of sulphides and metals to cause acid rock drainage (ARD). The abundant locations of aggregate pits near population centres of volume waste generation make them more suitable as MSW sites. Metallic surface mines are fewer in number, generally more distant and less well-located with regard to population centres, thus making them of lesser import for solving the urgent need of MSW landfilling sites. However, the hauling distance to MSW landfilling sites has become of less economic import. As full restoration of all types of non-fuel surface mines is advocated herein by possible landfilling with waste, treatment of all surface mines' problems of AMD is touched upon. Furthermore, since unreclaimed surface coal pits have been mentioned herein as successful examples of MSW landfilling to full surface restoration (see Ch. 10 §7 — Using Abandoned Surface Coal Mines for Landfills), even the problematic AMD from coal mines appears to be surmountable, at least for landfilling purposes.

AMD study and research programmes have been underway in recent years in attempting to neutralise acid drainage from metallic and coal mine and mill waste dumps. One such programme, Mine Environment Neutral Drainage (MEND), with joint industry / government aid in British Columbia, has been active in the study. Studies at "the University of Saskatchewan have also been underway to monitor and model the moisture profile in soil covers over waste rock dumps. High moisture contents, if maintained, will significantly reduce the rate of oxygen transfer and thereby the rate of acid production. Optimum materials for covers can be selected. In addition, the use of water cover on tailings (subaqueous disposal) has been studied over the past six years as several sites in Canada as part of a MEND project. Work has successfully demonstrated that this method prevents sulphide oxidation and subsequent ARD problems in tailings". (MEM, 1995, p. 8).

Other hopeful areas for treatment of and removal of contaminants from landfill leachates, and metals from mine and mill dumps and their water drainage are by bio-remediation methods. For discussion of treatment of leachates, see Ch. 11 §3.2, Peat — The New Waste Treatment "Wonder" Material.

Another prominent and highly hopeful vegetative method is phytoremediation which utilises plants that have affinities for metals and natural attributes for collecting them in their systems and removing them from polluted soils and waters. Of particular interest in this area are plants, shrubs and trees with high absorptive qualities ("metal up-take") exhibiting the process, hyper-accumulation. Hyper-accumulator plants are those species that have the ability to accumulate excessively high concentrations of metals, e.g., cadmium, lead, zinc, cobalt, nickel, copper, chromium, manganese, selenium. Such plants can tolerate 10 to 100 times higher metal content than normal crops. These contained metal elements exist in ranges higher than trace elements, with concentrations exceeding 10^3 mg/kg (1%) in the plant dry matter, or 10 to 20% in the ash. Phytoremediation-agriculture offers a distinctive and possible solution for decontamination and removal of high metallic content of soils. (MEM, 1995a).

Farming or cropping of hyper-accumulator plants also offers a new, speculative possibility of economic metal recovery. The U.S. Bureau of Mines recently completed a preliminary study on farming hyper-accumulating plants as an environmentally-friendly method of extracting nickel from contaminated soils at mine dumps and wastes from industrial operations.

Research on the potential farming of metal hyperaccumulators gives indications that they could remove enough metal to decontaminate soil, making hay from the biomass, recycling the metals, and repeating the process over a period of years. Researchers in this area are suggesting that "phytomining" might be a viable, economic process with the plant ash becoming an ore that would at least offset the costs of decontaminating the soil. It is also noted that phytoremediation is purported to be a lowest cost method to decontaminate soils, being not only lower than engineering alternatives, but by far, more bio-friendly (MEM, 1995b).

Wetland treatment and use of microbial mats are other new methods being currently investigated. Authors of an article on wetland treatment of discharge waters from the abandoned Wheal Jane tin mine at Cornwall, U.K., have noted that "Some of the processes occurring in natural wetlands, such as the uptake of metals by plants, are relatively insignificant as mechanisms for metal removal".

However, they further note that "There are, in principle, a wide range of physical, chemical and biological processes operating within a wetland environment which can promote the removal of metals from acidic metal-rich mine drainage. These include the oxidation of dissolved metal ions and subsequent precipitation of metal hydroxides, bacterial reduction of sulphate and the subsequent precipitation of metal sulphides, the co-precipitation of metals with iron hydroxides, the adsorption of metals onto organic substrates, and metal uptake by growing plants. *** The conditions required for some processes are incompatible with the conditions required by others, e.g., conditions for precipitation of most metal hydroxides are incompatible with conditions for sulphate reduction, *** and the pH in most wetland environments rarely rises to the 7.5 pH level necessary for the precipitation of zinc hydroxide".

At Wheal Jane, the simulated shallow wetlands comprised a pre-treatment of lime dosing and anoxic limestone drain pre-treatment. A series of aerobic cells and anaerobic cells followed. The authors concluded that "Passive treatment using wetlands can have a significant advantage over conventional water treatment processes". (MEM/Dodds-Smith, 1995c, pp. 22–24).

In "Biotreatment of Mine Drainage", the authors note that "Microbial mats are being used to promote metal removal for mine drainage in Alabama and Colorado, U.S. These mats are a microbial consortium of blue-green algae and bacteria which are highly tolerant of toxic metals and harsh environmental conditions". (MEM, 1995d, p. 25).

In a practical application for treating mine drainage, the West Glamorgan County Council and the National Rivers Authority co-sponsored a joint project to develop four wetland treatment systems for AMD from four abandoned coal mine sites in West Glamorgan, Wales. Iron-rich discharges are being made into the River Pelenna. Reportedly, each wetland has an artificial or clay liner, peat-free compost and reeds. The more acid discharges are channelled through anoxic limestone drains. Iron reduction is reportedly about 90–95% (MEM, 1995e, p. 28).

7.4 Canada — Environmental Impact Assessments

As the Canadian provincial and federal governments have been confronted by jurisdictional overlapping of environmental problems, they have begun to realise that some environmental issues, controls and laws for ecosystems cannot be confined within geographic boundaries. As noted by Canadian researcher Kennett at the Canadian Institute of Natural Resources Law, "Two events in the first half of 1992 added impetus to interjurisdictional concerns. First, the Supreme Court of Canada's decision in *Friends of Oldman River Society v.*

Canada (Minister of Transport) [1992] 2 W.W. R 193 (S.C.C.) (hereafter *Oldman River)* confirmed that environmentally significant projects will significantly trigger both federal and provincial EIA jurisdiction. Second, the enactment of the Canadian Environmental Assessment Act moved intergovernmental cooperation towards the top of the EIA agenda. According to the Canadian Council of Ministers for the Environment (CCME), 'the focus should now be on negotiating bilateral federal-provincial agreements as soon as possible to ensure effective joint environmental assessment procedures' ". (Kennett, 1992, p. 2).

Consequently, Canadian provinces in 1995 were still undergoing revisions of their environmental impact assessment (EIA) processes. "Treatment of interjurisdictional arrangements for EIA in provincial legislation presently varies considerably. Manitoba's Environment Act sets out a general 'equivalency' standard and specific requirements for joint assessment processes. More commonly, provincial legislation contains little or no explicit mention of interjurisdictional agreements and joint assessments. The Saskatchewan legislation merely gives the Minister authority to enter intergovernmental agreements. Conversely, in New Brunswick, there is no statutory reference to interjurisdictional co-operation on EIA.

The reaction by the Minister of Environment for Quebec was that the act was "totalitarian". *** British Columbia's legislation recommends that the EIA process 'should enable the province to work cooperatively with the federal government, neighbouring jurisdictions and local government' ". (Kennett, 1992, pp. 1–3).

The following subsections are discussions concerning the current Canadian Provincial attempts to revise their environmental acts.

7.4.1 *Ontario*

The current law is the Environmental Assessment Act, R.S.O. 1980, particularly §5(3) which sets out the statutory requirements of an environmental assessment document, pursuant to sub §1, to be submitted to the Minister and shall consist of:

"(a) a description of the purpose of the undertaking;

(b) a description of and statement of the rationale for:

(i) the undertaking

(ii) the alternative methods of carrying out the undertaking, and

(iii) the alternatives to the undertaking

(c) a description of;

(i) the environment that will be affected or that might reasonably be expected to be affected, directly or indirectly,

(ii) the effects that will be caused or that might reasonably be expected to be caused to the environment, and

(iii) the actions necessary or that may reasonably be expected to be necessary to prevent, change, mitigate or remedy the effects upon or the effects that might be reasonably be expected upon the environment, by the undertaking, the alternative methods of carrying out the undertaking and the alternatives to the undertaking; and

(d) an evaluation of the advantages and disadvantages to the environment of the undertaking of alternative methods of carrying out the undertaking and the alternatives to the undertaking".

7.4.2 *Alberta*

The *Oldman River* case, as noted above in §7.4, was an especially contentious one in which feelings ran high between proponents of a dam being constructed in the late 1980s on the Oldman River in southern Alberta by the provincial government, and an environmentalist group opposing its construction.

The purpose of the dam was to impound water for storage and as a management facility for relief of an adjacent region subject to drought. Opponents contended that the provincial government had not obtained the proper federal permit for construction, and had not adhered to the environmental screening process required by the Environmental Assessment and Review Process Guideline Order. Whether the Order had the force of law was a question at issue. Opponents were concerned over flooding destruction of fish habitats. The controversy is reminiscent of the U.S. *Snail Darter* case, except that no species was endangered. Another distinction was that dam construction was not held up for years by the on-going litigation and appeals by opponents.

Construction of the dam was continued by Alberta throughout the litigation, although enjoinment of construction was sought and denied.

Oldman River was heard by the Supreme Court of Canada. Its decision affirmed the construction of the dam even though handed down in the year after the construction was completed. The Court did find that a section of the Navigable Waters Protection Act and of the Fisheries Act were sufficient to trigger the requirements under the Guidelines Order. The court justified the decision-makers' discretion to consider socio-economic effects of such a project, e.g., relieving a drought, as well as the biophysical environmental effects on navigable river waters, for without such consideration no dam would ever be approved.

Alberta's current environmental assessment (EA) programme for non-energy projects followed in the same year of the *Oldman River* decision. The Environmental Protection and Enhancement Act (EPEA) 1992, S.A.c. E-13.3, is a multi-stage process and provides for an initial review, a screening report, and a more comprehensive environmental impact assessment report (EIA). When the Natural Resources Conservation Board Act (NRCB), 1990, S.A., c. N-5-5., applies to a project, public hearings *may* also be held (i.e., discretionary). The procedure for determining how far a given project should progress through this process is central to the operation of EA.

The following pertinent information, central to the thesis herein, concerning the EPEA and NRCB process was given by Ms. Susan Blackman, an attorney-researcher for the Canadian Institute of Natural Resources Law at the University of Calgary in Alberta:

> "When the question is utilisation of 'old quarries as municipal landfills, the municipal landfill would normally be approved by the local government, with appeals of that decision going to a public hearing before the very busy Development Appeal Board. The provincial environment department and the provincial public health department might render advice on the proposal to the local government and might appear at the hearing. The Natural Resources Conservation Board (NRCB) and the full brunt of the Environmental and Enhancement Act (EPEA) would not normally be brought to bear on the project unless the project was large, involved more than one government, involved

hazardous wastes, or, perhaps generated a serious public outcry. So I think that most landfills would be exempt from environmental assessment under the EPEA, but would usually be subject to a Development Board Appeal public hearing. Arguably, the EPEA process is redundant and the issue is more properly considered by the local authorities rather than the provincial authorities …

"The following appear to be the most relevant parts of the regulations and the NRCB Act:
Alberta Reg. 111/93 (under the EPEA) includes under Schedule 1 for mandatory assessment: (b) a quarry producing more than 45,000 tonnes per year; … (g) a surface coal mine producing more than 45,000 tonnes per year; … (i) an oil sands mine … (aa) a landfill that accepts hazardous waste from an off-site source.

"Schedule 2 for exempted activities includes: '(a) construction, operation or reclamation of … (vii) a sand, gravel, clay or marl pit that is less than to hectares (5 acres) in size'.

"The Natural Resources Conservation Board Act (S.A. 1990, c.N-5-5) states: 4. The following are subject to a review with this Act and the regulations: … (c) metallic or quarryable mineral projects; … (e) any other type of project prescribed in the regulations; (f) specific projects described by the Lieutenant Governor in Council". (There are no regulations yet.) Also, "6 In conducting a review under this Act the Board may (a) make inquiries and investigations and prepare studies and reports; (b) hold hearings or other proceedings; and (c) do anything that it considers necessary to carry out the purpose of this Act.

"The following section is also important: '2. The purpose of this Act is to provide for an impartial process to review projects that will or may affect the natural resources of Alberta in order to determine whether, in the Board's opinion, the projects are in the public interest, having regard to the social and economic effects of the projects on the environment'.

"If landfilling a quarry is considered 'reclamation', then a small quarry would be exempt from environmental assessment. And, this exemption may exist partly with the municipal landfill in mind (see Development Appeal Board above)". (Blackman, 1996).

The EIA report is the most detailed stage of review under EPEA. An EIA report is mandatory for certain classes of activities specified by regulation.

For non-mandatory activities, statutory decision-makers have considerable discretion to determine whether this level of EA scrutiny is required. The extent to which legal constraints guide the exercise of this discretion is therefore an important issue for the EA regime.

An order to prepare an EIA report has two significant consequences: (1) it results in a detailed project review which must address the issues listed in §47 of EPEA, unless otherwise excepted by the Director. These issues include potential environmental impacts, including cumulative, regional, temporal and spatial considerations; the significance of these impacts; and planned mitigation measures; (2) for several classes of projects, it triggers application of the NRCB Act.

The decision not to order an EIA report therefore precludes a public review. The Director is entitled to terminate the EA process if public comments are frivolous and vexatious (Kennett, 1995, op. cit.).

7.4.3 *British Columbia*

In recent years, British Columbia has had numerous, bitter environmental disputes over the utilisation of its well-endowed natural resources for commercial purposes versus disturbing its grand scenic topography and diverse ecological systems. The disputes have centered around the logging and mining industries, e.g., logging of old-growth forests at Meares Island, Lyell Island, the South Moresby archipeligo, et al; and for mining, expropriation of mining claims in Strathcona Park (see Ch. 5 §5.1, Mineral Land Takings By Environmental Regulation and Expropriation cases of *Casamiro Resource Corp.* and *Cream Silver Mines Ltd.*), and more recently, the environmental regulatory taking of a planned world-class copper-gold deposit in a proposed provincial park in the Tatshenshini-Alsek region, northwest of Skagway, Alaska. (See Ch. 12 §5).

Canadian Institute of Natural Resources researcher, Steve Kennett, has posed the question, "Is British Columbia leading the way in natural resources management?" by virtue of its greater experience and resolution of such disputes. As part of the environmental resolution, British Columbia has established the Commission on Resources and Environment (CORE). CORE has the "overall responsibility for developing a provincial land use strategy and reforming the processes for resources and environmental decision-making in British Columbia. ***

"CORE was established in 1992 *** and established a new approach to land use planning and decision-making, 'one that will put and end to valley-to-valley conflicts'. *** 'CORE's role is not to resolve directly a multitude of local resource disputes, but rather to 'design new processes for the future'".

The commission began its work by declaring an 18-month moratorium on logging on Vancouver Island which had become a top priority for CORE. The moratorium allowed the Commission time to study and formulate a regional land-use plan. Subsequently, it used the same procedure to study the Windy Craggy mineral claims in the Tatshenshini-Alsek region.

CORE is a giant step in the right direction for land-use planning. However, the undertaking of land-use planning for an expanse of territory as large as British Columbia is staggering. British Columbia has an area of 366,255 square miles (948,600 sq.km.). Compared with the United Kingdom's area of 94,226 square miles (244,058 sq.km.), British Columbia is 3.89 times the size. The population densities are equally startling: 9 per square mile for British Columbia, compared to 660 for the U.K.

Nevertheless, as argued in Ch. 5 §4.7, Mineral Land Planning, supra, forethought or foresight is virtually unheard of with regard to mining and minerals in land use planning, that is, to set aside and reserve known land areas that contain certain valuable and unmined minerals. Forethought to reserving land areas for mining should be given to avoid future conflicts with other types of land uses, particularly recreational, park lands and scenic areas. Mineral deposits must be mined where they are found, but the ecology of mined areas can be restored.

7.4.4 *Yukon Territory*

Since the Yukon is not a province but a territory, its powers are limited to those granted by federal legislation. The Yukon Environment Act (YEA), Yukon Stats., 1991, c.5, was assented to May 29, 1991, but will not be fully in force until all parts are proclaimed in 1996. YEA is comprehensive covering many new areas, e.g., solid waste management, reduction and cycling, release of contaminants and spills. It also provides for integrated resource planning and management; development assessment process for close integration with the permit stage, to be applied under a forthcoming Yukon statute, the Development Assessment Process Act; and, finally, it provides for citizens suits to be brought against any person "likely to impair the natural environment "or the government of the Yukon "for failing to meet its responsibilities as trustee of the public trust to conserve the natural environment". The Act is generously armed with provisions, giving it "teeth", to pursue alleged violations, e.g., offering protection for employees against retaliation for "turning in" their employer, and for monetary rewards in keeping levied fines upon violators after bringing successful private prosecutions.

7.5 A Negative Legislative Response — Withdrawal of Potential Mineral Lands

The Shrinking Mineral World: As a result of the strong tide of conservationism and the governmental responsive legislative actions, one of the major problems that currently threatens activities of the mining industry, especially in the developed nations, is the governments' withdrawals of huge land areas and acreages from mineral prospecting. These massive areas of potentially mineralised lands withdrawn from exploration, or placed under highly restrictive regulations severely limiting access for exploration and mineral development, are on lands set aside for national, state and provincial forests and parks, wildlife habitats, fish and game preserves, world heritage areas, native lands, seashores, coastal management zones, recreational and primitive areas, et al. The volume of "environmentally sensitive" lands steadily mounts with the public clamour for cultural preservation and conservation without regard to the severe reductions being placed on the area of mineral-base lands from which future mineral production for society might come. As said before, minerals must be mined where they are located. When the world's base-area is greatly reduced by massive land withdrawals from mineral prospecting, the future potential for mineral production is similarly greatly reduced; or said another way, as the untouchable lands grow in number and area, the world's potential mineable mineral deposits correspondingly shrinks.

This argument is supported by a document published by The Australian Mining Industry Council entitled "Shrinking Australia" which treated the problem of access to land, and announced that "29% of the area of the State (of South Australia) is either closed or severely restricted to mineral and petroleum exploration". In 1988, the Council predicted that "about half of Australia's land area will be severely restricted or prohibited to exploration and mining unless the current growth in conservation areas, particularly protected wilderness and aboriginal lands, is reduced sharply, or existing restriction applied on these lands are relaxed". It has been noted that cultural affluency of the developed countries has passed the point where untouchable lands have far greater consideration and priority than does the economy and mineral development within their borders. (Aston, 1993a — in an article, "Enviromining Law" referring to an article "Environmental Policies Towards Mining in Developing Countries" by Professor Thomas Walde, Editor of the International Bar Association's *Journal of Energy and Natural Resources Law*, where Professor Walde points out that in the developed countries of the North, "** mining is no longer automatically assigned precedence over other land-uses and environmentally (socially or culturally) important land is no longer available for mining".)

Governmental withdrawal of public lands from mineral prospecting and mining activities has been prevalent in Canada and the U.S. Of consequence to this work is the concern that surface mining as a part of mining will be reduced, not only severely reducing the supply for minerals, but thereby reducing potential voids for landfill sites during their reclamation process. This concern is predicated on the hypothesis that substantially all open-pit mines could and should be fully reclaimed by infill of wastes before restoring the surface to it original contours and surface. If the natural state of the worked out mining pit is found to be suspect of having natural leaks into the groundwater, the thesis herein is predicated on modern technology's ability to repair it; that it can be made leak-proof and safe by proper lining methods to hold solid wastes. Further contention is that it can be made safer than any andfill placed in soil or overburden. Treatment of this mining-landfill process n preserved land areas is made in Ch. 10, infra.

7.6 Conclusions and Comments

The two major approaches of the application of environmental law to effect he desired environmental improvements of this thesis are: (1) that of the British, visually, a discretionary and persuasive manner of obtaining compliance for established law and new regulatory developments, and (2) the U.S. approach of a "command and control" system for established law and new regulatory developments.

The Canadian approach in the past has been understandably more like that of Britain, employing government discretion and persuasion, than of the U.S. There is a notable change, however, in recent years, in which environmental egulation appears to be patterned increasingly after U.S. environmental egulations which allow the public to have a larger input and influence through public hearings. However, as part and parcel of emulating the American way, ncreased contention and litigation over environmental issues have also accompanied the increased Canadian public participation. In support of this changing Canadian attitude toward environmental law are the following statements.

In an article entitled, "Environmental Law and the Greening of Government: A Cynical Guide", a book reviewer states "One conclusion *** is that the constant in the development of environmental law in Canada has been the dogged retention of discretion by government. As environmental laws have been dramatically altered and the public increasingly included, the influence of the regulated community has been diluted for there is by no means a real sharing of power over decisions".

Another, similar article entitled "Environmental Groups and the Courts: 1970–1992" is one in which the author complains about the lack of input by he public. In it, the author traces the number and type of public interest environmental law suits brought in Canada since 1970. The reviewer commenting on the author's article states, "There has been a relatively low level of such litigation, as compared to the United States, although it has bee increasing since the end of the 1980s. Even though most of this activity ha focused on the actions of governments rather than those of business, *** man in business in Canada feel the rate of public interest litigation is 'already out o hand in Canada and any further growth is unacceptable' and are working 't impede the ability of citizens to bring [such] suits'. Examples of such strateg include the efforts *** to revoke public funding for Elgie's group, the Sierr Legal Defence Fund, and the avoidance of litigation written into the new Ontari Environmental Bill of Rights". (Wood, 1995, p. 469).

While both authors of the preceding articles complain there is a lack o environmental litigation in Canada, and that the public has only illusory inpu into the environmental "say-so" of their country, the statement in Ch. 3 §3. by a Canadian natural resources attorney explains the scarcity of Canadia litigation by commenting that "Canadians are far less litigious than Americans and where there are such contentious matters, Canadians tend to resolve the without court involvement". Though other conclusions may undoubtedly b drawn, a suggested one is that the British and Canadians place more confidenc in their elected leaders' direction, "discretion" and judgement, and are mor willing to follow them without litigating. Americans, on the other hand, ar more dissenting, litigious and prone to squabble. Environmental litigation i the U.S. is so prolific that environmental progress is actually impeded. I keeping with arguments here, it proposed that "command and control" regulation apply to permission, permitting, and environmental regulation fo surface mining and landfilling. The statement of the University of Leicestershir Senior (Environmental Law) lecturer Hughes' (see Ch. 8 §3.1) that Britis "Planning law has not historically been generally over-concerned with givin third parties rights at the development control stage" would seem to coalesc with command and control environmental regulation.

The determination of which is the better system is questionable and arguable Each has its merits and its drawbacks. Arguably, to allow too much publi input in the approval process leads to prolonged public hearings, results i increased contention and conflicts, impedes environmental progress, and i generally not conducive to serving the general public's welfare. Useful an beneficial projects to the public may be thwarted or easily defeated by a voca and militant minority. The argument is here injected that the public hearing process for approval should be minimal, and not required in all cases of public and private development. Public reliance should be placed on the technically-trained environmental agencies that have been emplaced by the government to protect the environment, public welfare and health, not in placing reliance on the non-technically educated lay public that is often subject to unfounded fears, misinformation and hysteria, as in the NIMBY syndrome.

Section III

TRANSITION FROM PRESENT TO FUTURE

CHAPTER 8

TODAY'S ENVIRONMENTAL REGULATORY STRENGTHS AND WEAKNESSES FOR TOMORROW'S NEEDS

.1 Introduction

wo of the principal problematic regulatory areas affecting the irreputed "bad eighbour" industries of concern here are: (1) the planning approval, public quiry process (PIP) and licencing permission; and, (2) mineral lands ithdrawal reducing the mineral base.

2 The Public Inquiry Process (PIP)

he PIP is principally encountered after the land use, or zoning, has been pproved, for general industrial use. A host of publicly sensitive areas precede, occur during, the applicant's licencing process, as aesthetics (visual trusion), road traffic, noise, dust, injury to natural beauty and scenic areas, d to minor degrees, historical, aboriginal and cultural areas. These problems e the most difficult to deal with in persuading and assuaging the lay public at their fears are exaggerated or unfounded and can be dealt with in a tisfactory, environmentally-safe manner. Although the engineering profession ows that environmentally sensitive problems affecting groundwater, drinking d surface water sources, air contamination, landscaping, road traffic, et al, troduced during the public hearing process, can be treated successfully by odern technology, the general public remains skeptical. Apparently little trust placed by the public in governmental environmental regulation and

engineering feats for industrial sites. Their fears frequently succumb to th
NIMBY syndrome. The problem for the so-called "bad neighbours" is that th
NIMBY syndrome is becoming universal. As Justice Dooley of the Vermon
Supreme Court so aptly stated concerning the NIMBY syndrome, "We are i
serious danger of expanding 'not in my backyard' into 'not anywhere' ". (Se
In *Re. Meaker,* 588 A.2d 1362 (Vt. 1991), infra.)

Once an area's land-use planning, or zoning, is resolved with intelligen
foresight and the public accepts the fact that excavations for minerals an
waste disposal are essentials for mankind's welfare and for the public's good
health and well-being, the remaining problems for keeping the environmen
safe can be overcome and dealt with through modern technology and enforce
by properly trained governmental agency technicians and personnel.

The weaknesses of the land-use and permitting procedure, prior to actua
permission, permitting or licencing of a mining or landfill project, are th
frequently incurred and prolonged delays by public hearing appeals, excessiv
or unnecessary administrational and litigation costs. This often results in defea
for a much-needed site for the general public's benefit, which, in turn, mean
that the expensive public process has to be repeated for each new propose
site.

It is therefore strongly advocated that the general public's participatio
should largely terminate, or be greatly reduced, with the land-use permittin
process. Once the public has participated, aided and approved in determinin
that certain land areas must necessarily be reserved for certain uses, includin
permissible areas for excavations for minerals and the depositing of MSW, th
deployment and monitoring of such excavations in the permissibly approve
areas should be left to the determinations and supervision of scientists, engineer
and technologists skilled in environmental protection. Further involvement b
the lay public and placing reliance on the unscientific whims as generated b
unbased or misinformed fears of environmental damage and the NIMB
syndrome only serves to vexatiously delay essential excavation projects. The
incur an added and unnecessary public inquiry process, superfluou
environmental protection costs, and generally compound the problems an
conflicts. In the case of urgently needed landfill space, the delay by prolonge
public hearing process only serves to exacerbate the urgency.

Ultra-expensive, ultra-extensive and intricate scientifically-based :nvironmental programmes have been emplaced in the Anglo nations for the ›ublic's protection. For the greater part, the environmental protection ›rogrammes are administered by well-informed governmental agencies placing eliance on scientists and technology for environmentally-safe procedures. \gencies are in a far superior position to the lay public to determine, assure ınd resolve environmental protection for the public. Hence, reliance by the ›ublic should be placed in them and on their decisions without further public nvolvement, agitation and prolonged conflict.

.2.1 *A Prime Example of Disruptive Public Hearing*

\ prime example of disruptive public interference based on unfounded nformation, continuing long after extensive natural resource planning by the :entral government for the public welfare, is illustrated by the U.S. Tellico River Dam site and construction controversy about 1971 in Tennessee. Environmental fears fanned by environmental groups through the media swept he U.S., incurred extremely high costs to the public for years of litigation up o the highest court of the land, and finally required resolution by the U.S. Congress. The end result was that the dam was finally extricated from the :nvironmental controversy by the government overruling the environmentalists' ›bjections, put into service, and the environmental fears evaporated upon liscovery that they were unfounded.

The environmental controversy arose when a federally approved and 'inanced dam was being constructed in a rural, mountainous area of East Tennessee at the confluence of the Little Tennessee and Big Tennessee Rivers. The area was one of great natural beauty and minor historical importance. The ırea to be flooded contained several ancient sites of Indian villages, whose ırcheology had not been explored, and Ft. Loundon, established in 1756 as England's southwestern outpost in the French and Indian War. An area of 16,500 acres (6,677.46 ha), some of which was valuable and productive 'amland, was to be inundated by damming of the rivers to form a reservoir 30 miles (48.28 km) long.

Between 1967, when Congress first appropriated money for the projec and every year until 1972, the dam, although virtually completed, never operate due to a tangle of lawsuits and administrative hearings brought about b environmental groups attempting to stop the dam. An injunction was finall granted in 1972 to halt the dam, and remained in effect until 1973 when federal court of appeals lifted the injunction on finding that the EIS for the si was in compliance with NEPA. [*Environmental Defense Fund v. Tennesse Valley Authority* (TVA), 371 F.Supp.1004 (E.D. Tenn, 1973); aff'd.492 46 (6th Cir.1974)]. Unfortunately, that was not the end of the controversy for th Tellico dam, and the conflict continued for nearly six years more.

Just prior to dissolving the injunction, a University of Tennesse ichthyologist while exploring the Tennessee River waters discovered previously unknown species of perch, the snail darter, a 3-inch (7.62 cm) lon fish whose numbers were estimated to be in the range of 10,000 to 15,000, an one of some 130 known species of darters. Four months after the snail darter' discovery, Congress passed the Endangered Species Act (ESA) of 1973. Th ESA authorised listing of species of life that are either classified as "endangered" or "threatened", and to protect their habitats from destruction. In January 1975 the anti-dam environmentalists resumed their opposition to the dam' completion by requesting a rating of "endangered" under the ESA for the snai darter. Having been listed as an "endangered" species with its habitat describe as critical, the Secretary of Interior found that "The snail darter occurs only i the swifter portions of shoals over clean gravel substrata in cool, low turbidit water. Food of the snail darter is almost exclusively snails, which require clean gravel substrata for their survival. The proposed impoundment of wate behind the proposed Tellico Dam would result in total destruction of the snai darter's habitat". Thus, the small fish became a *cause célèbre* for the opponent to the dam. The dam's relief was short-lived with the lifting of the injunction only to be impeded a year later by further litigation. In 1976, a permanen injunction against dam completion was issued [*Hill v. TVA*, 549 F.2d 1064 (6 Cir. 1976)] "halting all activities incident to the Tellico Project which ma destroy or modify the critical habitat of the snail darter". (id. at 1075). The Court of Appeals "directed that the injunction remain in effect until Congress by appropriate legislation, exempts Tellico from compliance with the Act, o the snail darter has been removed from the list of endangered species, or its critical habitat re-defined". (Ibid). A final appeal was made to the U.S. Supreme Court, 437 U.S. 153 (1978), which affirmed the Court of Appeals' decision that the ESA as written by the Congress, made no allowance for exceptions and the Courts had no alternative than to halt the federal dam project. The Supreme Court said, "It is clear that TVA's proposed operation of the dam will *** have the effect of eradication of an endangered species. Concededly, this view of the Act will produce results requiring the sacrifice of the anticipated benefits of the project and of many millions of dollars in public funds". (Actually, $100 million had been expended in dam construction costs up to the injunction, not inclusive of litigation costs in defending the federal project's continuance.)

"As a result of the *Snail Darter* case, Congress *** in 1979, legislated an exemption for Tellico Dam and the project was completed". (Schoenbaum, 1985, p. 399).

Ironically, the near-decade of delay in completing the dam and the millions of dollars spent on litigation was, in the end, found unnecessary and based on erroneous environmental fears and false reports. The snail darter's occurrence and habitat at the dam site was not the only one in existence. Since the initial discovery of the snail darter, several other populations and locations away from the proposed dam site were found. The fact is that the snail darter was never an "endangered" species. The U.S. Fish and Wildlife Service announced that it was removing the snail darter from the "endangered" species list. Professor Schoenbaum poses the question, "Was the environmentalists' primary purpose in the *Snail Darter* case to save the fish or to stop the dam"?

The *Snail Darter* conflict is exemplary of costly problems created by leaving determinations to hearings with the lay public; so-called environmental problems that frequently have already been investigated and determinations made by environmental and scientific professionals.

8.2.2 *Other Examples of Public Protest Interference*

A December 1995 editorial in Mining Environmental Management, London, treats the subject in more detail:

PUBLIC MISUSE OF THE MINERAL REGULATORY SYSTEM

"There have been a number of events, recently, which indicate a trend whereby project permitting no longer follows the path laid down by the legislation of that country. An extreme example of this happened earlier this year when Shell U.K. tried to implement a plan to dispose of the derelict Brent Spar storage buoy in deep water off northwest Britain. Activists, led by Greenpeace, mounted an extremely vocal opposition to the proposal, insisting that Brent Spar be disposed of on land. It had taken over three years of research and negotiations with the government to finalise the details of the plan. All along, Shell insisted that its proposal was by far the safest. Media exposure reached fever pitch until Shell capitulated and said it would rethink the options. Greenpeace has since apologised publicly, admitting that it had miscalculated the quantity of potential pollutants.

"In a separate incident, Ready Mix Concrete applied for an extension of its Eldon Hill Quarry in the U.K.'s Peak District and, in return, would have implemented a rehabilitation programme over previously quarried areas outside its lease. Although there was public opposition to the scheme, it was being considered by the local authority through the proper channels. The BBC televised an extremely biased report against the project, and shortly afterward the application was turned down.

"Similar interference with the legal permitting system can be seen at Crown Butte Resources' New World gold project near the Yellowstone National Park in Montana. Objections from local associations and environmental groups are being accommodated within the regulatory system, but the project has attracted the attention of President Clinton and the U.N.'s World Heritage Committee, which has proposed the establishment of a buffer zone for the park. This would preclude many types of activity and goes counter to federal laws that try and keep public lands open to a range of activities.

"Another type of pressure against a country's regulatory system can be seen in recent attempts to take a company to court, but not in the country of its operations. The most infamous case is the action filed against BHP in a Melbourne court. The compensation claim for $A4,000 million was brought by an Australian law firm representing P.N.G. inhabitants in respect of alleged damage to the Fly River by the

Oki Tedi mine. The P.N.G. Government objected to a foreign court being asked to judge on its internal affairs, and has drawn up legislation to restrict the amount of compensation payable.

"Similarly, Delaware-based Southern Peru Copper Corp. (SPCC) is being sued by some 700 residents from Ilo in Peru, but the action was filed in a district court in Texas where some SPCC shareholders are based. The plaintiffs are seeking compensation for alleged damage to their health from SPCC's Ilo smelter and refinery. Not only is historic pollution not attributable to SPCC as it only purchased the facility in June 1994, but it is already conducting a $US151 million project to improve environmental conditions. Interestingly, the attorney may now face sanctions for 'judge shopping' after the same lawsuit was filed 17 times before it was assigned.

"All these cases have something in common: the company and the local authorities follow legislative procedures but adverse public opinion has created the environment where politicians and lawyers act outside the established practice. It is also possible that companies are placing too much weight on scientific argument rather than seeking to counter the negative image being generated by environmental groups who appeal to public emotion". (Mining Environmental Management, 1995f, p. 3).

8.3 Land-Use Planning — In General

Land-use planning in the U.K. under its Town and County Planning Acts is possibly the most advanced of any nation in the world. Its provisions and considerations are without doubt most far-reaching, extending to most all human activities. Although land-use planning is not mainly environmental in nature, but managerial, its designs perhaps make environmental regulation and control more systematic, and consequently, more efficient and manageable.

The U.K. appears to be well-advanced over the U.S. and Canada in environmentally regulating regional areas as evidenced by its Town and Country Planning Act, and as amended. Perhaps its advanced regional land-use planning may be attributed to the much smaller size of the nation, in that closeness of population forced stricter control and planning of land uses in an attempt to making living in close-quarters more comfortable and compatible and with less friction. With the comparative sizes of the U.S. and Canada being about 38 times that of the U.K., "closeness" or crowding has not yet been an overall compelling consideration for detailed land-use planning in the two larger countries. However, crowded conditions or more dense populations in the eastern, Atlantic and Maritime coastal areas of both the U.S. and Canada, and more recently in the growing Pacific coastal areas of both nations, are seeing the need for megalopolis and regional land-use controls. However, as noted in Ch. 7, at §4.3, British Columbia, with an area 3.89 times the size of the U.K, has undertaken a staggering land-use plan through its Commission on Resources and Environment (CORE).

Especially in the U.S. and Canada, several areas of environmental regulation are in much need of overhauling with respect to need of larger area authorities, as opposed to state and provincial control, particularly in the areas of land-use planning, navigable rivers, surface and groundwaters, solid and hazardous waste landfills and surface mining pits as disposal sites.

8.3.1 *Land-Use Planning — U.K.*

Planning of land use in the U.K. has been controlled by the Town and Country Planning Act (TCPA) since 1932 and its various updated amended acts over the decades, with the latest act in force being that of 1990. The Act's very title word "planning" implies that plans are not irrevocable, but serve more as a methodical guideline for intended best land use and its development. In fact, reviews must be periodic and constant. There is a general limit of between 10 and 15 years on approved plans, which must then be re-appraised or re-planned.

Under TCPA, three development plans are statutory in nature: (1) unitary development plans, which apply to metropolitan areas; (2) structural plans, which apply to "shire" counties and districts; and (3) local plans, which also apply to "shire" counties and districts. In addition, non-statutory plans exist to fill temporary needs, but only until a statutory plan is passed. It should be noted that "local plans must not contain policies on winning and working minerals or mineral waste deposition, save for those plans relating to National Parks, nor similarly, any policies on waste or refuse deposition". (Hughes, 1992, p. 113). Under Ch. 8, §37 of the TCPA 1990, mineral planning authorities are required to prepare any Minerals Local Plan (MLP) (other than within National Parks) containing the formulated and detailed policies of winning and working minerals or the deposition of mineral wastes. §38, similarly, deals with the making of Waste Local Plans (WLP) with respect to refuse and waste deposition. These plans, filed with the Secretary of State, must reflect central government guidance in the form of planning policy guidance (PPG), mineral planning guidance (MPG) and regional planning guidance (RPG) (RPG's generally extend over a 15-year period).

Public participation for commenting during the proposal-making process is provided for by sections of the TCPA, and objections are considered by the Secretary of State, who has very wide discretion. The Secretary may hold an examination in public on any specified matter. As noted by Hughes, "The examination is not a public inquiry in the full sense*** for no person has a right to be heard***". Hughes also observes that, "Planning law has not historically been generally over-concerned with giving third parties rights at the development control stage. *** Nevertheless, there are publicity and notification requirements in the 1990 Act". (Hughes, 1992, p. 112, 126).

A noted difference between the nature of British and U.S. concern for meeting environmental problems is in the setting of pollution standards for planning consents. As stated by the U.K.'s Undersecretary of State for the Environment in March 1992, "The adverse effects of a working can be significant even though the operations conform with a standard, are within the terms of the planning consent, or conform with a code of practice. A standard is a compromise between the costs of control and the benefits to be gained; a standard rarely seeks to avoid any adverse impact. Planning consents are likely to take account of local conditions and sensitivities". (op. cit., p. 5).

From a U.S. viewpoint of environmental protection, that statement evidences a difference in attitude for environmental standard-setting and compliance to obtain improvement of quality for the environment. In the U.S., pollution standards under its EPA are rigid and applied uniformly nationwide. Ultimate, obtainable standard goals were set to obtain zero pollution. Where they could not be reached at the time of setting of the standard because of lack of technology or for extremely different conditions from the national norm, intermediate standards and time limits were set, but only to be considered as a time-limited step to the ultimate zero-pollution goal. For example, under the USEPA air pollution standards, nationwide standards were applied, but temporary exceptions were only made where approved progressive time schedules were approved in certain metropolis areas affected by intense smog conditions, as in Los Angeles, et al, not being the national norm.

To state that "*** a standard rarely seeks to avoid any adverse impact" is arguable under the USEPA. An adverse impact, large or small, by a proposed project subject to an EIS in the U.S. is likely to meet with defeat. Such statement is evidenced by the Tennessee Valley Authority's nearly completed dam at Tellico, Tennessee, in 1972, which came very close to being a multi-million dollar monument to a falsely-reported endangered species of fish. Moreover, it is highly debatable whether the snail darter was of sufficient socio-economic consequence compared to those benefits of the dam, particularly when it is considered that the snail darter was only one of 130 unendangered species of the darter (see the review of the TVA "*Snail Darter*" case, supra at §8.2.1).

Under the TCPA 1990 §53, "development" is described "as the carrying out of building, engineering, mining or other operations in, on, over or under land, or making any material change in the use of land ***". There are two divisions of development, viz., operational development, and making material changes of use.

As defined by the courts, operational development generally requires physical alteration of the land with some degree of permanence (see *Parkes v. Sec. of State for the Environment* [1979] 1 All E.R. 211).

"Material change" development is much more elusive and difficult to define. The courts have had difficulties in deciding whether there has been a "material change" in a land use. It may involve merely an intensification of the present use as illustrated in the case of *Brooks and Burton Ltd v. Secretary of State for the Environment* [1978] 1 All E.R. 1294, where the annual production of concrete blocks quadrupled from 300,000 to 1,200,000. In other cases, constant but small progressive intensification of a use is difficult to detect when it reaches a point of change of character.

Under §55(2)(f) of the TCPA 1990, Use Class Order 1987 SI 1987/764, six broad categories with subdivisions of land uses are given. Change within a use class will not be considered an act of development. However, a change across classes may be considered development if it is "material". No categories are specified for excavations, either mining or landfilling, but certain ancillary mineral processes are, such as mineral crushing, cement production, brick or lime burning, and smelting works.

A reading of British cases litigating whether there has been a "material change" constituting development frequently results in a feeling of "much ado about nothing" or that there is much nit-picking and wasting of time over seemingly irrelevant aspects of land already in use. Nevertheless, the statutory approaches of British landuse commendably include collaborative planning with trained, mining and waste-knowledgeable personnel for consideration of future area mineral and waste excavation requirements.

8.3.2 *Planning: Voluntary Versus Mandatory Compensatory Measures*

The Planning and Compensation Act 1991 provides for developers as mine and landfill operators to make unilateral undertakings. By way of further example, if allegations are made by neighbours at a public hearing for an applicant's new site, or for extension of an existing site, to loss of water in their wells, the developer might well offer to provide piped-in water, or alternatively to drill a new, deeper well, if such a water loss occurred. Such conditional guarantees by the applicant developer of correcting a nuisance or a wrong before the event occurs are less onorous for both parties than the alternative of "going to court" over an actual incident, and at the same time alleviates the friction at the hearing between the developer and the public.

As noted by the U.K. Commission in its 1992 Research Report on Environmental Effects of Surface Mineral Workings "Compensatory measures are seen by some operators as a valuable approach but, whilst there seems to be some willingness to compensate communities as a whole, there is a reluctance to compensate householders directly, e.g., by money, sound-proofing (a home), an offer to buy. It is recognised that operators inevitably make a contribution to the local economy in the form of rates and the provision of employment. In some cases their workings remove existing dereliction. *** Guarantees, e.g., to provide piped water if wells dry up, are more conventional. These conditional offers are likely to be less onorous for an operator than providing facilities before the event. They have the drawback that problems have to occur first, but can ensure that a remedy is provided without too much argument and delay". (id., pp. 114–115).

As an illustration of the difference between the British voluntary style of pre-mining negotiations between operators and community residents are the following two U.S. case occurrences. A distinguishing difference between the two acts, i.e., Planning and Compensation Act 1991 and the U.S.'s SMCRA is that the British act encourages voluntary interaction, while the U.S. act is mandatory adjustment.

In the following case, voluntary compensatory amelioration would have been preferable to the ensuing and costly litigation.

REPLACEMENT WATER SUPPLY MANDATED BY SMCRA

"The operator of a Pennsylvania surface coal mine was required, pursuant to a mining permit, by the Surface Mining Conservation and Reclamation Act (SMCRA), subsection 119, which mandates: "the operator of any mine which affects a water supply by contamination, pollution, diminution or interruption shall restore or replace the affected water supply with an alternate source, adequate in water quality for the purposes served by the supply ..." on a permanent basis.

The adjacent landowner's water source had been a spring. Carlson's surface mining disrupted its flow and under enforcement proceedings, Carlson installed a drilled well, and later, a water treatment system. The owner's costs were increased by $200 per year, which the operator must pay.

DER required Carlson to establish an irrevocable letter of credit (LC) to cover the permanent increase in operation and maintenance. If Carlson failed to pay the owner's monthly bill within 30-days, it would forfeit the entire amount of the LC to the landowner. Carlson argued that DER was without statutory authority to devise such method.

In *Carlson Mining Company v. Department of Environmental Resources* (DER) (Pennsylvania), 639 A.2d 1332 (Pa.Cmwlth. 1994), Carlson petitioned the court for review of decisions of the Environmental Hearing Board, which had approved the proposal of DER of a method

> of funding the increased cost of operating and maintaining the replacement supply of well water for the adjoining property owner.
>
> The Pennsylvania Commonwealth Court held that: (1) the operator was required to provide for increased maintenance costs of adjacent property owner's replacement water supply on a permanent basis; (2) increased operation and maintenance costs of replacement water supply were sufficient to require operator to compensate the property owner *ad infinitum* (from now on); and (3) DER could require an operator to create individual trust or escrow accounts to provide for future payments of additional costs in the event that an operator's company should dissolve, change ownership or enter bankruptcy." (Aston, 1995b).

In a second illustration under the U.S. SMCRA, duplicated by the West Virginia state version, the Surface Coal Mining and Reclamation Act (SCMRA), a similar claim of loss of a private water supply was litigated in *Russell v. Island Creek Coal Co.*, 389 S.E.2d 194 (W.Va.1990).

Russell had conveyed the surface rights for coal mining on a 60-acre tract adjacent to his homesite. During 1983, the 60-acre tract was mined and Russell alleged contamination of his spring supplying his home with water. For several years, Island Creek voluntarily attempted to rectify the problem for Russell by drilling a new well and by installing a purification system, both of which failed to correct the contamination. In *Russell*, the West Virginia Supreme Court of Appeals held that the provisions of the WVASCMRA require a coal operator to replace the water supply of an owner of interest on real property whose water supply has been affected by contamination, diminution, or interruption approximately caused by a surface mining operation.

As an aside in *Russell*, it should be noted, however, that the court denied liability of Island Creek for the contamination of the spring, since Russell had signed a waiver of surface damages to Island Creek, and, additionally, Russell had been found culpable in causing the pollution of his own spring since he had previously mined coal on his own land even closer to the spring before conveying the mining rights of Island Creek Coal Company. His actions thereby mitigated liability by Island Creek in causing the water damages.

It should be noted that in the *Carlson* case, the compensatory act of replacement was not voluntary by the mine operator, but required by environmental regulation and involved expensive litigation cost and time. Certainly the voluntary act by the operator would have been a less expensive alternative. In the latter *Russell* case, voluntary correction was attempted by Island Creek Coal, but the damage had already been done to the water source by the owner himself.

8.3.3 *Planning: Consideration of Private Property Rights*

Are private property ownership rights being lost to over-regulation?

It has been noted earlier in the Introduction of Ch. 5, that environmental regulations, particularly in the US, are of a "command and control" nature which encroach on previously-held areas of private ownership and self-control of individual property rights. For example, an owner of mineral-bearing property, whether an individual or a mining company, could previously mine any place on its property, or as many acres as it wished. After planning, zoning, permitting and licencing controls were in place, a landowner could not freely excavate or move about on his own property without the approval and licencing of authorities. At times, even public notice hearings were involved before approval could be given for exercising excavation of the property owner's minerals on his own land. Clearly, the property rights of the owner have been severely encroached upon, limited, restricted, and even taken away from him. This is a result of the "command and control" type of environmental regulation. However, to obtain the desired results for the good of the public, the new "command and control" system proscribed environmentally "harmful" acts and effects requiring specific control measures to produce the governmentally and environmentally-desired benefits for the general welfare.

8.3.4 *U.K. — TCPA Environmental Assessments*

EC's Directive 86/337 EC, effective July 1988, caused the Environmental Assessment (EA) to be incorporated into British planning law. The EA ensures that the probable effects on the environment of any planned development will be taken into consideration. Planning authorities are no longer limited to just land-use considerations, but must now also consider issues of pollution control with development. In certain, specified cases, an EA is mandatory, while in others an EA will take place if the project is anticipated to have "significant effects on the environment". Under Sch 2 of SI 1988//1199, relevant planning projects listed under the category "Extraction" and requiring EA's are: extracting peat, deep drilling, mineral extraction of sand, gravel, shale, etc., coal, petroleum and natural gas, extracting ores, open-cast extraction, ancillary surface installations, coke ovens, cement manufacture.

An interesting statement by the Infrastructure Policy Group of Institute of Civil Engineers published in a paper on "Pollution and Its Containment" pointed out a weakness in the provisions of the Control of Pollution Act — "though site licences have enabled stricter controls to be imposed and, together with stringent planning conditions, they go far in reducing the environmental impact of waste disposal on landfill sites. However, this is, to some degree, treating the symptoms rather than the disease. The disease must be treated as well as the symptoms".

8.3.5 *Land-Use Planning — U.S. — Narrow Jurisdictional Versus Regional Needs*

A recurring U.S. problem in both land-use planning and environmental regulation is the oft-occurring conflict and interference of local, narrow jurisdictional controls and planning with larger, regional controllers and interests.

Hopefully, local planning at the county level is becoming obsolescent to some degree as it either increasingly interferes with planning goals for the larger concerns in regional planning, or it may not conform with regional needs, and occasionally even conflicts with the larger, regional areas for environmental controls, e.g., clean water for larger rivers and their tributary drainage systems under area water authorities, waste disposal and landfill requirements for region-wide areas that are unhampered by narrow jurisdictional lines, etc.

Local planners are properly concerned with strictly local environmental matters. However, as the adage states, they cannot "see the forest for the trees". The regional "forest" is of greater concern, which is the perspective taken by state agencies and regional authorities. Regional authorities are better situated and informed as to the needs of greater areas and larger populations, enabling them to coordinate efforts more effectively than the unconcerted individual local efforts.

As others have observed, geological provinces do not conform to political subdivisions and boundaries. It is the geological formations that control physical and topographical features of the earth's surface which include rivers and drainage patterns and a host of other features affecting the environment on the Earth's surface. The larger English county or shire, the Canadian province, and the U.S. state political division boundaries do not conform to the geological provinces. Administrative environmental controls within these larger political subdivisions may not suffice for efficient and best control of the environment, e.g., river and water control, clean air control.

Land-use controls in the U.S. have always been considered the sacred domain of the states and their subdivisions. Consequently any encroachment on this sensitive area reserved to the states has been jealously guarded and considered inappropriate for the federal government to attempt to administer or become involved in to even a slight degree. However, the federal government has made some successful efforts through its grants-in-aid powers to cultivate area planning that is unconstrained by narrow political boundaries. The grants are exercised under the federal spending power as an inducement to the states to employ state-wide planning and regional approaches requiring mandatory controls in intergovernmental efforts, e.g., in solid waste disposal, interstate highway planning, interstate air-pollution control regions, and region-wide waste effluents under the federal Water Pollution Control Act (§208), and in the largest of all regional acts, under the Coastal Zone Management Act of 1972.

The Coastal Zone Management Act (CZMA) was an early federal law passed for national land-use planning. There are no direct federal enforcement powers involved, except for the greatest indirect power, being the power of monetary funding which will be withdrawn for lack of state compliance with the federal guidelines. Congress's purpose in the CZMA is basically for "more effective protection and use of the land and water resources of the coastal zone to encourage and assist the states to exercise effectively their responsibilities in the coastal zone through the development and implementation of management programs to achieve wise use of the land and water resources of the coastal zone giving full consideration to the ecological, cultural, historic and esthetic values as well as to needs for economic development".

Penetrations into local planning domains are gradually being made under developing regional authorities for environmental and pollution control of rivers, water and air, and on a lesser scale for regional waste disposal sites.

An example of the narrow controls of local zoning boards was recognised by a judge in an Ohio case where a sand and gravel operator had been issued a conditional use permit by the local zoning board. The Board later, upon the operator's application for renewal, subverted their former decision by claims of possible contamination of their water supply. In *Barrett Paving*, supra, at Ch. 5 §4.6, the Ohio Court of Appeals stated that "the purposes of local zoning and environmental regulations are inherently different *** are complementary but wholly independent of one another. The EPA is solely concerned with the environmental protection and protection of human health from pollution and improper waste disposal. A local zoning board *** is primarily interested in land usage *** affecting the development of the community".

The point of the argument is to limit approval for mine permitting to a state agency. Once local authorities have reached their decision to approve mining activity by zoning, the control over mining that may affect the environment, health and safety of the community should be left to the state EPA and mining agencies which are professionally trained in such matters. The purpose of state agencies for natural resources and environmental protection is to administer environmentally protective regulations (see Nimby Syndrome Continues for Ohio S & G Operator, supra at 5.4.6).

If local communities find they are not being adequately protected environmentally from quarry operations, they may bring suit against the state agency for failure to perform their duties. [See also, the case of "Concerned Citizens Question Vulcan Quarry Permit", supra at Ch. 5 §4.7 (iv)]

8.3.6 *Land-Use Planning — Canada*

Planning regulation in Canada is of far less concern than in the more populated lands of her cousins, the U.K. and the U.S. However, mineral exploitation and other natural resources on Crown lands is one of the federal Canadian government's areas of land planning. In all jurisdictions, the common law rules dealing with surface rights are put aside for activities under mining legislation "These statutes provide for the conditions under which a miner can enter private land in pursuit of Crown minerals and the manner in which the miner is to pay compensation to the surface owner and occupier. The Ontario and British Columbia statutes allow the miner to enter private land without seeking permission, while those of the older provinces and territories require the miner to obtain the permission of the owner or an order from the tribunal or officer that the statute designates". (Barton, 1993, p. 192). Where both surface and mineral rights are both privately owned, the mining legislation, above, does not apply.

As noted in Ch. 7 §4.3, British Columbia is leading the way in natural resources management. As part of its environmental resolution, in 1992 the provincial government established the Commission on Resources and Environment. CORE has the overall responsibility for developing a provincial land-use strategy and reforming the processes for resources and environmental planning and decision-making in British Columbia. CORE's role is not to resolve directly a multitude of local resource disputes, but rather to design new processes for the future.

8.4 United Kingdom — General Environmental Protection

Prior to passage of the Britain's Environmental Protection Act 1990, there was no overall, well-defined national environmental policy within the U.K. No central body existed, having entire responsibility for development, oversight and operational implementation for the loose assemblage of environmental regulations in force. Various governmental bodies, authorities and agencies shared the duties and supervision of implementing environmental directives and regulations authorised under an assortment of Acts. The lack of a central environmental organisation frequently led to duplication of supervision and control, which in turn led to confusion, less effective environmental control, and certainly greater administrational costs.

The British system of enacting regulations and carrying them out seems to be, both historically and presently, (1) lacking in centralisation; (2) lacking in strict mandate for their application and enforcement, being permissive and voluntary; and (3) lacking in simplicity, being complexly intertwinned with other acts, regulatory bodies and agencies. Offered in corroboration of those weaknesses are quoted statements by the eminent British environmental lawyer, David Hughes, Senior Lecturer in the Faculty of Law, University of Leicester and Honorary Consultant to the National Housing and Town Planning Council, in his 1986 book *Environmental Law* which follow:

> "Between 1848 and 1872 a multiplicity of enactments covering issues such as nuisances, sewage and sanitation ... general public health ... were put on the statute book. The essential basics of modern public health law were created in this period, but, sadly, in a confused and tangled manner which was beyond the comprehension of even trained minds ... The division of responsibility made the law unnecessarily unwieldy and difficult to know and interpret. Added to this complexity was the fact that most of the law was permissive and not mandatory. These basic defects made the law unworkable ... the twentieth century has also signally failed to create either a comprehensive system of environmental law or machinery to enforce it. The legacy of past divisions of responsibility lies heavy on us. (p. 5).
>
> "*** the Public Health Act 1936, Part III, (which) deals with nuisances and offensive trades ... was generally applied to London by the London Government Act 1963, ¶40(1), **but the law is made irritatingly complex** because ... boroughs, London and the Temples retain an inherited independence with regard to certain public health issues. (p. 6) (emphasis added).
>
> "Environmentally, the most important matters are noisome or unhealthy accumulations and deposits of trace or manufacturing dust of effluvia ... These provisions are enforced by centrally appointed inspectors ... Much more reliance is placed on informal liaison between local environmental health officers and the various central pollution inspectorates who ... prefer to use persuasion rather than the powers of enforcement in their functions. (p. 8).

> "A further restriction on the powers of local authorities to take action in respect of certain nuisances arising from coal or shale mining, is contained in the Clean Air Act 1956 §18(2), whereunder §92 of the 1936 Act does not apply to … the combustion of refuse deposited from such mines and quarries *** (p. 8).
>
> "It is apparent from what has been said above that the public health legislation is hedged about with technicalities which limit its usefulness. A further limitation is to be found in §109 of the 1936 Act which states: 'Nothing in this Part of this Act shall be construed as extending to a mine of any description so as to interfere with, or obstruct the efficient working of, the mine, or extending to the smelting of ores, and minerals, to the calcining, puddling, and rolling of iron and other metals, to the conversion of pig iron into wrought iron … so as to interfere with or obstruct any of these processes' ". (p. 14).

Hughes (op. cit) in quoting Charles Webster in *Environmental Health Law* (1981) page 17, who writes: "... it is suggested that a new look at the rather archaic provisions relating to offensive trades might be appropriate. Control by means of the development control system under the town and country planning legislation is likely to be more efficient than control under the public health legislation. **In any case, a single unified form of control is less likely to produce confusion and injustice**". (emphasis added).

In discussing the use of planning law to effect environmental control over developments, Hughes states:

> "It is this large measure of discretion which principally makes planning law so imperfect an agent of environmental control. Planning law, however, has other defects such as the lengthy and cumbersome nature of its procedures, the principal that an authorised development or activity can only be halted following the payment of compensation, slow moving and somewhat ineffective enforcement systems, and a failure to rationalise the legislative overlap between planning law and other *ad hoc* industrial safety, environmental and public health measures that have grown up around it in recent years. The last named defect is readily visible in the division of responsibility for environmental issues between a wide variety of central and local agencies. Frequently there is no legal

> obligation on these agencies to consult and cooperate with one another, though informal collaboration has become rather more common of late. *** the present author believes that the current system of planning law is too much subject to the manipulations of central government, and as such, is not a truly effective system of environmental control". (op. cit., p. 25).

In Hughes' writing, he makes a plea that planning law, in its present form (1986), is not "as effective a system of environmental control as it might be". He quotes from Malcom Grant's *Urban Planning Law* at page 432: 'So far as new development is concerned, the broad powers of planning authorities ... are a significant means of preventing development which may have an adverse environmental impact. *** But there is a strong case for arguing that a more formalised procedure ought to be built into development control to encourage special weight in decision-making to be attached to environmental damage'.

In support of Grant's point, Hughes writes, "Such a system of 'environmental impact statements' is well established in the U.S.A., and the Commission of European Communities has been working on a similar system for its member states for some time. The introduction of environmental impact assessments would undoubtedly give much more weight to environmental considerations in the planning process as against technological, social, and nowadays the pre-eminent, economic considerations. **On the other hand, it is hard to see how such a system could be made to easily fit into the British system of planning which is so very much based on fluid discretion**". (Hughes, 1986, p. 23, emphasis added).

Hughes argues support for this position by stating: "As we said in '10 years in Europe: Success — and setbacks — in cleaning up the environment'. (Supplement to *Europe '82 No. 12, December 1982*):

> "One characteristic of British legislation, at least in the environmental field, is that it is normal to place a duty on an authority (perhaps a local authority or a central government agency) to achieve certain ends, to give that authority powers to carry out the duty, but then to leave to it a great measure of discretion as to how that desired end is to be achieved. British legislation also has a tendency to caution with only small steps

> being taken at a time, and with care being taken that demands are not made that cannot be realised. *** For the system to work well there has to be a consensus between the government and the governed and close collaboration between them: this has always been the essence of the British democratic tradition. **In other countries it is more common for legislation to be used to force the pace, sometimes making demands that cannot be achieved, but with timetables for doing so**". (emphasis added).

The emphasised words in the preceding statement accurately describe the environmental programme of the U.S. In fact, the American legislative environmental programme has been described as oppressive with long-standing individual rights succumbing to the regulations. Agreeably, the point of Hughes' argument is that for environmental regulation to be successful, it must be forceful; its administration and enforcement powers should not be distributed, or in Hughes' word "fragmented", amongst several departments, each doing their uncoordinated, independent bit. The efforts of all must be concerted under the direction of one environmentally responsible department. To this point, it is notable here there is no legal obligation of the various agencies to consult and cooperate with one another over environmental matters and regulations.

Hughes points out that in planning for mineral extraction, "plans may not be over rigid". (p. 244; p. 250/92 book). § 29 of the 1971 Act (§70 of the 1990 Act-p. 254) imposes conditions on mining for the hours and manner of working, screening operations, hours of permitted maintenance works, phasing extraction and restoration on a progressive basis. In view of the infinite amount of specified planning controls for mining, the statement is ironical. Policies that also regulate hours of work, production output, control stockpiling and erection of ancillary plant and machinery are no better than government-owned and operated mining operations which leave little management to private owners and certainly less incentive to entrepreneurs. Past lessons in private enterprise versus government-owned and operated mining operations, as in the British coal industry, and in mining investments for third world countries requiring 50% or better ownership-control, should demonstrate that detailed government controls over the inner workings of a mine should be very limited, not too "rigid", and left to free enterprise management. Certain specified controls for ecological protection have been accepted by the mining industry as necessary, but interference with the actual operation of mining in government planning will be a deterrent to successful environmental management mining.

The following short item from Mining Journal, April 7, 1995 (p. 258) purports to corroborate the cost-effect on U.S. mining of over-control by government.

U.S MINING REFORM COSTS JOBS

> "More than 60% of the $US3,640 million invested by U.S. mining companies over the past two years was spent outside the U.S.
>
> The study by the Mining Resources Alliance said that more than 16,000 U.S. jobs in mining and mining-related industries have been lost as a result. The study by two professors at the U. of Nevada-Reno said, "It is no coincidence that the decline in expenditures began when draconian mining law reform proposals were introduced that would discourage domestic mineral development". The Gold Institute president, John Lutley, said in a statement that the results of the study were a call for Congress to enact mining law reform to reverse the trend". (The Mining Journal, 1995d).

8.4.1 *The European Community*

The 15-member nation European Union / Community occupies only 3.5% of the world's land area, but contains approximately 10% of the earth's population. It furnishes one-third of the world's gross domestic product and captures about 50% of the world exports. A credible part in previous decades has been its large degree of self-sufficiency in mineral production. After many centuries of mineral products ranging from ferrous and non-ferrous ores, coal, industrial minerals, and construction minerals, mineral production has decreased rapidly since the 1970's, and in some cases has virtually stopped. Metal mining has decreased greatly, with none at present in France and Germany. Iron ore mining in 1994 has been reduced to one-third of its production in 1974. Zinc production has dropped 50%, lead 40%, and all copper mining has ceased except for two remaining mines in western Europe. In total, the E.C. is more than 80%

dependent on imports for most of its metal needs. However, it continues to be a leading producer of potash, magnesite, kaolin, fluorspar, and construction minerals.

Total expenditures by European-based companies for mineral exploration invested in Europe during recent years has been about $40 million out of a total $450 million, or less than 10%. Dr.-Ing. R.K.F. Nemitz, president of the German Mining Association, in a paper presented at Leeds on January 25, 1995, attributes the European decline mainly to richer mineral deposits abroad which are easier and cheaper to exploit, but also because planning and environmental requirements are less onerous due to fewer environmental constraints imposed by governments of the developing nations.

8.5 United States — Weaknesses of Surface Mine Permitting Procedures

Until quarry and surface mine permitting is placed solely under the control of state mining agencies and removed from local control, surface mining operations will continue to suffer irrational and illogical local mining regulations by nonprofessional, layman local authorities. Several formerly reviewed cases, as well as a 1992 New Jersey Supreme Court decision, illustrate the point.

In the State of Washington, the court supported state regulation of quarries stating that "Such (*mining*) regulation is vested exclusively in the State's Department of Natural Resources (DNR)". In *Fjetland v. Pierce County,* Wash.App. No.12448-3-II, Div. Two, (unpublished) (1990), a Washington court supported the state's DNR agency as being exclusively in control by denying county limitations on mined area and annual tonnage because DNR had issued the permit. In spite of that ruling, Washington county governments still exercise limited authority over quarrying that severely controls or hampers mining. For example, inconsistent with the earlier *Fjetland* decision was the later case of *Meridian Minerals v. King County*, 810 P.2d 31 (Wash.App.1991), where the local authority's denial of quarry enlargement and increased production was upheld by the court because the quarry was operating under a locally-issued nonconforming use permit. Even worse and more illogical was the decision in *Valentine v. Kittitas County*, 753 P.2d 988 (1988), where the court ruled that

DNR had the exclusive right to regulate mining under the State's Mining Act, but site processing was not state-regulated and, thus, left to local control. As a result, Valentine was allowed to quarry stone, but could not crush it on the quarry site because local regulations prohibited stone crushing. Such a ruling is not only inconsistent, but idiotic and virtually leaves mining control still in the county despite rulings that DNP is in control. Legislators should be able to visualise that roads cannot be paved with large quarry-blasted size rock, but need crushing. Mine-permitting and regulation over mining and mineral processing needs to be solely controlled by one, professionally-staffed, state-wide mining agency.

Split or joint governmental control over the mining process only leads to chaos, expensive and numerous litigation, delays and frustration of the mining economy.

In the 1991 Vermont case, *In Re. Meaker*, 588 A.2d 1362 (Vt. 1991), local controls were upheld in preventing a mining permit to a sand and gravel operation because the already locally deteriorated roads could not accommodate the extra trucking anticipated. (An aggregate operation could have cured the local bad road conditions.) A dissenting justice of the Vermont Supreme Court aptly stated that under such a decision, **"We are in serious danger of expanding 'not in my backyard' into 'not anywhere'"**. (emphasis added).

In the New York case, *Hunt Bros. v. Glennon,* Park Dir., 585 N.Y.S. 2d 228 (1992), when local state park authorities tried to regulate mining activities and interfere with a state-approved mining permit, the court properly found that the local agency had no control over mining within the park and that regulation was vested in the State. For further supportive arguments to eliminate local controls over mining operations see *Bernardsville Quarry v. Bernardsville Borough*, supra at Ch. 5 §4.6 (vi).

In the New Hampshire *Wolfeboro* case, supra at Ch. 5 §4.9.1- Grandfather Act, the three-pronged test designed by its Supreme Court as to whether there was intent to quarry all of the land from its inception of acquisition, illustrates the lack of understanding of the judiciary of mine planning and operation. An article in Pit & Quarry that responded to the *Wolfeboro* decision and its three-pronged test is re-produced following and entered into evidence of the weakness of the *Wolfeboro* decision and for any states that would adopt *Wolfeboro* as case law.

Wolfeboro decision sets a dangerous test for allowing quarry expansion in New Hampshire

A QUARRYMAN'S REPLY TO WOLFEBORO

(Pit & Quarry, March 1990)

by R. Lee Aston, J.D., LL.M., Legal Editor

"The *Wolfeboro* decision by the Supreme Court of New Hampshire may prevent quarry owners who should have the right to expand their operations from doing so.

The decision set a three-pronged criterion test for allowing quarry expansion under the grandfather clause of the state's 1979 surface mining law. The test is ill-conceived, ill-devised and grossly unfair. In defence of mining operations across the nation, the Court is urged to revise this test before other states mistakenly adopt it as good case precedent.

The Case: Smith began operating a sand and gravel pit on a 35-acre tract of land in Wolfeboro, N.H., in 1950. Twenty-nine years later, the state legislature passed a law granting "municipalities the authority to cope with the recognised safety hazards that open excavations create" through zoning and permitting. The law was designed to regulate mining operations that started production after its effective date of August 24, 1979.

As with most licencing laws, the act contained a grandfather clause, which allows prior mine operators to continue their operations. In Wolfeboro, the local zoning board took the untenable position that only the land currently in use on the law's effective date was exempt under the grandfather clause. This meant operators could not expand their operations or even change pits on the same property without a permit, once the law took effect.

The Answer: The Supreme Court rejected that argument saying many operators would find they could not continue mining much longer, if at all, under the zoning board's reasoning. The Supreme Court said the question was whether lateral expansion constitutes a permitted continuation under the grandfather clause.

In answering its own question, the Court said it was not aware of any other jurisdiction that ruled continuing an excavation included only vertical — and not horizontal — expansion.

Up to that point the Court's reasoning was supportive of mines that had operated before permitting laws were passed. But from that point, the Court created a new obstacle to operators who supposedly were grandfathered in under the act.

The Catch: In cases where the quarry operator only used part of his property before passage of the act, the court questioned whether he had the right to expand his excavation without permits.

In the Wolfeboro case, Smith still had 10 acres on unmined reserve on his 35-acre tract. The question was whether Smith's entire 35 acres was grandfathered in, or if he needed a permit to excavate the last 10-acres.

As all quarrymen know, operators acquire as large a tract as possible to insure a long life for the mine. Mining is not a short-term venture; it is predicated on lasting as long as the mine's reserves — the greater the reserves — the longer the mine's life.

In its impractical attempt to decide whether all land held before the act's passage was grandfathered in, the court devised a three-pronged test:

(1) The operator must prove that excavation activities were being actively pursued when the law became effective;

(2) He must prove that the area he desires to excavate was clearly intended to be excavated, as measured by objective manifestations, and not by subjective intent; and

(3) He must prove the continued operations do not and will not have a substantially different and adverse impact on the neighbourhood.

The Ruling: The court found that Smith failed to demonstrate an objective intent to quarry the entire original tract. Hence, he was required to apply for a use permit from the Wolfeboro zoning board to mine the remaining 10 acres of his land.

The Mistake: From a quarryman's standpoint, the three-pronged test is ill-conceived, ill-devised and grossly unfair because it is economically not feasible, and nearly impossible to meet and pass.

It unfairly and unreasonably sets requirements in 1989 for quarriers to have met before 1979 without knowledge that there were any requirements to prove at a date years later.

The first prong: "He must prove that excavation activities were being actively pursued when the law became effective."

This may be the only prong that quarriers can reasonably meet. However, the test fails to clearly explain to what extent "he" must prove the excavation activities were under way before the law took effect — over the entire acreage owned, or merely over some part of the property?

The second prong: "He must prove that the area he desires to excavate was clearly intended to be excavated — as measured by objective manifestations, and not by subjective intent".

According to the Court, the quarrier cannot merely demonstrate that he bought X number of acres to start his operation on, and that he intended eventually to mine it all. That would be subjective. The Court requires objective manifestations.

To qualify objectively, the quarrier must have carried out some act of exploration, evaluation or assessment over the entire purchased acreage before 1979.

For example, to qualify he would have to drill, strip, test overburden depth or cut timber over the entire tract, in order to signify preparation for stripping. Drilling, stripping and testing overburden depth are all costly and time-consuming operations. A quarry operator cannot and should not use capital funds to develop acreage many years in advance of actual development and excavation. Smaller operators with less capital could not afford such a large outlay of money to start overall exploration and mining. As for cutting timber, legal ramifications exist if the operator leases the land. Even if the operator owns the timber rights, cutting timber off the entire tract would eliminate a natural barrier for quarry noise, dust, and vibrations and air concussion. Yet the second prong requires manifestations on all of the tract.

Not only would it prove unwise to make such advance moves simply to satisfy the second prong of this test, but in 1979 quarry owners had ways of knowing they would have to prove their intentions 10 or more years later. It has, in effect, created an *ex post facto*, judge-made law, depriving quarriers of their right to mine all the land they obtained for that very purpose.

The Court's objective test should center on what are sound financial and economic mining principles. It should accept that when a mining operator acquired land prior to 1979 for mining, he intended eventually to mine it all.

The third prong: "He must prove that the continued operations do not and will not have a substantially different and adverse impact on the neighbourhood.

The *ex post facto* criticism, above, also applies to this third prong of the test.

With regard to the "different and adverse impact" phrase, the Court stated in its arguments that "a great increase in the size or scope of a use has also been considered to be a factor in determining whether the character of the use has been changed". In other words, a quarrier, by Court-made law, cannot anticipate or be allowed to share in business growth as any other business can.

In an era of great national expansion after 1979, when quarriers were needed to supply stone for highway and building construction, the quarrier alone cannot expand or enlarge his business or else he loses his standing for a continuous exemption — a penalty for growth! For the court to penalise responsive growth by the quarrier creates a judicially established Bill of Attainder.

Bills of Attainder, or laws which inflict punishment for past or future conduct without judicial trial are expressly prohibited by the Constitution. This test may not only be ill-advised, but may be unconstitutional as well.

The quarrier's subjective intent before 1979 when he bought or leased the land should qualify the entire acreage for exemption from the 1979 law under the grandfather clause.

The challenge: There are several legal grounds for overturning this decision. The costly and prohibitive objective manifestations created in 1989 as criteria for pre-1979 acts smack strongly of *ex post facto* laws. And the penalty of requiring permits when a quarrier fails the 1989 test implies a judicially devised Bill of Attainder.

Should these claims of unconstitutionality fail, quarry owners should challenge the test on the grounds that it runs afoul of the Fifth Amendment's right of due process. It appears there may be substantial and solid grounds to contest the three-pronged test of *Wolfeboro*". [Aston, 1990 (Pit & Quarry)].

However, contra to the 1989 decision in *Wolfeboro,* in New York's 1992 *Fletcher* decision, and in California's 1995 *Hanson Brothers* decision, both courts took the minority zoning position in supporting expansions for quarries under the grandfathering acts.

8.6 The Detrimental Cost-Effect of Currently Prevalent Zoning Denials

The current, prevalent, uncoordinated actions in the U.S. for planning of land use without included consideration for mining of construction materials, or other essential minerals, can only have the unnecessary effect of causing higher costs.

Through personal communication in 1995 with the Northmont Sand & Gravel Company, supra, Ch. 5 §5.4.6, the continued denial of several successive zoning hearings over five years since closure of their worked-out pit, to allow the operator to open a new tract of land for sand and gravel mining, has adversely affected the supply of materials for concrete and road construction and maintenance in the local area. Builders and contractors are having to pay higher costs to bring the materials in sufficient quantities from further away.

This increased cost-effect of the NIMBY syndrome is dramatically borne out by a current and similar example in the Denver, Colorado, area: "along the booming eastern flank of the Rockies in Colorado — where sand, gravel and stone demand is approaching 12 tons / a. per person (*the average American uses 9 tons of aggregate per year — the equivalent of a 50 lb. bag every day*), no new aggregate mine has been permitted in the past twenty years. In the suburban areas surrounding Denver, one of the nation's fastest-growing cities, vocal citizens have killed three proposed quarries in three years.

"Meanwhile, the Denver area is already coming up short of aggregate. Rubble for the new Denver-International Airport had to be brought by rail from Wyoming. People, who do not want pits in their neighbourhood, say the rubble should be brought in from elsewhere. But experts say the cost doubles every 30 miles that is transported by truck. Consequently, in urban areas that are running out of gravel, costs are skyrocketing, and the high prices are drawing sea-faring shipments from Mexico and Canada. While in smaller cities gravel costs as little as \$6/ton, the Philadelphia street department, for example, is paying \$11.70 a ton to have gravel delivered". (*Wall Street Journal*, March 1 1995).

As Justice Dooley of the Vermont Supreme Court has said, "To allow the general public to prevail with the NIMBY syndrome will be to place the nation in serious danger, of expanding 'not-in-my-backyard' into 'not anywhere'". The point is further supported by a comment from the National Stone Association Digest of March 3, 1995, on Martin Marietta Corporation's Chairman Augustine's article, "On Permitting Delays" which appeared in the *Wall Street Journal*, March 1, 1995: The article on Norman Augustine "does an excellent job of identifying why aggregate prices are climbing, due to a reluctance of local government authorities to expedite approval of permits for new aggregate operations, because of complaints usually generated by overly zealous environmental activists".

The "rule of thumb" frequently stated by executives in the aggregate (crushed stone) business for obtaining permitting of a new surface mine is minimally one year, and more commonly two years, provided there are no environmental obstacles and opponent reactions and prolonged public hearings by citizen's groups. However, a current example of obtaining permission for re-opening a former gold mining operation in California now stands at two years without receiving approval for a county Use Permit. Emperor Gold Corporation, after two years' work to overcome environmental obstacles to reopen its former mine near Grass Valley, California, received the county's approval in November 1995 of the company's final EIS. Approval of the EIS is an important preliminary hurdle prior to receiving a Use Permit. Even then, the purpose of the Use Permit is only to allow the company to do drilling and further exploration of the old mine.

8.7 Canada

A limiting aspect of Canadian environmental law lies in that there is lack of strong federal or central control, i.e., laws are not uniform as each province is responsible for its own jurisdiction. The federal government allows far more local autonomy for its provinces than the U.S. does for its states, thus following the governing pattern of the U.K. Environmental regulatory matters are largely left to the provincial legislative bodies, with legal interpretations and decisions left to the provincial courts. The main environmental concerns of the Canadian federal system are for regulation of Crown lands owned by the central government, for coastlines and international agreements.

Canada's constitutional division of powers is generally non-prescriptive in policy terms, enumerating and allocating powers without requiring that they be exercised in a particular way. The currently proposed Environmental Management Framework Agreement (EMFA), now in its second draft submitted in May 1995, is "quasi-constitutional in the sense that it specifies functional areas of responsibility, but leaves substantive policies to be determined by the provincial government, or in some cases, through an inter-governmental process. *** Although the EMFA addresses the environmental management responsibilities of both levels of government, it is the resultant federal role that will generate the most controversy. The objective is to clearly redefine federal priorities, focusing on areas where the federal government has exclusive authority or which require a national perspective, while ceding responsibilities in other areas to the provinces. *** The EMFA's comprehensive approach to environmental regulation is also seen as a step forward from the current pattern of largely uncoordinated bilateral agreements and informal arrangements among governments. The goal of reallocating responsibilities between the two levels of government is to be accomplished through administrative agreement, thereby avoiding the quagmire of constitutional reform". (Kennett, 1995, pp. 1–3). Kennett notes that the first draft of the EMFA in 1994 "was criticised by environmentalists as a wholesale federal abandonment of substantial areas of environmental regulation".

The EMFA "adopts a functional approach to the detailed allocation of government roles. Environmental management functions are divided amongst the schedules, eleven of which are specified in the Framework Agreement. These functions are: monitoring; environmental assessment; compliance; international agreements; guidelines; objectives and standards; policy and legislation; environmental education and communications; environmental emergency response; research and development; state of the environment reporting; and pollution prevention". (Ibid.)

Under the heading of "Accountability", Kennett writes of an argument of the environmentalists in the first draft of EMFA, that "it is argued that the negotiation and implementation of federal-provincial agreements and the role f intergovernmental bodies [in this case the Canadian Council of Ministers of ıe Environment (CCME)], can remove important decision-making from public crutiny and political accountability".

This is counter-argued herein (see 8.0– 8.1) as a progressive step forward) the reduction of: conflicts in public hearing and litigation over environmental ıatters; excessive and unnecessary costs of promulgation of rules and egulations and ensuing litigation. As previously argued, for the greater part, ıe environmental protection programmes are administered by well-informed overnmental agencies placing reliance on scientists and technology for nvironmentally-safe procedures. Agencies are in a far superior position to the ıy public's to determine, assure and resolve environmental protection for the ublic. Hence, reliance by the public should be placed in them and on their ecisions without further public involvement, agitation and prolonged conflict. nticipatory fears in environmental regulatory matters frequently turn out to e uncalled for as evidenced in the U.S. *Snail Darter* case, and the U.K. Brent par occurrence, supra.

.8 Legislative Weakness of Land Withdrawal from Mining Activities

'he inevitable conflict between conservationists / environmentalists and the ıining industry for use of public lands for park development is noted by rofessor Barry Barton, "Mineral exploration and mining are regarded as recisely the kinds of activity against which protection is needed, no matter ow little land is taken up with actual mining. *** No satisfactory process is in lace for reconciling or arbitrating these different needs for mining and parks ecisions, be they to create new parks, to open parks to mining or reduce levels f protection for parks, or to adjust park boundaries. Decisions are made by an xercise of governmental discretion without any settled procedure or pportunity for input from the affected parties". (For discussion and argument oncerning the decision-discretion of elected officials, see Ch. 7 §6.0, ummary — Conclusions and Comments, supra; and, Ch. 13 §4, Deciding the rocedure for Permission of Projects, infra.)

The concern in this work is for excessive land withdrawal for numerous arks, forests, wildlife preserves and habitats, and other preservation projects,

in its limiting of the potential land-area mineral base necessary for future miner
prospecting and production. In turn, the number of potential future surfac
mines, or pits, are similarly reduced, thus affecting the potential of future site
for landfills in the full restoration and reclamation process.

The inevitable conflict between the conservationists and the mining industr
is that mining disturbs the natural ecology during the mineral removal an
until the operation is worked out, leaving a scar on the natural beauty of th
area. Mineral removal should be viewed as only a temporary period
minimised interruption of the ecology that is necessary to provide the mineral
for mankind's way of life. As so often said, minerals must be mined wher
they are found. Parks and forests can be located elsewhere, but not miner
bodies. Further, with proper planning of land use, particularly by government
for large areas of the public domain or Crown lands to be set aside for park
habitats and recreation, those lands should be first be reviewed and evaluate
for potential mineral contents. Where bodies of minerals are suspect in th
areas to be preserved, an opportunity should be provided for the mineral industr
to explore and evaluate the mineral bodies, even exploiting them if economicall
called for and reclaiming the surface before the land is set aside and classifie
as "untouchable".

The thesis herein calls for a compromise that allows mineral productio
from environmentally sensitive lands during a relatively short period of yea
for the mineral removal, accomplished with modern mining consciousness f
the environment that minimises injury to the ecology. On depletion of th
mineral body by surface mining, land reclamation would immediately follo
with full restoration of the surface to its original state. This can be done b
placing solid waste infill back to replace the volume of the ore body remove
An additional, but necessary, disruptive feature required for mineral remov
and infill for restoration in such areas is the road and power lines infrastructur
Such roadways are frequently in place in national forests as requisite for timb
cutting and mining where permitted. Construction and maintenance of thes
roadways are already strictly supervised by the Forest Service. In regard
national forest mining, roads must be viewed as temporary features until th
surface mine is fully restored, whereon, the road access to the restored mi
and landfill may also be removed and reclaimed to its natural state. This positi
is given more attention in the Ch. 12, under future trends.

8.9 The Future Development and Enforceability of Environmental Law

Avoidance of extremism, moderation or temperance, i.e., utilising environmental management, is herein the strongly recommended path for future development of environmental law in contrast to extremism. As Hughes points out concerning stewardship of the earth and the polarity of the various environmental stewards, those "Dark Greens" who espouse "*** the need for considerable restraint on activities (both procreative and economic), while "Lighter Greens" are prepared to adopt a less strict approach. *** seen in legal systems between those which stress environmental protection and those who favour environmental management.

"It is not always easy to know what the environmental consequences may, at some unknown future date, flow from the particular uses of the environment and its resources, or from particular industrial, or agricultural processes, nor the ways in which they may happen. The law could require:

(a) cautious progress until a process is judged "innocent";
(b) ordinary progress until findings of "guilt" are made; or
(c) no progress until intensive research has been conducted into a proposed process.

"It is (c) which most fully represents a "precautionary principle" in law corresponding to the economic notion of 'anticipatory action'. It is (b), however, which represents the general British response to the problem". (Hughes, 1992, p. 13, 17).

The EC's "polluter pays" principle has been debated as to whether it is adequate. A recognised problem is that many industrial polluters may absorb the cost of fines for environmental violations, or costs of remediation, as overhead or operating costs, then pass the costs on to the customers as increased production costs. There are also those industrial polluters that may have an inability to pay high fines as a deterrent to polluting. More effective alternatives to sanctions against corporate and industrial polluters have been considered, e.g., corporate dissolution, disqualification to take government contracts, stock dilution to pay fines, etc. Whether imprisonment for environmental violations, or "crimes", is morally acceptable is still debatable.

8.9.1 *Criminal Penalties for Environmental Violations*

Until about 1990, the U.S. was the leading proponent of punishment b imprisonment for higher magnitude "criminal "violations of environmenta laws. Canada had revised and implemented its federal EPA in 1985 by addin imprisonment as an alternative to fines. U.S. terms for imprisonment wer between six months and a maximum of five years. The Canadian Province followed suit about 1989 to 1990 by increasing the amounts of fines and / o imprisonment. (R.S.Ontario and R.S.Saskatchewan). The U.K. joined the U.S and Canada in April 1992 by amending its Environmental Protection Act 199 with the addition of imprisonment sentences to supplement fines for crimina violations. Nevertheless, the U.K. and Canada have remained far less aggressiv in applying and meting out imprisonment terms as sanctions for environmenta violations, generally employing only fines. According to a natural resource attorney in Canada, "To the best of my knowledge, nobody has ever actuall been imprisoned under the sections, at least nobody connected with any oi gas or mining facility". (Blackman, 1996).

8.9.2 *U.K. Environmental Crime Punishment*

For example, in the applicable sections of the U.K.'s EP Act concerning lega obligations for waste disposal, Section 34 imposed a "duty of care" and wen into force in April 1992. Section 33, the Prohibition on Unauthorised or Harmfu Deposit, Treatment or Disposal of Waste, went into force in April 1993. Bot sections proscribe penalties for criminal offences, but only Section 33 proscribe imprisonment terms for violations.

Essentially, §33 replaces the former waste disposal licence with a wast management licence. Under the new section, all waste disposal, by any perso carries with it a responsibility and liability for the safe disposal of controlle waste and they may not knowingly cause or permit waste to be deposited excep through a waste management licence holder.

Under the U.K.'s former Control of Pollution Act 1974, only waste disposa was regulated. §33 goes beyond final disposal by proscribing violations fo environmental harm resulting from any pre-disposal treatment of waste, e.g., recycling, or during waste storage, e.g., as of recyclable materials (glass, tyres, etc.).

As with the U.S. law for environmental violations, provisions are made under §33 for citizens to report observed roadside tipping. §34 also provides for public "self-policing" to encourage reporting by citizens of observed environmental crimes and states in ¶39, "Compliance with the duty of care will be secured mainly by waste holders checking on each other in connection with the transfer of waste". This latter section provides, in American environmental law terminology, "cradle to the grave" protection for disposal in the waste chain. Previously, there was no link in the waste disposal chain for regulation of waste carriers and transfer of waste. Hauliers and waste transfers are now controlled by transfer records. Authorities now have investigative, as well as search and seizure powers where crimes have been committed.

Under §33, the provisions for a person committing an offence shall be liable:

(a) on summary conviction, to imprisonment for a term not exceeding six months or a fine not exceeding £20,000 or both; and
(b) on conviction on indictment, to imprisonment for a term not exceeding two years or a fine, or both.

For persons committing an offence under the same section but for special wastes on indictment-conviction, the maximum imprisonment term is five years.

In emulating U.S. environmental law regulation, §34 goes so far as to provide for citizens reporting on each other when violations have been observed.

§§35 to 44 of the Act tightened up regulations for licensing waste treatment and disposal activities. §39 in particular, prevents licence holders from simple surrendering of their licences and walking away from the site as they were able to do in the past. Surrendering of site licences now requires an application, investigation of the site, and satisfaction that the site is left in a safe condition whereby the land is unlikely to cause pollution or harm thereafter (Orlik, 1992).

It appears from the above summary of environmental criminal punishments, that British "discretion" and "persuasion" are slowly giving way to American style "command and control" regulation.

8.9.3 *U.S. Environmental Crime Punishment*

In 1990 the U.S. Department of Justice (DOJ) reported that criminal sentences for environmental offenses were on the increase with successful convictions being 95% of the indictments. Jail sentences were given to more than half of those convicted and large fines were made on corporations. The vast majority of criminal cases were brought under the Clean Water Act and RCRA. A reason attributed to the rise in prosecutions and convictions is that an increased number of environmental crimes were re-classified from misdemeanor status to felony. Prosecutors are generally more willing to pursue felonies than misdemeanors.

In 1994, the EPA referred 220 cases (a 36% increase over 1993) to the DOJ. Criminal charges were brought against 250 individuals and corporations (a 40% increase over 1993); $36.8 million in criminal fines were assessed (a 19% increase over 1993); and jail sentences imposed totalled 99 years (a 25% increase over 1993). (EPA Memorandum, 1994).

Title VII of the 1990 Amended Clean Air Act made criminal the release of hazardous pollutants into the air by a Knowing Endangerment clause. An individual violator is subject to 15 years imprisonment, and a corporate violator fined up to $1 million. Lesser, misdemeanor imprisonment up to one year and fines may be made for negligent pollution releases. In addition, the amended Act authorised a greater number of environmental inspectors ("police") for stepped up enforcement. Additional provisions have also been made to coordinate the CWA with violations other environmental statutes. Knowing violations of reporting and monitoring requirements are punishable as two-year felonies. Five-year imprisonment and fines may be imposed for knowing violations of implementation plans, new source performance standards, et al. Primacy states may also impose their own punishments. The California case of *People v. Martin,* 211 Cal. App. 3d 699 (1990) held that criminal sanctions were also applicable to persons who "should have known" that their act of unpermitted hazardous waste disposal would cause injury and harm. The California Code provides for imprisonment on conviction for 16 to 36 months in state prisons.

A policy known as The Responsible Officer Doctrine was approved in the case of *U.S. v. Dee*, 912 F.2d 741 (4th Cir. 1990). The case serves as precedent in case law and empowers a federal judge to instruct a jury to convict a manager of an operation for a felony based on failure to prevent or detect an environmental violation. Numerous cases in 1990 illustrate the serious nature of court-imposed jail sentences and fines for managers and corporate officers that were responsible for environmental violations.

8.9.4 *Canada Environmental Violations Punishment*

The following case examples illustrate the less severe penalties meted out in Canada for environmental violations.

8.9.4.1 *Ontario*

Non-compliance with mine cleanup leads to conviction Owners of an inactive surface copper-gold mine at a remote location in northern Ontario Province were charged by the Ministry of the Environment with non-compliance of an order to clean up the site where inspectors found eight transformers with polychlorinated biphenyl (PCB) staining the ground at their bases. After 20 months of failure to get the mine owners to voluntarily perform a cleanup, the EPA issued an order requiring cleanup and securing the area in seven days. The owners failed to comply and EPA levied fines against the owners and the owners appealed.

At trial in *Regina v. Consolidated Mayburn Mines Ltd.*, [1992] 73 C.C.C.(3d) 268 (Ont.Prov.Div.), the owners challenged the validity of the EPA orders and argued a defence of due diligence by non-compliance of unreasonable orders. The owners were able to show the court that all but one of the procedures required by the EPA were environmentally unsound and that non-compliance would better protect the environment than compliance. The court then severed those unsound counts from the EPA order. The remaining ground not complied with, and unproven as sound procedure, was for the owners to remove the PCB contaminated soil. For its non-compliance, one defendant individual and the corporation were convicted and fined.

8.9.4.2 *Newfoundland*

A fluorspar producer was found guilty of severe water pollution under the Fisheries Act by a Provincial Court. Damage to the environmental life of fish was found to be extensive. The company had been warned in the past, but was conditionally allowed to continue its operations provided it kept effluents from waste water to a minimum. Despite repeated warnings to correct increasing levels of effluents, the producer did nothing.

In *Regina v. St Lawrence Fluorspar Ltd.* [1989] 80 Nfld. & P.E.I.R. 171(Nfld.Prov.Ct.), the government fined the mining company C$15,000.00 on three charges. The Court stated that "Pollution was a serious crime and the sentence was therefore appropriate".

8.9.4.3 *British Columbia*

In *Regina v. Jack Crewe Ltd.* [see supra, Ch. 5 §5.3.5], the Canadian court fined a sand and gravel operator that allowed silt, sand and clays as deleterious contaminants from its washing plant operation to enter a stream, being harmful to a fish habitat.

8.9.4.4 *Yukon Territory*

In *Regina v. United Keno Hill Mines Ltd.*, [see supra, Ch. 5 §5.3.6], the defendant company operated an open pit and was fined C$1,500 for discharging mine waste effluents into Yukon waters in excess of effluent contaminants discharge limits prescribed by an Industrial Water licence issued by the Yukon Territory Water Board contrary to the Northern Inland Waters Act.

The *United Keno* decision (see Ch. 5 §5.4(vi) and Ch. 7 § 1) and arguments for penalties for corporate violations of environmental law is reviewed.

Section IV

TRENDS AND FUTURE NEEDS

CHAPTER 9

PRESENT AND FUTURE MINERAL AND WASTE TRENDS

9.1 Introduction

As nations grow and populations increase, so does the construction minerals mining industry. With the great surge of world growth since World War II, the demand for building minerals (crushed stone, sand, gravel, limestone, clays, gypsum, et al) has been tremendous.

Presently, there is a continuing need for more minerals from the earth to sustain Mankind's quality of living for his growing numbers. Again, the two needs are mutually dependent and cyclic in location occurrence. As referred to in the Introduction, Ch. 1, mining provides minerals and materials for Man's goods; Man creates waste; wastes need to be disposed of; mining creates the voids for waste depositories. "... for dust thou art, and unto dust shalt thou return".

9.2 Present and Future Global Mineral Needs

On the threshold of the third millennium, the global mining industry faces a range of challenges, not least, a fast-depleting resource base and economic and ecological threats in augmenting the mineral production volume to meet the rising demand of a burgeoning global population. The size of the task ahead can be gauged from the current global output of 20,000 Mt of mineral raw materials for a population of 5,200 million. Of this staggering volume, the share of different minerals includes: 8,000 Mt of energy raw materials; 550 Mt of iron and ferrous metals; 160 Mt of non-ferrous metals; 500 Mt of industrial minerals; over 3,000 Mt of quarried stone and dimension stone; and 8,500 Mt of sand and gravel.

For extracting this volume of mineral raw materials, the mining industry globally has to excavate and move some 200,000 Mtonnes of earth and rock, 90% of which is waste or 'residual' and needs to be disposed of in an ecologically sustainable manner. The conflict between mining and environment can be best understood from the concept of a support square, proposed by Skinner. This is defined as "the scrap of land that must supply all the resources that an individual uses throughout life and that must fulfill the same purpose for others who follow. Somehow that space must also absorb most of the solid waste left over.

"The support square has been inexorably shrinking from about 300 m^2 / person in 1890 to about 160 m^2 at present, and it may shrink to less than 100 m^2 by the year 2090". For sustainability, the mining industry has not only to produce more, but also to manage the environment in the face of a shrinking support square. (*Mining Journal*, London, November 11, 1994).

Further support of this fact of shrinking mineral base is indicated in the statement in Blakeman's 1976 report to Environment Canada: "Because of increased demands in conjunction with transportation costs, development of pits and quarries concentrated in the rural townships surrounding the Toronto-centred market. Consequently, the rural residents (often former urbanites) and municipal councils in these areas not only voiced strong opposition to pit and quarry development, but in many instances local councils prohibited new extractive activities through official plans and zoning by-laws. This action effectively removed from the resources base extensive areas containing potentially available aggregate resources".

9.2.1 *Future Aggregate Site Alternatives*

As an example of future planning in England, Mineral Planning Guidelines 6 (MPG6-1994) states a future aggregates objective is to reduce the proportion of supply from primary-won sources in England from 83% in 1994 to 74% by

2001 and 68% by 2006. Planning of this nature is a result of environmental and land-use pressures making the siting or location for new inland quarries difficult. An alternative trend, to move away from inland, won aggregate sources in the developed countries, which began about 1985, was to establish coastal super-quarries. Super-quarries are broadly defined as those capable of producing at least 5 m.t./a., and with reserves of at least 150 m.t/a. Locating these new large-size coastal operations offers two opportunities to quarry developers, viz., (1) with the sea fronting on at least half of the property, ocean-side sites offer fewer neighbours with nuisance and annoying complaints against the operation; and (2) it provides deep-water loading for long distance, large load transportation by vessels at increasingly interesting lower rates as opposed to escalating trucking and rail haulage rates. It could also reduce the ubiquitous site-hearing complaints of "increased traffic" problems.

Super-quarries were reputedly pioneered by Foster Yeoman Ltd. in 1986 at its Glensandra quarry on the west coast of Scotland. Glensandra was closely followed by Vulcan Materials Company's (USA) joint venture on Mexico's Yucatan Peninsula and Newfoundland Resources and Mining Company's (NRMC) super-quarry at Lower Cove on the Port au Port Peninsula shipping to its terminal in New York City, U.S.

A new coastal marine quarry in Ireland, the Wimpey Fleming Adrigole Quarry, started production in 1993. Annual production for 1995 was about 1.2 M tonnes and production was expected to reach 2.0 M tonnes by the year 2000. It is the second large-scale operation of its kind in Europe. Outbound shipping will transport the aggregate. The markets of the U.K. and northern continental Europe are expected to provide backhaul cargo.

New coastal super-quarries are being developed and planned, e.g., a 5 M tonnes per annum operation at Jossingfjord in Norway; one at the south coast of Sweden intended to supply the Berlin and the German Baltic coast markets; and one on South Harris Island, Scotland. NMRC plans to establish more terminals along the U.S. eastern seaboard to take delivery from its super-quarry aggregates operation in Newfoundland. In Nova Scotia, plans for a major coastal marine quarry are delayed pending an environmental review at a deep-water site north of Sydney on Cape Breton Island.

9.2.2 *Seabed Aggregate Mining*

Seabed mining or dredging of marine aggregates offers another alternative for avoiding the increasing environmental and zoning constraints for inland quarry siting and permission. In the U.K., it is already an important alternate source of aggregate for the construction industry. By 1989, mined seabed sand and gravel made up 18% of the 20 m. tonnes consumed in England and Wales. In Japan, seabed sands account for about 40% of their total domestic production of fine aggregates used in concrete. Concern for the marine environment, disturbance to fisheries, and coastline damage is considered in U.K. licencing for marine dredging. Concentration of seabed site licences is avoided as are known, sensitive seafloor areas. Seabed mining by the U.S. has been delayed due to the drafting of regulatory laws that would hopefully alleviate major environmental concerns.

(i) **Remoteness:**

Remoteness from population areas, i.e., moving into rural, sparsely populated areas, placing distance between the excavation site and people, is an alternative that has been avoided for decades prior to the green movement. Previously, the low-cost, high-bulk product could not tolerate the addition of costly freight to construction sites and the market. Proximity to market, i.e., the population centres, for site location of an aggregate surface mine, has long been a critical, strategic and economic matter for pit operators. The ability to find a good stone deposit and move in closer to the urban area than a competitor has been a tactic that frequently "weeded out" or hurt the competition. Thus, the ability to ship to market at a fraction of a dollar cheaper per ton than the competition often meant the success or failure of an aggregate mining operation.

It appears more and more strongly that the public will accept the addition of higher hauling or transportation costs to the low-cost bulk construction minerals in order to prevent scenic intrusion of excavation in their area. However, it must be realised that construction materials mine operators can neither physically nor economically afford to pick and move to a new pit location every generation. As the population centre continues to grow and expand, moving outward, it encroaches on the surface mine that was located in a rural area just a generation ago, thus, ultimately forcing the mine operator to move further away again. A scenario like that might fit shopping centres and malls in an ever-expanding metropolis, but does apply to mining of minerals. It bears repetition, "Minerals must be mined where they occur". Surface mine operators cannot simply "pick up and move" further into rural areas with each successive encroachment of people complaining about the unsightly surface mine and its "nuisance". Re-location of a mine is dependent upon the local geology and the occurrence and extent of construction minerals in the earth. All rocks and sands are not suitable for making aggregates, cement, mortars, brick, glass, wallboard, insulation, roofing granules, et al.

As a consequence of this state of continuing and mounting conflict between mining and public environmental and aesthetical complaints, an alternative trend is for mining to move far away from population centres. Remaining close to the construction market will not always be possible if the geology of the earth does not conform or provide deposits of suitable building minerals. The increased cost of transportation for construction minerals will have to be borne by the consumer.

However, if deposits of suitable minerals can be located in a remote area, the costs of transpiration will be the main concern, i.e., finding the lowest means for delivery to market. Obviously, the further from market, the higher the truck haulage cost. Railway movement can become competitive with road movement at distances greater than 50 miles. Large volume rail movements of coal in "unit trains" are well established for the coal industry. Unit train shipments for aggregates, sand and gravel, and other lower-priced minerals are beginning to be of interest as viable considerations for the mining industry. Witness the super-quarries, supra, §9.2.1, with volume shipments by water. Blakeman noted in 1976 that Great Lake "Boat haulage into Southwestern Ontario has increased substantially since 1972, amounting to 2.5 million tons (2.25 million metric tons) in 1975". (op. cit., p. 23.) Long distance haulage of processed construction minerals will require storage and distribution yards in the major urban centres.

In addition to the price increase for transportation from remote quarry sites, an important detrimental environmental concern for consideration is the increased demand of energy consumption for diesel fuel that will be made to accommodate remoteness of surface mines. The fuel consumption costs of moving as much as 50 million tons (45,359,000 metric tons) over 100 miles (160.93 km) would result in an annual increase in diesel fuel consumption of approximately 12% over present levels (Blakeman, 1976, p. 24). In 1996, the cost and consumption would be even greater. Can this fuel and energy consumption be justified with environmentally advocated sustainable development for non-renewable fuels?

(ii) Transportation Costs: Long Distance Refuse Haulage vs. Long Distance Aggregate Haulage:

The construction industry has long argued it requires proximity to the construction growth and population centres because the low-priced minerals cannot tolerate an added high freight cost. Contrasted with virtually valueless MSW, long distance haulage of MSW and other wastes to distant landfills is already environmentally justified. It has become a beginning reality and practice (see discussion in Ch. 11 §3 (ii), infra).

Similar with the justification of high haulage costs for MSW, higher hauling costs will be environmentally justified for aggregates in the near future. Inroads on the added costs for long-distance hauling of aggregate minerals are already taking place in the construction industry, e.g., ocean hauling from super-quarries, hauling from seabed dredging points. The days of proximity for many aggregate sites to the market will wane as aggregate quarriers find remoteness offers less permitting and operating contention. Furthermore, backhauling of waste to landfilling quarries on unit trains can offset the costs of aggregates to market as does ocean back-hauling freight for super-quarries. Nevertheless, the cost of construction minerals will necessarily increase to satisfy the demands of environmental aesthetics and contention. Thus, the trend of increased costs of construction aggregate materials for long-distance haulage may be similarly environmentally justified as are the higher costs for MSW to remote landfill sites. As always, the increased costs will have to be borne by the public.

In keeping with the thesis of utilising surface mines for waste depositories, the long-distance haulage of aggregates, and other minerals, to population centres to distribution terminals can be more readily justified and compensated by a return haul load of MSW to the surface mine used as a landfill. Unit trains can be assigned for surface mine-landfill haulage, and are already being assigned to certain landfilling sites.Unit trains for mineral haulage are not uncommon, e.g., coal and clay in North America and Britain. Unit trains for aggregate haulage up to 80 miles to stockpile terminals for the Washington, D.C., metropolitan market are already in existence.

9.3 United Kingdom

In the U.K., domestic mining of metals is now virtually non-existent, and reliance is principally on imports with very minor relief from recycling. In the last decade, the only significant metal mining remaining has been of ironstone for the winning of iron and it has all but stopped. Production of iron drastically decreased from 10,228,000 tons in 1971 to 731,000 tons in 1981, to 59,000 tons in 1991, ceasing in 1992 with 29,000 tons. Thus, surface mining of non-fuel minerals has been reduced to mainly that of industrial minerals or construction materials and clays. With the exception of gypsum (2 million tons in 1995), the U.K. is self-sustaining for the production of construction minerals.

9.3.1 *U.K. — Current and Prospective Mineral Production*

The British Geological Survey (BGS) reported production of construction minerals (crushed stone, sand and gravel, limestone, sandstone, dolomite, shale, common clays) for the U.K. as follows:

1991 .... 300 million tonnes
1992 .... 294 million tonnes
1994 .... 300 million tonnes ..... (with a value of over £1,240 million)
1995 .... 304 million tonnes ..... (estimated)

Projections of aggregate demand by the BGS are for a 55% increase by 2011.

In the U.K., environmental and land-use pressures have resulted in relatively less construction minerals production from inland quarries and more production from coastal super-quarries. Entertainment of the idea for coastal shoreline super-quarries in the U.K. is developing with the worldwide trend. At present, there is only one in operation, located in Scotland.

U.K. Mineral Lands: In 1994, BGS reported 106,900 ha. of England's and Wales' land, or less than 1% of the land area, was permitted for surface mineral workings. For the same year, the Confederation of British Industry (CBI) reported some 96,000 ha of U.K. surface land was affected by mineral workings, amounting to 0.75% of the land area.

The Department of Environment reported in its Survey of Derelict Land that in 1988 only 60% of surface mining lands with planning permission "was covered by satisfactory restoration conditions". (Hughes, 1992, p. 245). Nevertheless, presently, about "59% of reclaimed mineral sites in the U.K. are subsequently used for agricultural purposes, and 4% for forestry, while nearly 30% is used for increasingly popular recreational purposes, e.g., golf courses, nature reserves and water sport-facilities". (*The Mining Journal*, 1995a, p. 89).

9.3.2 *U.K. — Mineral Land Planning and Permitting Trend Indications*

The planners of future mineral extraction have been forced to consider more seriously and include systematic reviews of environmental effects as a result of U.K. regulations and procedures, e.g., TCPA Regulations (Assessment of Environmental Effects) 1988 and Dept. of Environment / Welsh Office, Environmental Assessment — A Guide to Procedures, 1989, and the EC Directive on Environmental Assessment, 88/337/EEC 27 June 1985.

Whilst mineral extraction in the U.K. is carefully regulated by planning authorities to minimise ensuing harmful environmental effects, planning authorities are also charged with taking into account local and national demands for minerals, especially where few or no alternative sources exist. It has been similarly recognised in Britain that "geology dictates where minerals are found", and this has been a crucial factor in the U.K.'s industrial development. According to the CBI, local and regional planning authorities identify current workings and resources, reportedly indicating areas which may be needed for future minerals extraction.

The 1990s have brought increasing demands through population growth for new construction. Under the pressures for increased construction minerals production accompanied by environmental controls, the British central government planning has matured to the point where local and regional planning authorities are directed to identify mineral-bearing areas which may be needed for future mineral extraction. Thus, policies for land development include planning for the extraction of minerals. Guidance for the planning of development and use of mineral lands is set out in the statutory development plans drawn up under the TCP (Minerals) Act 1981 and the TCPA 1990, as amended by the Planning and Compensation Act 1991. Mineral Planning Guides (MPG) are periodically published by the Department of the Environment (see Ch. 5 §4.7 for introduction of the "Green Book" / MPG in 1951). After the 1960 revision of the first MPG in 1951, the following MPG's and subjects were published:

MPG1 — (1988) the general principles and national considerations of mineral planning with specific advice on the development system;
MPG2 — (1988) planning applications for minerals development; planning permissions; and the imposition of planning conditions;
MPG4 — (1988) review of mineral working sites, including the compensation implications;
MPG5 — (1988) aspects of the general Development Order of special relevance to mineral interests;
MPG6 — (1989/1994) guidelines for aggregates provision in England;
MPG7 — (1994) The reclamation of mineral workings.

Highlights of MPG6 (1994): MPG6, first published in 1989, was cancelled and superceded in 1994 with regard to England (not Wales). Arguments throughout this work are stressed in MPG 6's Policy and Objectives, to wit:

"9. The Government wishes to see indigenous mineral resources developed within its broad objectives of encouraging competition, promoting economic growth, and assisting the creation and maintenance of employment. The Government believes that for the economic well-being of the country it is essential that the construction industry continues to receive an adequate and steady supply of aggregates so that it can meet the needs of the community and foster economic growth.

"10. At the same time, the Government recognises that aggregates extraction can have significant environmental impact and often takes place in areas of attractive countryside. *** (it) stresses the importance of combining economic growth with care for the environment in order to attain sustainable development. ***

"12. The aims of this Guidance Note are: (i) to provide guidance how adequate and steady supply of material to the construction industry may be maintained at the best balance of social, environmental and economic cost ***; (ii) to provide a clear framework within which MPAs can develop aggregates policies in development plans and carry out development control; (iii) to serve as a national framework for the Secretary of State, et al; (iv) to help reduce the number of planning appeals; ***

"67. Safeguarding: Planning authorities should make every effort to safeguard resources of all types of construction aggregates which are, or may become, of economic importance, against other types of development which would be a serious hindrance to their extraction.

"68. It will usually be necessary to consider the need for aggregates over a longer period than for most other land-use planning issues. When considering the need to extract the mineral as opposed to letting surface development proceed, *** where it is possible to extract minerals prior to some other more permanent forms of development, this should be encouraged ***.

"70. forms of (mineral) development in National Parks, the Broads, New Forest, AONB, et al, should not take place save in exceptional circumstances, and

"71. *** should normally include an (environmental) assessment ***

"91. Environmental assessment: *** Where proposals for mineral development are likely to have significant effects on the environment, applications will need to be subject to EA under the TCP (Assessment of Environmental Effects) Regulations 1988. Whether, or not, a particular mineral development will warrant an EA will depend upon such factors as the sensitivity of location, size, working methods, proposal for disposal of waste, the nature and extent of processing and ancillary operations".

Highlights of MPG7 (1994): The Reclamation of Mineral Workings

The objectives and aims of reclamation and after-uses are:

"1. England and Wales are rich in minerals, but in many areas they are also densely populated. There are pressures on land and competing claims on space for housing, industry, commerce, waste disposal, agriculture, forestry, recreation, nature conservation and other uses. *** and with the concern for protecting the environment, it is very important that land worked for minerals should not become derelict or remain out of beneficial use for longer than is absolutely necessary. In seeking to reconcile the winning and working of minerals with other claims on land, one of the main aims of planning control is to ensure that land taken for mineral operations is reclaimed at the earliest opportunity and is capable of an acceptable use after working has come to an end.

"2. Unlike most other forms of development, mineral extraction is an ongoing activity as a result of which the land can, and should, be **recycled**, either to its former use or to a new and acceptable one". (emphasis added).

MPG7 is concerned with imposing reclamation conditions for new permissions, e.g.,

"8. In granting planning permission for mineral working, mineral planning authorities should always carefully consider the applicant's proposals for reclamation of the site***".

The intended after-use of the reclaimed land is stressed, and how it fits into planning for the local area. MPG7 notes that agriculture is the most common and appropriate after-use for reclaimed mineral land. It notes that "There is now much more consideration of nonagricultural after-uses, particularly for forestry or amenity". The use of surface mines for MSW disposal sites is treated lightly with the emphasis on restoration for agriculture, forestry and amenity and some industrial uses. An illustration for MSW landfilling is included, but not discussed in great detail.

19. The preparation of an EA is again recommended where there is likely to be a significant impact on the environment.

MPG7 gives detailed consideration and information concerning the orderly procedures for pre-mining stripping, storage and saving of soils for the after-uses requiring the growth of vegetation. It also details backfilling with mine overburden, mining and milling wastes, and other types of waste materials, including MSW, in preparation for the various types of after-use.

In addition to "imposing reclamation conditions for new permissions", MPG7 also proposes that "new or improved reclamation conditions" may be imposed "on existing permissions and workings". The 1981 Act introduced new powers, in effect 1986, to enable mineral planning authorities to revise older permission conditions which may have been considered adequate at the time of granting, but are now considered ineffective, or in some of the older case, non-existent conditions.

One broad objective of MPG6 notes is to reduce the proportion of primary land-won aggregate sources in England. U.K. planners have noted a trend in aggregate consumption over the 12-year period, 1977 –1989: "there has been a consistent proportional shift from sand and gravel to crushed rock. Proportions for the National primary aggregate consumption in 1989 were around 44% for sand and gravel and 56% for crushed rock, which is now the largest, single source of primary aggregates used in England and Wales". (MPG6, 1994, p. 60123).

9.4 United States

In the United States for 1993, the U.S. Bureau of Mines estimated that the mineral industry provided employment for 1.86 million people; that the value of domestic mineral production was $32 billion, whilst processed materials of mineral origin were worth $326 billion and exports $40 billion.

According to a 1994 report of the U.S. Bureau of Mines, during one person's lifetime, he will "use more than 2 million pounds of minerals and metals, including 1.2 million pounds of sand and gravel, 360,000 pounds of iron and steel, 27,000 pounds of clay, 26,000 pounds of salt, 500,000 pounds of coal, 800 pounds of lead, 28,000 pounds of phosphate and potash, 3,200 pounds of aluminum, 1,500 pounds of copper and 840 pounds of zinc.

9.4.1 *U.S. Aggregate Production — Potential Landfill Sites in the Making*

Crushed Stone: For the year 1994, the U.S. Bureau of Mines reported total crushed stone production of 1.23 billion metric tons (1.36 billion short tons) with a value of $6.62 billion. The given tonnage was produced by 1,600 companies from 4,000 active quarries. Additionally, about 8 million metric tons (8.82 million short tons) of aggregate stone was imported. It is noted that aggregate importation has slowly increased from 5 million metric tons in 1990 to 8 Mm tons in 1994. Source countries for importation are Canada (49%), Mexico (29%), Bahamas (12%), other (10%).

The types of stones quarried for aggregates shows that 71% was limestone and dolomite; 15% granite; 8% traprock (diabase/basalt); and the remaining 6% were shared, in decreasing order, by sandstone and quartzite, miscellaneous stone, calcareous marl, shell, marble, volcanic cinder and scoria, and slate.

About 58.1% of the production was used as construction aggregates, mostly for highway and road construction and maintenance; 9.8% was used for chemical and metallurgical uses including cement and lime manufacture; 1.4% for agricultural use; 0.7% for special uses and products; and the remaining 30% went to unspecified uses.

Sand and Gravel: For the year 1994, the U.S. Bureau of Mines reported total construction sand and gravel production of 891 Mm tons (982 million short tons). The given tonnage was produced from 6,020 active pits. 42.8% of the production came from operations with an annual production of between 100,000 m t and 499,999 m t; 25% from operations between 500,000 tons and 1,000,000 tons / a.; 19.3% came from the largest operations producing over 1 million m t / a. The balance was produced by smaller pits.

The distribution of uses was: 35.5% for unspecified uses; 27% as concrete aggregates; 15% for road base and coverings and road stabilisation; 9% asphaltic concrete aggregates and other bituminous mixtures; 8% as construction fill; 1% for concrete products (blocks, bricks, pipe, etc); and the balance for assorted miscellaneous uses, e.g. snow and ice control, plaster, roofing granules, water filtration, etc.

Transportation for sand and gravel was by trucking, by far the major means. Waterway and rail served a very minor mode of transport.

Projected Volumes of Production: The following U.S. projections, as noted, between 1995 and 2000 were reported by the editor of *Rock Products* magazine and the National Stone Association (NSA).

year	crushed stone (billions s/t)	sand and gravel (billions s/t)	source (1 short ton (s/t) = 1.0165 tonnes)
1995	1.42	1.01	Rock Products
1995	1.4	–	NSA
1996	1.45	1.04	Rock Products
1996	1.45	–	NSA
1997	1.5	–	NSA
1998	1.58	–	NSA
1999	1.6	–	NSA
2000	1.61	–	NSA

The U.S. Bureau of Mines in its November 1995 Annual Review for 1994 on Construction Sand and Gravel reviewed major concerns for the sand and gravel segment of the mining industry:

1) implementation of the Clean Air Act Amendments of 1990 and its complex legal and technical provisions;
2) the amended Federal Water Pollution Control Act of 1977, the Clean Water Act, §404, dealing with "wetlands" and the associated "no net loss of wetlands" policy;
3) the Storm Water Pollution Prevention Program;
4) the Occupational Safety and Health Administration's (OSHA) Hazard Communications Standards regulating the use of products containing more than 0.1% crystalline silica; and
5) the provisions of the Federal Endangered Species Act.

Agreeably, these are areas of over-regulation creating difficulties for the sand and gravel segment of mining, as well as other mining segments.

9.5 Canada — Current Mineral Production

According to Natural Resources Canada (NRC), "The mining and minerals industry is a vital industry to the Canadian economy. In 1993, this industrial sector contributed over C$20 billion to the Canadian economy, an amount equal to 4.2% of the Gross Domestic Product. Mining and minerals-related industries directly employ 335,000 Canadians.

"Canada is the third largest mining nation in the world. *** Almost 80% of Canada's mineral production is exported, making it the world's largest exporter of minerals.

"Ontario (30%), Quebec (17%) and British Columbia (16%) account for more than 60% of Canadian mineral production. Alberta accounts for 6%. Producing mines are located in all provinces except Prince Edward Island.

"The value of production of structural (construction) minerals, including clay products, sand and gravel, stone cement and lime increased 7.7% (C$200 million) to reach approximately C$2.5 billion in 1994.

"Less than 0.03% of the land area of Canada has been used by mining since mining began more than 125 years ago. This intensive use of a small area produces the minerals used every day.

"Mining represents a temporary land use, disrupting relatively small areas of land for a specific (usually short) period of time. Once the ore deposit is depleted, the land is cleaned up or reclaimed for other uses including recreation".

Production from surface mining operations pertinent to this work follow. The Mineral and Metals Sector of Natural Resources Canada reports the following statistics:

Sand & Gravel — Stone Production in 1992–1994 and average for period 1990–1994

(in tons)	1992	1993	1994	1990–94, avg.
Sand & Gravel	240,616,000	238,137,000	238,110,000	235,351,000
Stone	89,338,000	89,361,000	91,053,000	93,783,000
Total, s & g-stone	340,154,000	327,498,000	329,163,000	329,134,000

(Source: Natural Resources Canada, 1995)

A comparison of Canadian projections made in 1976 for Southwestern Ontario with present Canadian national volume figures for the production of sand and gravel and crushed stone is interesting. Although based on the figures used for a populous part of Ontario Province and not the nation as a whole, it shows a sizable shortfall, or an underestimate of considerable tonnage for those two construction minerals.

Comparison of Sand & Gravel — 1976 Stone Projections with 1994 Production

(tonnes)	projected for 1995 (Blakeman, 1977, p. 9)	1994 production (NRC, 1995)
Sand and Gravel	153,800,000	238,110,400
Crushed stone	66,300,000	91,035,000

As readily seen, both construction materials were underestimated by very large amounts, viz., actual sand and gravel production was 35% more in 1994 than predicted in 1976 for 1995; actual crushed stone production was 27% greater in 1994 than predicted in 1976 for 1995.

Applying the same amount of error from the 1976 projections for the year 1995 to the 1976 projections for the year 2000 yields a corrected projection for 2000 for sand and gravel and stone as follows:

ALL CANADA

(tonnes)	1976 projection for 2000 ÷ correction = corrected 1976 projection for 2000 (Blakeman, 1977, p. 7) factor
sand and gravel	183,200,000 ÷ .65 = 281,846,153
crushed stone	79,000,000 ÷ .73 = 108,219,178

The new projections are made on the assumption that the same rate of growth for the period 1975 to 1994 will continue for the five-year period between 1995 and 2000.

Blakeman's 1976 report (based on work by Redfern & Proctor, consultants) predicted that "the Eastern Ontario Region could face a shortage of sand and gravel prior to 2025, and, in fact, Proctor and Redfern (1975) predict that the region will exhaust its sand and gravel supplies between 1985 and 2010". (Blakeman, 1976, p. 10). However, it is also noted that "tremendous resources of stone are available in Northern Ontario, and that there are (remote) extensive resources of sand and gravel, their utilisation being dependent upon the development of efficient transportation ***".

The implication is that production from remote deposits will effectively raise the price of the construction minerals at the market site, and the cost of construction projects will rise correspondingly. The price tag for eliminating visual intrusion for near-urban areas will be an expensive one for the consuming public.

Although unit values of construction minerals continued to increase through 1994, keeping pace with the rate of inflation in Canada, selling prices varied considerably depending on the proximity to consumers. According to the Canadian Minerals Yearbook 1994, Minerals Sector author, Oliver Vagt, writing on Mineral Aggregates, states "a broad indicator of demand for most primary construction minerals are housing starts". He notes that 'housing starts' dropped in Canada from 168,000 units in 1992 to 155,400 in 1993 and remained the same, 155,000 for 1994. *** The level of residential construction was expected to remain the same in 1995 with about 156,000 housing starts, according to the Canada Mortgage and Housing Corporation. "A related statistic from the same work states that "the construction of single-family homes triggers an overall demand of about 30 tons of aggregate per unit, while apartment construction requires 50 tons per unit". (op. cit.)

9.5.1 *Canadian Future Construction Mineral Trends*

It is noted in the Canadian Minerals Yearbook 1994 that "As existing land-source sources are depleted, there is a growing potential for the economically viable marine dredging of sand and gravel in Canada. Offshore sand and gravel resources in Canada have been used to meet special job requirements in the Beaufort Sea (off-shore of NWT and the Yukon above the Arctic Circle), the Prince Rupert area of British Columbia, and at Roberts Bank port facility near Vancouver. In Atlantic Canada (Maritime Provinces), it has been established that there is a good possibility of defining sufficient quantities of sand and gravel for marine dredging".

Although stated, supra, that super-quarries were pioneered by Foster Yeoman Ltd. in 1986 at its Glensandra quarry on the west coast of Scotland, NRC states that "large ocean transportation facilities have been used for many years in British Columbia to supply high-quality aggregates or high-calcium limestone, e.g., limestone producers on Texada Island, situated about 100 km northwest of Vancouver Island in the Strait of Georgia, supply raw material to cement and lime producers on the lower mainland and in the State of Washington. Holnam West Materials Ltd. and its predecessor have been shipping from Texada since 1957. Road-base material and rip rap for use in the lower mainland are also important products shipped to ports as far away as Alaska or northern California". (op. cit.)

In recent years, "granite aggregate from a quarry near Port Hawkesbury, Nova Scotia, has been transported to markets throughout the region, and when favourable backhauls could be arranged, 50,000–60,000 ton shiploads have been made as far away as Houston, Texas". (*Canadian Minerals Yearbook* 1994).

Thus, the trend for greater volume shipments by sea transport appears to becoming more firmly established and economically feasible.

9.6 Conclusions and Comments

The natural process for reclamation of surface mines and pits is by backfilling as landfills for municipal solid waste. CBI reported that in 1994 about 0.75% of U.K. surface land, or approximately 96,000 ha (237,216-acres), is affected by mineral operations. Planning authorities believe most of it will be restored to other beneficial uses, "*** **and that waste disposal may be usefully incorporated into the restoration phase**". (emphasis added).

As argued in this work, the location of aggregate and cement limestone quarries and their production reflect the population growth and centres. This is supported and evidenced by Canadian mineral statisticians using 'housing starts' as a broad indicator for predicting aggregate consumption trends. Additionally, "an analysis by the Industrial Minerals Division of Natural Resources Canada's Mining Sector has confirmed a high statistical association between cement shipments on the one hand, and 'housing starts' and one- and five-year mortgage rates on the other".

A possible fallacy and potential weakness of the trend for siting super-quarries on coasts and shorelines is the probable furor that will be raised by the environmentalist sector over the potential harm that may done to coastal marine ecology. That battle is presently being fought in Scotland over the proposed super-quarry planned by Redlands Aggregates Ltd. on South Harris Island, Scotland. Apparently, the planned quarry site is in a natural scenic area causing a conflict over environmental concerns. One may ask, "With all the world declared 'a natural scenic area', where will mining go"?

A prospective positive outlook for coastal super-quarries is for their anticipated reclamation and surface restoration by MSW landfilling. In addition to being remotely located, any potential contaminating seepage from landfilling would more than likely be into the sea which would have a far greater ability of absorption than land sites. The leachate-contained dreaded toxic metals for ground waters would simply join the already high metal content of the saline waters of the seas.

Attention is again called to the difference in attitude between the British "way" and the American for applying regulations. Persuasion, influence, suggestion and discretion is conspicuous in the foregoing MPG6 and 7. It stands out in marked contrast with the American style of "command and penalties" for disobedience or non-compliance.

CHAPTER 10

CURRENT AND FUTURE TRENDS FOR WASTE DISPOSAL: THE URGENCY FOR LANDFILL SPACE

10.1 Introduction

There is no future long-range indication that the critical need for more void space in the earth's surface for depositing Man's wastes will abate or slacken. The ever-growing population only indicates an increase in the volume of waste with the consequential increased requirement for disposal. In the Anglo nations, this can only mean an increased need for void space for deposition of solid wastes. Simultaneously and similarly, the demand for construction minerals is predicted to grow at an ever-increasing pace with the population growth.

The tenor of this work, specifically and in general, is to promote mandatory regulatory control for reclamation of worked-out, presently operating and future surface mines, quarries and pits to use their voids as depositories for MSW whenever the hydrology of the location is practical, and as part of the pit's complete restoration to original land surfaces. Surface mining voids created annually, let alone those in abandoned mined lands (AML), are far more than sufficient in volume to resolve the annual critically-needed space for waste.

Referring to the author's previously stated argument that surface mining pits for construction materials are bound to be located near population centres by virtue of (1) the demand for urban construction materials, and (2) construction materials must be delivered to the construction site for a low price, inclusive of haulage. To date, mined construction materials costs cannot tolerate a large increase in price for long-distance haulage. Consequently, their location near to population centres make them a natural possibility for a MSW site. However, there is noted a recent trend towards greater hauling distances for aggregates from more distant sites with the introduction of super-quarries and low-cost ocean transportation (see Ch. 9 §9.1.1, supra).

10.2 Current Methods to Relieve the Urgency for Landfill Space

An obvious solution to reducing the need for burial space of MSW is to reduce its volume. This would hold true for a constant population figure. Nevertheless, any reduction of the waste volume is helpful and important. The two prominent, current trends for MSW volume reduction, viz., recycling and thermal destruction (incineration, pyrolysis, gasification) are being pursued by most nations in the attempt to resolve the inadequate and acute space problem for landfilling. Those two methods of waste volume reduction unquestionably alleviate the landfill volume and siting problem. Until investigation in the U.S., it was supposed by many that items such as disposable diapers, fast food packaging and other plastic products placed a major load on landfills. After investigation, it was determined that diapers took only 1% of the landfill volume, fastfood packaging 0.1%, and all plastics 12%. The research further revealed that plastics were not as big a problem as supposed, but paper was a major problem in taking landfill space. Newspapers took up 10 to 18% of the landfill volume and all paper products combined occupied 40 to 50% of the space (Liptak, 1991, p. 24).

Obviously, recycling of such wasted materials as paper and plastics can have a great effect on conserving landfill space. However, in view of continued increasing population gains and products demands, recycling falls short of needed relief and has only a delaying effect of reducing current waste volume at landfill sites. When salvaged material from recycling is placed in new products-stock, a percentage loss of unusable salvaged stock material is inevitable. Recovery for re-use is never 100%. Thus, the entire original product material will eventually be used up and returned to waste, albeit bit by bit.

It should be noted that the principal reason for recycling glass is not to conserve natural resources, but to save fuel. Silica is the earth's most abundant material. As the principal ingredient of glass, cullet (broken / recycled glass) requires less fuel in its reuse to manufacture new glass. The same energy savings are also applicable to recycling of metals.

Incineration is a great volume reducer, but there are residues from burning which must be returned as waste for ultimate burial in a landfill. There is also particulate matter and high levels of nitric oxides emitted from these processes. However, technologies have been developed that are effective in reducing particulate and achieving acceptable levels of noxious gases, e.g., oxygen enrichment of combustion air, moisture content, advanced and controlled combustion, etc., thereby making the thermal destruction of solid waste environmentally acceptable.

An innovative and highly commendable economic solution for reducing the volume of MSW is in its use as a substitute fuel in the manufacturing of cement. Using shredded MSW as a fuel has great merit as opposed to incineration merely to reduce the volume before deposition of the residue in landfills, thereby saving space. Cement industry research continues to investigate cheaper fuels and using waste materials in cement kilns. The use of shredded rubber tyres is one illustration that has met with some success.

Cement research has found for Refuse Derived Fuel (RDF) about 70% by volume of municipal solid waste from post-recycled curbside garbage could be extracted for use by the cement industry. Obviously, this would reduce by two-thirds the volume of waste material for disposal in landfills. With certain qualifications, using RDF could affect reductions in requirements for fuels such as coal as high as 20–25%.

Suitable waste fuels for cement manufacturing are a desirable goal in the reduction of total energy consumption and costs. Pyro-processing accounts for about 30% of total production costs. In the U.S. and Europe, the use of waste derived fuels and spent organic solvents has grown. Very satisfactory waste materials already established as fuels are paints and coatings, surplus oils and greases, solvents, inks and cosmetics. Additionally, research is currently investigating the possibility of using fly ash as replacement in Portland cement. In the interest of conservation and sustainable development of non-renewable fossil fuels, the utilisation of post-recycled MSW should be strongly pursued for pyro-processing, not only in cement manufacturing, but in various other mineral processes, and as a fuel for generation of electrical energy and steam.

10.3 Current Landfill Information

After reviewing the world's history of waste disposal (Ch. 4, the burial method), particularly of solid wastes, sanitary land fills became the major accepted method for disposal. In the U.S., landfilling is still the dominant method for municipal solid waste (MSW) disposal. It was reported in 1986 that about 83% of MSW in the U.S. was being landfilled, and 10% being recycled. By 1990, landfilling had been slightly reduced while recycling increased 3% and incineration amounted to 14%. In Canada, in 1987, 95% of MSW was being landfilled. By comparison with Japan, with a far higher population density and where land is held more precious, landfilling amounted to a mere 15% and recycling was about 50% (Liptak, 1991, p. 24). Japan purports its reason for low usage of landfills as a shortage of space. This is somewhat questionable as Japan has much uninhabited mountainous land, but customarily refuses to use it for public use other than scenic beauty. Tokyo's main landfill was expected to be filled and closed by 1995. Tokyo and Osaka are reportedly considering extension of the shore line or creating islands from solid waste, which varies little from dumping in the sea. A major problem for worldwide recycling advocates is that the daily volume of MSW is generating faster than the rate of recycling.

10.4 United Kingdom

Minor, in Hughes' *Environmental Law*, 1986, reported that the "EC produces over 2,000 million tonnes of waste per year and the vast majority (70%) is disposed of on land". (op. cit., p. 122).

Contaminated land clean up costs were estimated in 1991 to be between £10 billion and £30 billion, and this figure may rise as the cost of eliminating groundwater pollutants is assessed (Hughes, 1992, p. 298).

It is observed in the U.K. that the number of landfill sites has been decreasing in number over recent years, but increasing in size of the sites. As Tomes notes, "Since 1974 when public waste disposal became a County Council function under local Government re-organisation, the number of landfill sites reduced significantly, as the traditional local village tip disappeared. When site licensing was introduced two years later, due to the requirement for higher standards and hence higher costs, even more sites, including private ones, closed down. The practice now is therefore fewer, but larger landfill sites ***. *** it is becoming less cost effective to operate sites accepting less than 25,000 tonnes per annum of waste". (Tomes, op. cit. p. 2).

Alternative methods for waste disposal

In 1989 incineration cost in the U.K. ran on an average of £12 per tonne. New EC directives since that date, and new U.K. environmental scrutiny on emissions have already meant, and will continue to mean, costly modifications of the waste incinerators to meet new emission standards, estimated at £1 million each. Tomes calls attention to the fact that "incineration reduces the waste volume by 90%, but only 65% by weight, with 35% of the products of combustion still requiring landfill at appropriately licenced sites. *** In general, direct landfill will still be in the order of 50% more competitive than incineration. *** Whatever options are chosen for waste disposal, there will always be a requirement for landfill, for either direct placement, or after some sort of pre-treatment". (Tomes, 1989, p. 18).

Incineration offers a keen interest in competition to recycling of MSW by virtue of its energy-producing ability as a fuel. The energy content of landfill waste in the U.K. has been estimated to be worth £1 million per day (Hughes, 1992, p. 247).

10.5 United States

In the U.S, EPA's goal of reducing landfilling to 55% and increasing recycling to 25% by 1992 fell considerably short. The total amounts of MSW being generated is rising faster than the rate of recycling. In 1992, the U.S. rate of MSW generation was reported at 4.3 pounds (1.95 kg) per day per capita. Thus, a municipal area of 10 million persons would require space coverage of 1,000 acres (404.69 ha), of 8-feet (2.44 m) compacted depth every year for its waste disposal (Liptak, 1991, p. 25). Based on a 4.1 lbs. (1.86 kg) daily per capita waste figure, or 1,500 lbs. (680.39 kg) annually, another interesting analogy is that "after compacting, the 180,000,000 tonnes (163,292,400 mt) of waste generated each year would cover a football field with a column of trash 38 miles (61.15 km) high." (Brown-Ferris Inds., 1991.) By comparison with waste generation in the Third World nations, at times referred to as the "have less" nations, per capita waste generation in Calcutta, India, is 1.12 lbs. per day (0.51 kg) and 5.0 lbs. (2.27 kg) per day in Chicago, Illinois.

Similar to the U.K., there has been a notable decline in the number of landfill sites. In the U.S. in the 1970s, there were reportedly 18,000 operating landfills, which dropped to 10,000 by 1980 (to 9,000 by 1986 according to Brown-Ferris Inds.), and to 6,500 by 1988. One study reports that all but one state is running out of suitable locations for landfills. Landfills reportedly have a life of between 10 and 20 years. Obviously, there is a great need for new landfill locations. Available landfill space has become a major problem in the U.S. Simultaneously, the costs of controlled landfilling has increased greatly due to stringent federal standards.

Utilisation of reclaimed surface mines and pits for landfill sites would obviously more than relieve the critical need for MSW depositories. In 1994, the U.S. Bureau of Mines reported that 625 acres (252.93 ha) of land is daily disturbed by mining, with 337 acres (136.38 ha), or 54%, being reclaimed daily. (Much of this daily "reclaimed" land would have to be coal-mined land since hardrock, non-coal land is generally not reclaimed, at least to the extent that coal land is reclaimed. In the alternative, the definition of "reclaimed" land might be questioned.) The reclaimed land is potential landfill acreage that is not being used for that purpose. Further, reclamation that satisfies environmental regulations, excepting coal strip-mined lands, is not complete restoration of the land to beneficial surface uses other than water-surface uses.

As in the U.K., the current trend is for larger volume landfill sites, and it is predicted that they will become even larger in the future with greater total capacities and longer lifespans as future demands increase. Even more desirable in reducing tipping costs is their relative proximity to high-density populations areas. Thus, creating large landfill sites on, or in, virgin ground becomes a costly and major excavation-undertaking in addition to being unsound economics and poor land conservation practice. The logic of resorting to the large voids created by surface mining for landfilling, particularly those pits located near the high-density population areas producing low-cost construction materials, becomes even more logical, and economically inviting.

Americans produce waste at ever-increasing rates. The volume of garbage has increased 80% since 1960, and is expected to increase another 20% by the year 2000. The average American disposes of 3.5 to 4.5 pounds (1.59 to 2.04 kg) of garbage per day. "The huge volume of waste produced has caused crises in many areas of the nation; municipal governments are running out of landfill space, and most states lack comprehensive plans for safe disposal of wastes (Schoenbaum, p. 375).

Landfilling has been the traditional method of waste disposal, The other technological options open at present are increased composting, combustion, source reduction, reuse and recycling.

10.5.1 *U.S. Landfilling Costs*

Solid waste disposal is a $20 billion industry in the U.S. 25% of that, or $5 billion, is spent on the operations of landfills. Costs of well-designed landfills in 1991, as a result of strict state and federal environmental standards, were reportedly approaching $500,000 per acre. As the availability of landfills decreases, tipping fees increase. In the northeastern U.S., the area with the greatest population density and greatest generation of waste, the average tipping fees increased from $20 per tonne in 1986 to $40 / tonne by 1987; and doubled again by 1991 to $80 per tonne. (Liptak, 1991, p. 26).

10.6 Canada

Citing a highly successful operation in Ontario as an example, and quoting from the 1987 Ontario case of *Walker v. CFTO Ltd.*, 59 O.R. (2d) 104, at 107, "a landfill site is a natural outgrowth in the quarry business". This may not be taken as an authoritative, professional statement, nevertheless, it bears general acceptance in truth and examination because it is the basis in fact of this thesis. The statement is supported by a similar one made by the manager of the Fred Weber, Inc., a highly successful quarry-landfill operation at St. Louis, Missouri, who states, "It's a natural business for us. Excavating rock for crushed stone creates pits. We fill the pits with trash, compact it, and cover it with soil". (Seeney, 1988, op. cit.)

Communications with various provincial waste disposal agencies revealed generally that the use of surface mines, pits and quarries have been used to some lesser degree in the past, and a few are being presently used as MSW disposal sites. One or two provinces did not know of any abandoned that had been used.

A reply from the Senior Policy Advisor for the Ontario Ministry of Environment and Energy stated, "The regulations do not specify minimum technology standards for the design of landfills nor do they specifically prohibit landfills from being located in pits or quarries. *** Presently, pits and quarries are being used as landfills in the province".

An Abundance of Pits in Canada for Landfill Sites: Blakeman's 1977 report to Environmental Canada, (op. cit.) inventories the number of sand and gravel pits and stone quarries, active, inactive and abandoned in eastern Canada, from the Province of Ontario, eastward, through Quebec and all of the Atlantic (Maritime) Provinces. The inventory is summarised as follows:

Eastern Canada Province	sand & gravel pits (active, inactive, abandoned)	quarries / pits (stone, mines)
Ontario	2,640	190
Quebec	1,405	88
New Brunswick	204	18
Nova Scotia	209	29
Prince Edward Island	114	0
Newfoundland	87	13
Eastern Canada Totals	4,659	338

The total acreage for all pits and surface mines, active, inactive and abandoned is about 84,300 (34,116 ha), of which sand and gravel pits occupy about 70,400 acres (28,491ha), and quarries and mineral pits occupy about 13,900 acres (5,625 ha). (Blakeman, 1977, p. 174). Of the total, it is roughly estimated that 60% are located close to suburban areas of population centres. Thus, the prospects of solving any critical shortage of landfill space in the most populous areas of Eastern Canada appear to be excellent with a prolific number of potential sites for waste depositories.

10.7 Reclamation of Old Refuse-Filled Mine Pits: A Beginning Trend for Safe Waste Disposal in Today's Mine Pits?

The use of abandoned surface mines to dispose of public trash and waste is not without precedent, probably anywhere in the world. History reveals that openings in the earth's surface have long tempted man to discard waste into abandoned pits and other topographic depressions. The number of such ancient sites will never be known. Various metropolitan areas have been reputedly built on landfilled areas. Probably, in a few instances, the folly of indiscriminant, uncontrolled depositing of waste in mined-out pits has returned to haunt residents of the area. Two completed site examples of required cleaning-up and remediation of such uncontrolled and misused former mining pit sites follow, both in England. A third contemplated site, located in Wales, known as "The British", is currently planned and expected to enter the PIP in 1996.

10.7.1 *Examples of Old Refuse-Filled and Derelict Mined Land Reclamation Technology*

Following are completed reclamation examples of mined land, some that have been polluted with old, uncontrolled refuse and tipping, are offered in evidence to prove that landfilled mining sites may be restored to environmentally-safe beneficial surface uses for the public by employing current environmentally-protective technologies.

10.7.2 *Bowmans Harbour Project, Wolverhampton, England*

See Figure 3, Map Location of Bowmans Harbour, page 337 and Figure 4, Aerial View of Bowmans Harbour Site Before Reclamation, page 338.

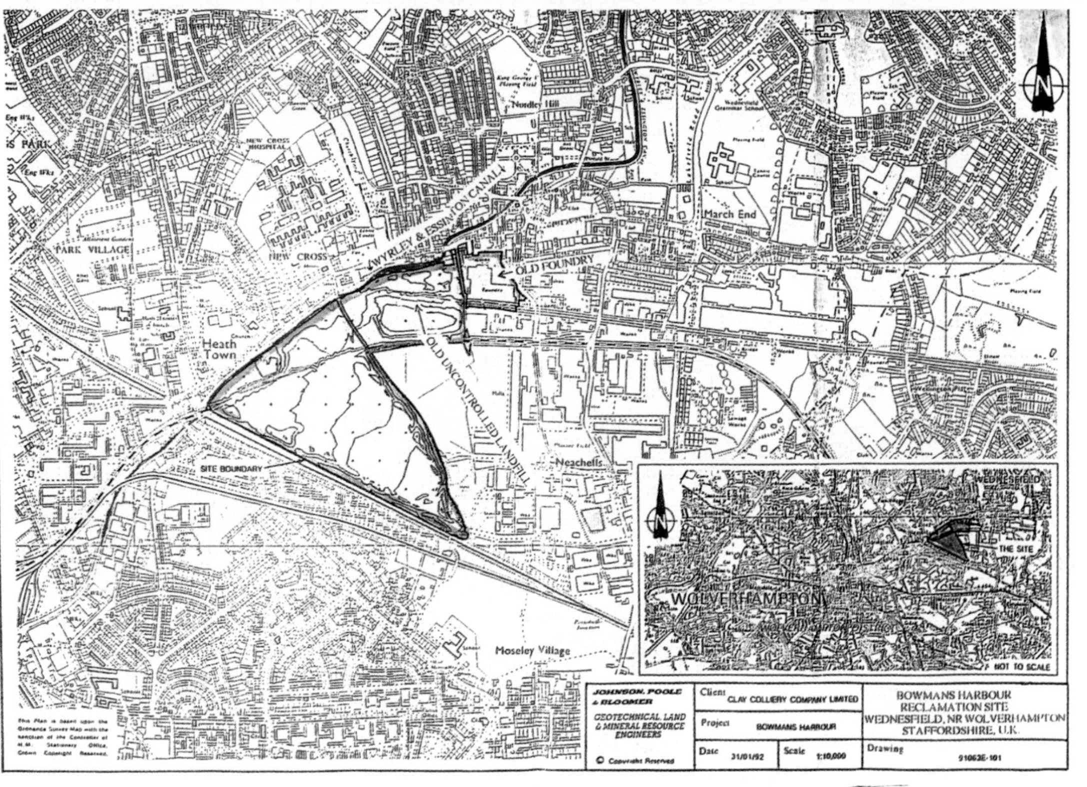

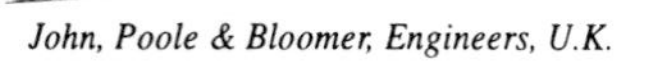
John, Poole & Bloomer, Engineers, U.K.

Figure 3 Map Location of Bowmans Harbour Reclamation

Clay Colliery Co.Ltd., Telford U.K.

Figure 4 Aerial View of Bowmans Harbour Site Before Reclamation

Synopsis of Project:

Site: 120 acres (48.56 ha) of derelict despoiled land including:
uncontrolled domestic landfill;
derelict metal works and metal recovery area;
former canal arms;
former railway embankments;
greater than 100 abandoned mine shafts and mine workings.

See Figure 5, Bowmans Harbour-Old landfill Surface Before Reclamation, page 340.

Objectives: To remove and contain waste and contaminated soils;
control of gas and leachate;
treatment of mine workings and shafts;
homogenisation of soil / backfill material and compaction to allow development of site for industry and residential uses with amenity land.

Methods: Open-cast coaling of residue minerals;
lining of void with low permeability compacted clay;
provision of drainage (gas and leachate) in void;
modern landfill of re-excavated former waste;
long term environmental monitoring;
treatment of mine shafts and workings by excavation and replacement by, or drill and grout techniques

Geology: Boulder clay overlying productive coal measures with dolerite intrusives.

Minerals: Coal, fireclay coal series, new mine coal

The Bowmans Harbour reclamation project started in 1992 and was nearing completion in the spring of 1996 as evidenced by the contruction of a new supermarket (a Safeway chain market; see Figure 7, page 348) on the reclaimed surface in late 1995. The project is illustrative of modern environmental-technology restoration that can be successfully accomplished with derelict mining land that has subsequently been used as an uncontrolled trash and refuse

Clay Colliery Co., ltd., Telford, U.K.

Figure 5 Bowmans Harbour — Old Landfill Surface Before Reclamation

dumping ground without concern for contamination and pollution of the local earth and the groundwaters beneath and adjacent to it.

Before reclamation, the derelict land contained old, shallow underground coal mine workings and a later-date land tip for refuse. Its refuse leachate had escaped into local ground water and the land had become a blight on the local scenery midway between Wolverhampton and Walsall in the northwest Birmingham metropolitan area (see Figure 5 for view of part of old surface of site, page 340). The project is situated in part of the formerly well-known Black Country, a highly industrialised area starting with the period of the Industrial Revolution and continuing well into the nineteenth century. The Black Country is dotted with old, abandoned shallow coal, iron and limestone mine workings, and iron foundries, all from the naturally-occurring ingredients necessary for the forging of iron and steel that made Birmingham an industrial centre.

The nineteenth century local sky was black with soft coal smoke and ash from hundreds of iron ore furnaces and thousands of chimneys of workers' homes huddled around the large centre of industry and employment. Described by the American consul in Birmingham in 1868, "The Black Country, black by day and red by night, cannot be matched for vast and varied production by any other space of equal radius on the surface of the globe".

The Bowmans Harbour site of approximately 120 acres (48.6 hectares) is one mile east of the centre of Wolverhampton on the West side of the City of Birmingham (see Figure 3, p. 337). The reclamation project was carried out by Clay Colliery Company Limited (CCCL), a mineral extractor, land reclamation and civil engineering company of Telford, Shropshire.

10.7.2.1 *The Project Environmental Assessment and Plan*

Quoting from CCCL's Precis of Proposals and from the Environmental Statement prepared by Johnson Poole & Bloomer, the site is "comprised of two distinct portions — the southern part, known as Bowmans Harbour, being largely grassland overlying a former landfill site, and the northern area being areas of former factories and industrial tipping. The whole area is visually of poor value and is unsupervised. These characteristics, coupled with their associated ground problems are leading to fly-tipping and general blight.

"It has been considered that the most effective way of reclaiming the land is to excavate the whole site and re-engineer it to create a stable and safe landform that will also be of significant environmental value in terms of visual amenities and after-use capabilities.

"The existing widely-spread domestic waste underlying Bowmans Harbour, some 500,000 cubic metres, will be excavated in a carefully controlled manner in accordance with the standards laid down by all the regulatory bodies, and will be re-interred and concentrated in a purpose-made modern engineered repository on the site. Some of this waste will be stored temporarily above the ground, again in a special contained and controlled cell, while the permanent repository is being prepared. During these excavation works, another form of land instability — the shallow coal seams — will be removed and transported to local power stations. The industrial wastes will either be put into the permanent waste facility or will be buried at depth as part of the normal backfill operations in accordance with the current codes of practice and standards. Any particular difficult or special works encountered may have to be taken off site.

"To ensure the above operations in a sensitive and environmentally acceptable manner the working site has been designed to ensure maximum screening value, minimal visual intrusion and proper noise, dust and odour control. Traffic, apart from the staff and service vehicles, will amount to some 30 loaded coal lorries leaving the site daily ***.

"The scheme is programmed to last approximately $2\frac{1}{2}$ years and will result in a comprehensive land reclamation exercise having resolved the problems being experienced in the area. The waste will have been contained and concentrated in a modern repository — thus all landfill gas and leachate will be properly controlled. The shallow mine workings and old shafts will have been removed creating a stable landform; and the restoration works will create a usable and useful area for public open space and development opportunities. Furthermore, the whole exercise will have been carried out in the most environmentally sensitive way — and shorter time scale — than had each portion of land been tackled independently.

"The Advantages of the Proposals are:

* maximisation of comprehensive reclamation for the widest range of uses;
* extraction of approximately 236,000 tonnes of coal- a valuable national mineral resource;

* removal and treatment of shallow underground mine workings, shafts and voids;
* treatment and complete containment of existing *in situ* wastes into a compact modern engineered cell;
* resolution of methane gas and leachate seepage problems from existing tip and coal seams (unlike other forms of treatment);
* substantial visual improvement to site and surroundings;
* employment creation in an area of above-average unemployment;
* no time scale penalty compared to alternative solutions;
* potential for any redevelopment to commence an earlier stage than with other solutions;
* environmentally acceptable means of land reclamation and waste treatment;
* preparation of proposed highway formation" (CCCL, 1992, pp. 1–2).

10.7.2.2 *The Non-Technical Summary of Bowmans Harbour Project*

The Town & Country Planning (Assessment of Environmental Effects), Regulations 1988, requires specific information to be provided on:

1. The nature of the development with information about the site, the facilities, design and size;
2. The date necessary to identify and assess the main effects which the development is likely to have on the environment;
3. A description of the likely significant effects, direct and indirect, on the environment of the development, explained by reference to its possible impact on: human beings; flora; fauna; soil; water; air; the landscape; the interaction between any of the foregoing; material assets; and the cultural heritage.
4. If any significant adverse effects are identified with respect to any of the above, a description of the measures envisaged in order to avoid, reduce, or remedy those effects.
5. A non-technical summary. "

10.7.2.3 *Environmental Assessment of Bowmans Harbour*

The non-technical summary of the Bowmans Harbour Environmental Assessment was prepared by consultants Johnson, Poole & Bloomer and pertinent parts are quoted following. The EA dealt with (1) the existing site and its environment; (2) description of the proposed development; (3) restoration and aftercare; and (4) environmental considerations.

(1) Existing Site and its Environment: The site is largely featureless scrub land, undeveloped and is crossed in a west to east direction by an abandoned railway line. Adjacent and to the north is the Wyrley and Essington Canal whilst at its southern boundary is the Bescot-Stafford railway line. There is a gradual fall only in levels from the northwest to the southeast. However, to all intent and purposes it appears flat and therefore well screened from view from most boundaries. Residential districts tend to be situated some distance from the site with the nearest being to the south and north. Slightly elevated views are available from high rise blocks to the northwest and northeast and also from the New Cross Hospital to the North.

(2) Description of the Proposed Development: Site operations: Operations to secure the land for further beneficial use comprise three main elements, site clearance and restoration of the area; excavation and re-compaction of shallow previously worked seams of coal ***, and the temporary and permanent placement of contaminated wastes into properly engineered repositories.

Site operations will allow for the control of gas and leachate, proper site management, environmental management restoration and aftercare.

Initial preparatory works will include placement of contaminated wastes currently scattered throughout the site into a single temporary but well-engineered repository. This will include proper containment measures allowing for cellular construction and lining of bases and sides. The temporary facility will have a capacity of the order of 125,000 m^3.

Minerals extraction will remove shallow previously worked seams of coal helping to pay for the proposed works. A permanent waste repository will be developed from the area of maximum void. This void will include lining measures, leachate and gas drainage systems and will take waste only from within the site as generated by the works. The restoration profile will be raised and left in a free draining domed shape as an open space for future leisure activity.

(3) Restoration and Aftercare: Environmental management will allow for the following main areas of concern: (1) restoration standards; (2) protection of surface water quality; (3) protection of groundwater; (4) landfill gas; (5) settlement.

Necessary measures to control against potential hazards of landfill gas and leachate will be incorporated into the landfill design. Ongoing monitoring of these elements and settlement of the compacted fills will continue after conclusion of the site operations. Other aspects of environmental management will include dust, noise, odour, vermin, litter, water, traffic and visual intrusion.

The site will be restored to a relatively uniform profile capable of further development for a range of uses from industrial to new-housing residential, and infrastructure development (see Figure 7, Safeway Supermarket construction on recalimed site, page 348). The permanent waste repository will be slightly elevated to introduce some relief to the finished landform and also in accordance with modern design criteria for restored landfill facilities. The landfill will be capped with a 1-metre blanket of clay and finally with a 2-metre covering of soil making material selected during site operations. Restoration will include soft landscaping appropriate to the medium and long term needs of the site (see Figure 6, clay sealing, left, on page 346).

(4) Environmental Considerations: In order to satisfy the requirements of the EA, the scheme has been rigorously examined and the following areas of potential concern evaluated: Visual intrusion, ecology, water, noise, dust, waste, odour, pest, vermin, mining stability, traffic, vibration, archaeology, socio-economic.

Summary of Assessment:

Visual intrusion: These will be alleviated by placement of screening mounds and over short length, close board fencing; landscaping by planting.

Ecology: the feasibility of relocation of the Southern Marsh Orchid and Bush Grass have been examined and in principle, confirmed.

The site will be one of the largest semi-wild open spaces in the borough and provides permanent or temporary habitats for a number of species of wildlife of varying degrees of interest (including) a potential habitat for the Little Ringed

Clay Colliery Co., Ltd., Telford, UK

Figure 6 Bowmans Harbour — Showing clay sealing (upper left) of Reclaimed Landfill and Final Stages of Coal Extraction (right)

Plover, a rare breeding specie in the country. The development will remove the necessary ecosystem to sustain this condition. However, the ecological salvage area may provide an alternative.

Water: The main potential impact of the landfill operations on water resources and water quality arises from the generation of leachate and its potential to migrate into ground water and surface systems. Migration measures include the provision of a leachate treatment and drainage system for both the temporary and permanent repositories, basal and side wall lining with impermeable clays selected specially for the task, capping with 1-metre thickness of clay overlain by 2-metres of soil-making substitute.

The scheme as proposed is unlikely to have significant effect upon surface hydrology. However, the impact of de-watering surrounding areas of ground water cannot be assessed with the present level of study. Ongoing monitoring of the ground water will be required during and after the proposed works. In the long term, it is unlikely that restoration will have a significant effect upon groundwater conditions.

(**Noise and Dust** are treated in the EA, but deleted here.)

Waste and Contamination: The site, if left in its present condition offers significant hazards in respect of contamination and gas emission. Research shows contaminated waste areas to be in contact with permeable horizons. Impacts deduced from study, likely, because of the proposed operations, potentially affect construction works and end users. Gas and leachate emissions will be controlled by means of appropriate venting systems, all excavations where waste is placed will be correctly formed and lined with impermeable clay liners. Post-reclamation environmental monitoring will be undertaken to assess the success of the engineering designs.

Landfill design will be to standards laid down by government guidelines.

(**Odour, Pests and Vermin** are treated in the EA, but deleted here.)

Mining Stability: Formation of new temporary and permanent earthworks, excavations below surface horizons and support to adjacent lands and properties will be undertaken within strictly laid down and controlled criteria. Regular inspection of the works for stability assessment will be undertaken. All back-filled areas will be supervised to conform with best working practices. Mine shafts exposed during operations and not removed by extraction of minerals will be further investigated and treated appropriately.

(**Traffic and Vibration** are treated in the EA, but deleted here.)

Archaeology: No impact is anticipated on archaeological resources.

(**Socio-economic** is treated in the EA, but deleted here.) (Johnson Poole & Bloomer, 1992).

10.7.2.4 *1996 Update on Bowmans Harbour*

Subsequent communication with Mr. Hugh G. Kent, Planning Manager for CCCL, advised that "Generally, the project has been successful in that it

Clay Colliery Co., Ltd, Telford, UK

Figure 7 Bowmans Harbour — After Use — Construction of a Safeway Supermarket on Site Reclaimed Surface

successfully re-deposited and contained the waste — subsequent monitoring by Johnson Poole & Bloomer has proved this. But, inevitably, the site has not been without problems, viz.: — (i) a larger quantity of waste was encountered than estimated, resulting in the need for enlargement re-design of waste repository and a consequent shortage of clean cover material; (ii) a problem with large inflow of water and need for pumping; (iii) the large number of different agencies involved and the sheer logistics of coordination; (iv) CCCl had to file for extra cost; and (v) some unexpected leaching at the surface.

"Despite these, the objectives have been (or shortly will be) met — i.e., containment of waste on site, control of gas and leachate, treatment of old mine workings and shafts, remediation of contaminated soils and compaction to allow hard development for industry and residential uses, with amenity land.

"There was no injury to health from contamination due to very stringent controls and checks. It cost more than envisaged (for various reasons e.g., inadequate initial information on which to calculate tender price, additional waste, etc.), and there was huge public opposition and the very tight involvement of the public agencies. *** It still cost a lot less than conventional land reclamation techniques not involving coal extraction.

"Some interesting figures for the project are:

- The total cost of the scheme was £12.4 million
- 272,220 tonnes of coal was recovered (see Figure 6, coaling on right, page 346).
- a £3.6 million grant from Black Country Development Corporation
- 5.3m m^3 of overburden excavation
- 1.2m m^3 of waste re-handled
- 30 mine shafts were capped
- 10,000m^3 of imported soils required
- £150,000 to treat old mine workings
- £1.4 million royalty paid to British Coal
- £0.2 million consultantcy fees
- £0.5 million for testing waste"

Additional information concerning the waste problem: Actually, the amount of encountered refuse turned out to be slightly over 1,000,000 m^3 of buried wastes belonging to a local authority landfill of yesteryears. Waste from the disused landfill site uncovered during the initial stages of overburden removal was transferred to a temporary surface repository, constructed with a drainage blanket underneath and a temporary seal, pending preparation of the new, engineered containment facility. Utilising glacial material excavated onsite, a 1 m-thick liner was compacted beyond minimum requirements to achieve a permeability as low as 1×10^{-11} m/s in parts, whilst for the side walls, a corresponding 1 m layer of clay was placed above a 5 m layer of slightly higher permeability material. With the coaling operation proceeding below the water table, it was required by the National Rivers Authority that the base of the repository be a minimum of 2 m above existing levels so as to offer added protection to ground water sources.

10.7.3 *The Poynter Street Reclamation Site, St. Helens, Merseyside, England*

This site formerly contained brick-clay mining pits, unstable shallow underground mine workings, brickworks with ovens / kilns, and a water-filled clay pit in a residential area. The abandoned clay pits had subsequently been used for uncontrolled refuse tipping.

The local Council granted three extensions during the project. "Despite initial fears, virtually no complaints were received and the Liason Committee organised by the Council did not meet for over two years as there were simply no problems to discuss". The land was stabilised and compacted for residential development (Kent, CCCL, 1995).

10.7.4 *Miscellaneous U.K. Reclaimed Old Mine Sites*

A tabulation follows of further completed reclamation projects by Clay Colliery Company Ltd (CCCL) is offered as evidence and proof that derelict mining land, whether coal, clay, or sand and gravel, with or without contained former uncontrolled tipping, located in densely populated areas, can be restored to beneficial, environmentally-safe surface uses (see Table 1, infra.).

Numerous, additional examples of reclamation of derelict lands in the U.K. with full environmental restoration to beneficial surface uses by Clay Colliery Company Limited can be given. The successful reclamation of surface mine voids using refuse as backfill without harm to groundwater has been proven. In the U.K., such use is a *fait accompli.*

10.7.5 *A Planned Reclamation — "The British" Site, Abersychan, Torfaen Borough, Gwent County, Wales*

The numerous, successful reclamation projects by Clay Colliery Company in the U.K. of derelict mined lands, which had subsequently been used for

Table 1 A list of other U.K. reclaimed old mine pits

Time Period	Site Name	Location	Description	Reclaimed Land Use
1988–1991	Swan Farm	Little Wenlock Shropshire	large unrestored open cast coal pit close to village	agricultural, wildlife meadows, conservation area, woodland lake; new rights-of-way net work completed
1990–1991	Croppings Farm	Shropshire	small, open-cast coal pit; shallow mine works; sealings flank adjoining public landfill	more favourable land contours made; final: agricultural, woodland and amenity land with pond completed
1991–1992	Coalmoor	Telford Shropshire	old coal working void; old refuse deposits; state of art engineered-redeposited landfill site; removal of remaining coal	restored to agricultural, woodland and amenity lands (completed)
1990–1991	Horsehay	Colwich Staffordshire	small open-cast coal pit with shallow mine workings removed; coal and clay removed	remaining land compacted for carriage way by-pass route (completed)
1988–1991	Ketley Brook	Telford Shropshire	old coal workings; coal and clay to be removed and site backfilled	451,000 tonnes of coal removed; several hundred thousand and tons of clay backfill compacted to accomodate housing development as part of city expansion (completed)

(Continued)

Table 1 (Continued)

Time Period	Site Name	Location	Description	Reclaimed Land Use
1988	Mount Pleasant	Buckley Clwyd Wales	two derelict clay to be reclaimed and backfilled	development of a well landscapeddistrict waste disposal site
1990	Bagillt	Clwyd Wales	currently operating sand & gravel pit with coal reserves	mining for minerals ontinues while land reclamation proceeds
1993–1995/6	Smelt Farm	Clwyd Wales	old, shallow mine workings; former steel works; former waste tip	remaining coal and clay to be removed; reclamation of uncontrolled former waste tip; restoration of two archaeological buildings and historical trail for Clwyd County Council; wildlife meadows, woodland with pond
1990–1996	Blinkbony	Gorebridge Scotland	underground coal workings	opencast excavation of coal reserves and reclamation to follow (ongoing)
1987–1989	Newdale Hall	Telford Shropshire	old, illegal, uncontrolled waste site in centre of town	recovery of minerals made while containing harmful solidwaste; restoration of surface for housing development and playing field (completed)

uncontrolled refuse tipping, have lead to a grander scale objective and plan to undertake one of the largest areas of dereliction in the Eastern Valley of Wales, some 16 miles northeast of Cardiff (see Figure 8, page 354).

The site is one of former collieries with shallow coal and ironstone mine workings, including some 70 recorded mine shafts and adits, mine spoil piles, colliery spoil piles up to 16 m thick in places, a former uncontrolled refuse tipping area, scrap yard, an abandoned quarry, a disused reservoir, abandoned buildings, and other small industrial operations. Parts of the site contain clinker, ash, furnace slag, masonry, concrete, metal contamination.

Waste from a previous consented landfill, from 1964 to 1968, and covering 6 ha (14.83 acres) is estimated to be 172,000 m^3. 'Waste' tipped contains nylon, polythene, household grate ash, metal, glass, and man-made fiber and occasional degradable material. Fly-tipping has also taken place in more recent years.

Some of the land allegedly contains low level contaminants, ground gas, and there is a generally inadequate stormwater drainage system causing flooding during heavy rainfall. Five reclamation feasibility studies were submitted to the Welsh Development Agency (WDA).

The feasibility studies considered the technical and safety aspects; environmental impacts (e.g., noise, dust, traffic, visual); time to completion; and cost. Options varied in time from 9 months (for landscaping only) to 36 months, and in cost from £0.5 million (£1.0 million for landscaping only) to £4.0 million. The WDA chose CCCL's open-cast option at a cost of £0.5 and 36 months. A monetary aspect that offsets the costs in the selected option of open cast restoration is that the coal recovery operations are anticipated to be largely self-financing. See Figure 9, page 355, for an idealised cross section of the planned reclamation process for "The British" by excavation and coal recovery.

It should be noted that the recovery of coal, clay, and sand and gravel in the above reclamation project is expected to contribute to a sizable reduction in the project's costs.

"The British" site cannot be offered into evidence because it has not been accomplished, and therefore, not proof. However, it is offered as evidence of technology planning for reclamation of derelict mined lands. A proposed reclamation scheme outline specification for treatment of The British Site has been compiled by CCCL following discussions with the Torfaen Borough

The Site Today

The site includes three distinct areas

Area 1 contains derelict buildings and the remains of old colliery spoil mounds. An old refuse tip occupies an area south of Big Arch. Some of the land contains low level contaminants, abandoned mine entrances and shallow mine workings.
The stormwater drainage system is inadequate and underground water courses are only partially effective resulting in flooding during heavy rainfall.

Area 2 contains unstable and unsightly spoil tips, together with dangerous derelict buildings, a disused reservoir and an abandoned quarry. Some areas contain abandoned mine workings which will require treatment.

Area 3 is not derelict and only requires minor improvements and management.

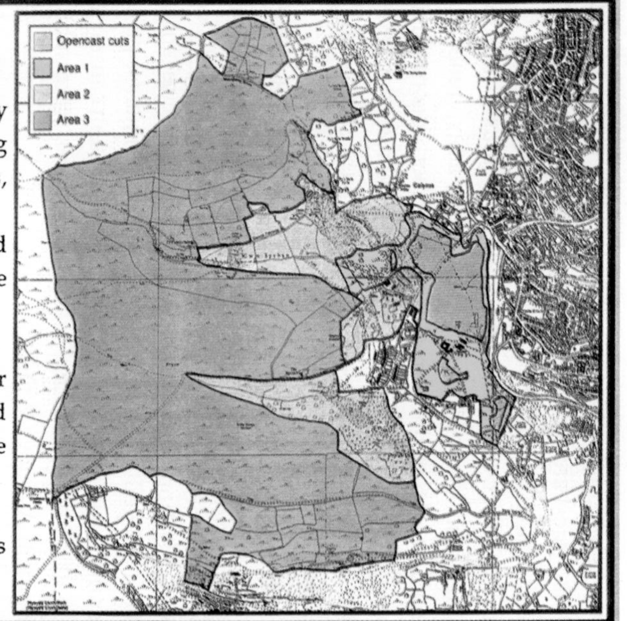

Clay Colliery Co. Ltd., Telford, U.K.

Figure 8 Plan for "The British" Reclamation Site, Wales

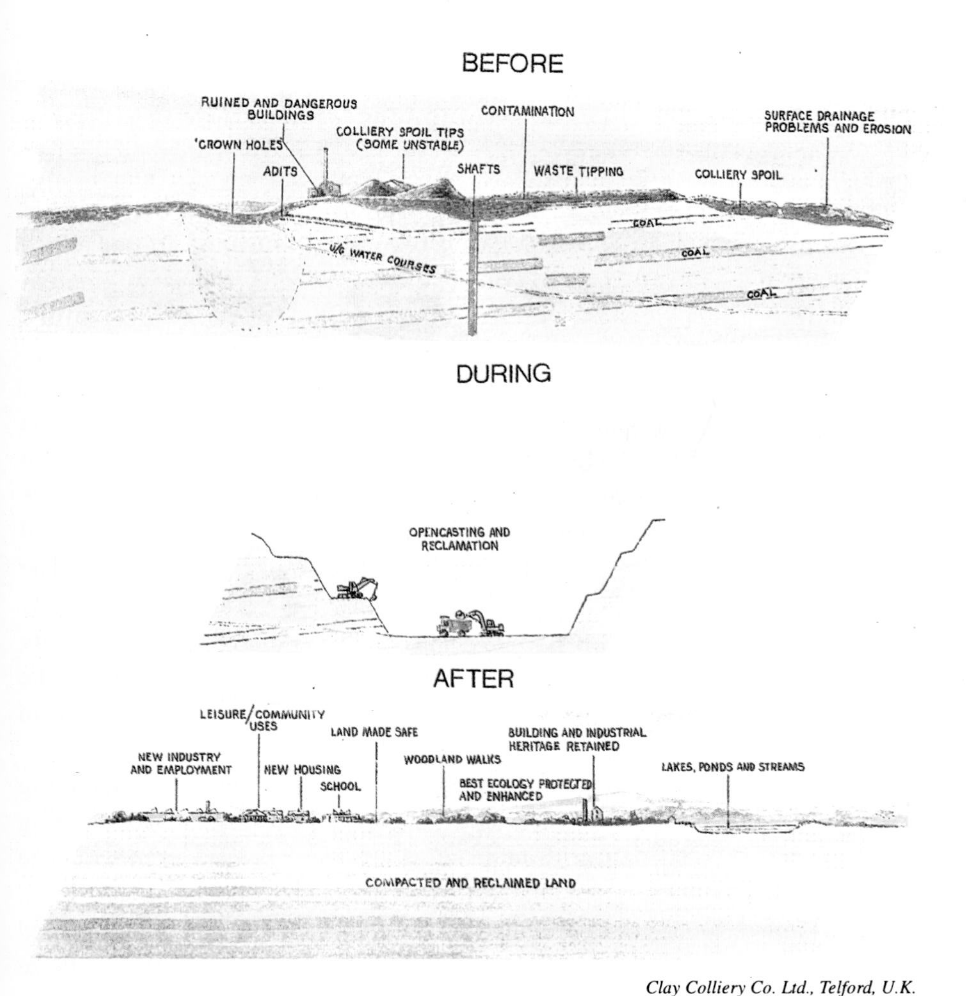

Clay Colliery Co. Ltd., Telford, U.K.

Figure 9 Idealised Cross Sections of "The British" Site

Council Environmental Health Department and the National Rivers Authority. The Scheme Outline is provisional and subject to their detailed approval.

10.7.6 *A U.S. Beginning Trend — Using Abandoned Surface Coal Mines* for Landfills

Closely approximating the U.K. trend of reclaiming coal reserves from derelict surface coal pits, including some that have been filled with uncontrolled refuse and industrial debris, followed by complete surface restoration, is an incipient trend of using abandoned surface coal mines in the U.S. as MSW landfills. The use of working aggregate and sand and gravel pits in the U.K. is also an incipient trend in the U.K., the U.S. and Canada (see examples in Ch. 11, infra.). A few years ago, this may have been somewhat startling news. Serious consideration amongst mining and geological engineering professionals in former decades has generally been lacking when the subject of using surface mines for MSW depositories was mentioned. The suggested use of sand and gravel pits or abandoned surface coal mines has generally been negative. The primary reason for rejection of coal pits in the past has been that they already present a hydrology contamination problem of acid drainage without creating a potentially additional source of polluting potential leachate seepage. The beginning trend is certainly still not universally accepted in the U.S., nor in Canada.

Opportunity for additional profit is at hand for the U.S. surface coal mining industry by reclaiming pits as municipal solid waste diposal sites. Reclaiming derelict opencast coal sites in Britain, examined supra, is a *fait accompli.* The British trend is being followed in the U.S. with successful results.

Abandoned lands from previous surface coal mining activity are located throughout the area of the Central United States Coal Field. Most of these lands have rough, irregular topography, little or no soil cover, and have revegetated under natural conditions. Many have final-cut lakes or other bodies of water associated with them. These lands are not productive, and in some cases, are the source of water contaminants that run off-site onto adjacent areas. There is little potential for most types of land-use development on these lands, but they do have unique characteristics which are well-suited for solid waste disposal facility siting.

A survey was made by Michael Owens and C. Dale Elifrits of the University of Missouri-Rolla during a graduate study by Owens. The focus of the survey was limited to the Interior Coal Mining District which can be divided into the Illinois Basin and the Western Interior Basin.

The following article is based on a 1995 Survey of the Use of Abandoned Surface Coal-Mined Land for State-of-The-Art Solid Waste Disposal Facilities. It was made within the Geological Engineering Department of the University of Missouri-Rolla, by Michael D. Owens and Dr. C. Dale Elifrits. It indicates a following of the English trend of using abandoned surface coal mines in the central U.S. for municipal solid waste (MSW) landfills.

A Summary of a 1995 "Survey of Use of Abandoned Surface Coal-Mined Land for State-of-The-Art Solid Waste Disposal Facilities"

by Michael D. Owens, M.S. and C. Dale Elifrits, PhD

A critical shortage of acceptable landfill space in Missouri is attributed to the stringent federal regulations governing disposal of MSW under the Resource Recovery and Conservation Act's (RCRA) Subtitle D. At present (1995), there are only a few disposal sites within the state that meet the federal criteria, thus, very little of the state's MSW is disposed of within its borders. Waste is consequently being hauled long distances from throughout the state at a high cost to local governments. The shortage is attributed to a lack of sites specifically designed to meet the Subtitle D standards, yet there is no shortage of potential sites that could meet the standards were Missouri to emulate the use of abandoned surface coal mined land (ASCML) as its neighbouring states have done.

At least eleven state-of-the-art landfills are operating in abandoned surface coal pits, meeting the standards of Subtitle D, in the adjoining Upper South states of Kentucky (3) and Kansas (2), and the mid-West states of Illinois (3) and Indiana (3); (See Table 2). The survey was limited to the Western Interior Basin and the adjoining Illinois Basin, having similar geology and past mining practice. Regional MSW regulators indicate that it is acceptable to use abandoned surface coal mines which are treated no differently than any other site. "The use of mine spoil for more than daily cover, specifically for liner construction material, is viewed without suspect as long as the 1×10^{-7} cm/s

Figure 10 Survey study area

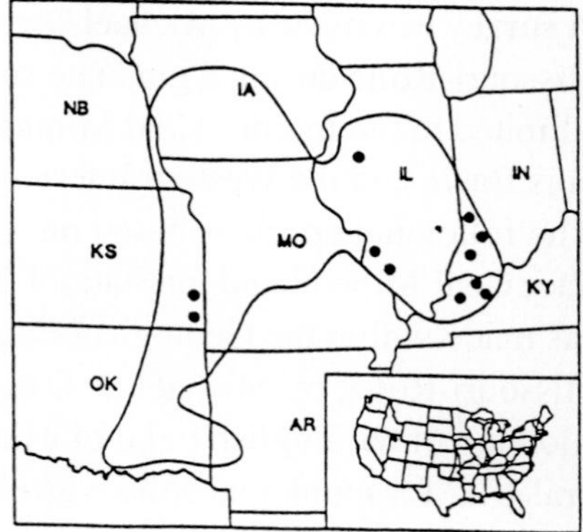

Drawings by M. Owens

Figure 11 Known coal pit-landfill locations in study area

Table 2 Municipal solid waste disposal facilities in former surface coal mine pits within the study area

State	County	Operator
Illinois	Fulton	Gallatin Balefil (BFI)
Illinois	St. Clair	Millstadt Landfill (BFI)
Illinois	St. Clair	Marissa Landfill (WMX)
Indiana	Greene	Worthington Landfill
Indiana	Pike	Rose Disposal (WMX)
Indiana	Vigo	Yawhill (Laidlow)
Kansas	Cherokee	American Disposal
Kansas	Crawford	Deffenbaugh Industrie
Kentucky	Davies	Daviess County
Kentucky	Hopkins	Bituminous Resources
Kentucky	Ohio	Addington Brothers

maximum hydraulic conductivity and compaction to 95% of standard Proctor density at 2% below to 4% above optimum moisture content are achieved. The use of mine spoil for daily cover is acceptable in all cases".

"**Base level construction:** Location of the base level of construction may be made within the spoil pile by standard penetration test or by establishing the top of the undisturbed underlying clay / fireclay. Spoil piles, located in the pits, must be excavated down to the base fireclay bed. This is followed by a controlled fill to "establish a sub-base or by direct placement of the constructed liner on the underclay. In the cases where a sub-base is constructed, non-selective spoil is used to form the sub-base such that the liner can then be constructed at a uniform thickness to conserve fine-grained materials. With this method, the controlled drainage for the leachate collection system is facilitated by the precision grading of the sub-base. *** dynamic compaction is being used in Kentucky to densify the spoil at depth.

"The underclay is in a naturally occurring shale unit directly below the mined coal beds. The underclay serves two functions: (1) as a strong bearing unit for subsequent loading, and (2) as a vertical hydraulic barrier. 'Re-compacted spoil forms the sub-base over the clay base.

"Bearing capacity is not a reported concern at any of the sites. The hydraulic conductivity of the underclay generally ranges between 1×10^{-8} cm/s for unfractured units to 1×10^{-5} cm/s for units with secondary permeabilities due to fracturing (by *in situ* packer tests).

Mine Spoil and Liner Construction: Mine spoil is utilised for daily cover at all of the sites, and for liner construction material at some. Spoil from the site, if it can be used for liner construction, reduces costs and avoids costly borrowing of material from adjacent undisturbed land or from off-site.

"An impermeable liner at one site was constructed of spoil with the following physical properties:

1) 50–70% passing the #200 sieve
2) liquid limit of 36.8%
3) plastic limit of 18%
4) plasticity index of 18.8%

5) standard Proctor density of 112.1 pcf at 14.2% moisture content
6) modified Proctor density of 121.1 pcf at 11.4% moisture content

"Subtitle D requires that a composite liner system be constructed to protect ground water. The liner requirements consist of a composite system using a minimum of two feet of compacted clay with hydraulic conductivity less than or equal to 1×10^{-7} cm/s which is covered by a 30 mil flexible membrane liner (FML), or 60 mil, if an HDPE liner is used. The clay liner system is constructed in 6 to 8-inch lifts, compacted to 95% standard Proctor density at requisite moisture content, and precision graded to facilitate drainage of the leachate collection system directly above. The recommended material properties for the clay are:

Liquid limit greater than or equal to 30%
Plasticity index greater than or equal to 15%
Greater than 50% passing the #200 sieve
Clay fraction greater than or equal to 25%

Liners using mine spoil consistently meet the hydraulic conductivity and strength requirements. Elimination of oversized particles in the upper 6 inches is important so that the integrity of the flexible membrane is not compromised.

Daily cover: Subtitle D requires a daily cover of the working face of at least 6 inches. The spoil heaps provide acceptable and ample daily cover material. Some alternative covers are being used, e.g. 4 mil thick Canvex plastic tarpoleans, over areas 100 feet × 75 feet. Use of tarps result in a 50% reduction of spoil material as daily cover. A spray-on bituminous emulsion is in use as daily cover at another site.

Underground Workings: At a few pit sites, underground coal drifts have been encountered. One at a Kansas site required extensive grout filling and sealing. Another site has used seismic surveys to search for unknown underground workings.

Water Management: Local acid mine drainage is exacerbated by disturbance of the spoil piles. Run-in water that come in contact with the active face, as well as diverted sheet-flow or run-off water that has not made contact, is collected in 25-year, 24-hour event detention basins. The water is tested and treated, if necessary, on-site or at publicly owned treatment works (POTW) to comply with the Clean Water Act. Sedimentation / detention basins that control sheet flow away from the active area can, after testing, often discharge directly into the natural drainage as long as no adverse impact results and according to the facility's NPDES permit.

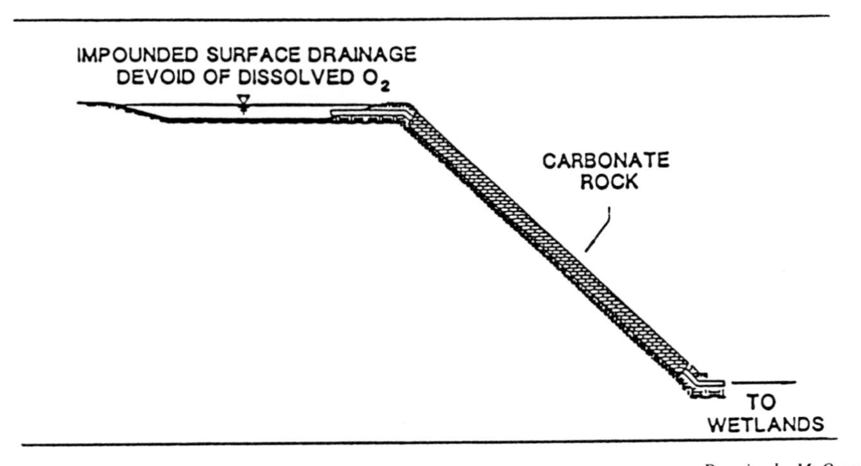

Figure 12 Schematic of anoxic drain system

A southern Indiana site employs an anoxic drain discharging into a constructed wetland. Surface water is diverted to a sump, then pumped to an elevated detention basin. Discharge from the basin is into a "closed filter blanket composed of crushed carbonate rock situated down-gradient toward the headwaters of the constructed wetlands". (See Figure 3) The included iron sulphides and deleterious matter are precipitated in the wetlands, and raise the pH of the water. The wetlands proceed down-gradient around the perimeter of the site in a series of basins with each having improved water quality. The pH values are improved from 3.5 to an exiting pH of 6.5 resulting in a better water quality than the former uncontrolled drainage from the undeveloped site.

Phreatic Water in Spoil: Hydogeologic testing of the Fulton County, Illinois, site found the hydraulic conductivity in disturbed spoil to be much higher than in the undisturbed adjacent land, viz., at 2×10^{-2} cm/s, yet almost impervious at the spoil-underclay contact.

Successful de-watering of ASCML depends on the elevation of the top of the underclay with respect to elevation of the local drainage system. After pumping, retained water in the spoil piles that were previously resting in pit water, can be a problem by continued drainage in the de-watered pit.

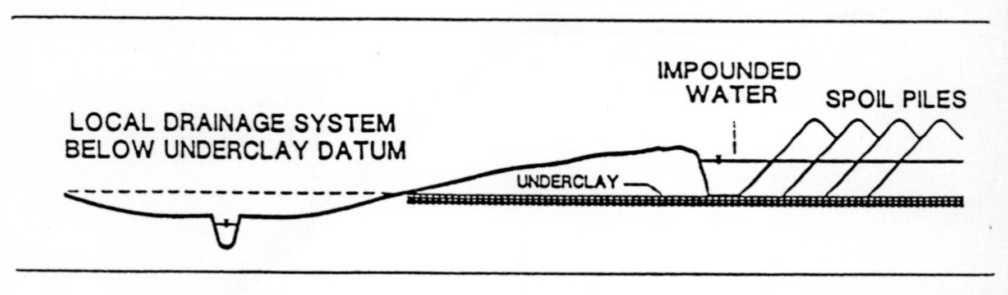

Drawing by M. Owens

Figure 13 Conditions necessary for gravity drainage

Ground Water Protection: Subtitle D requires ground water monitoring up and down gradient of the site, and must monitor the uppermost aquifer. Monitoring wells are completed in the spoil, within the underclay, and to depths below the underclay. Baseline conditions are established by a minimum of four independent samples collected quarterly for one year, for determining contamination. Missouri requires a minimum of four monitoring locations: one hydraulically up-gradient, and three down from the disposal site.

The following determine the number, spacing and location of the monitoring wells at a site: (1) aquifer thickness; (2) hydrogeologic properties of unsaturated and saturated materials; (3) hydrogeologic properties of the unit just below the uppermost aquifer; and (4) ground water flow rate and direction.

Final Cover and Closure: The final capping must be comprised of a low hydraulic layer similar to the clay liner, a 40 mil FML of very low density polyethylene (VLDPE), an optional drainage layer, and one to two feet of top soil.

Post-Closure Plans: Subtitle D, Sub-part F of RCRA requires post-monitoring, maintenance and performance for 30 years for the site with quarter-annually inspection of cover, semi-annual ground water sampling, methane gas monitoring, leachate disposal and monitoring, cover repair, clean out of swales and sedimentation / detention ponds, and record-keeping.

Owens and Elifrits estimate that the clay liner and constructed sub-base can conservatively exceed 5000 cubic yards per acre. Processing of spoil on-site is estimated at US$2 to $3 per cubic yard.

****

10.7.7 *A Money-Saving and Money-Making Solution for a National Environmental Problem*

Environmentalist groups, and principally the Mineral Policy Center (MPC) of Washington, D.C., are clamouring for the U.S. Congress to spend billions of dollars in reclaiming abandoned mined lands. "The cost of cleaning up more than 550,000 abandoned hardrock mine sites in 32 U.S. states could be between $32,000 million and $71,500 million" according to a report prepared by MPC. MPC calls for an Abandoned Mines Reclamation program to be enacted. Utilisation of the abandoned surface mines as MSW depositories is the solution, accomplishing not only the goal of MPC and environmentalists, but relieving the urgent need for waste disposal space. Furthermore, landfilling is a profit-making business. Reclamation of surface mines can be done at a profit rather than a high cost to the public. Additionally, the restored surface land has far greater value to the landowner than a derelict hole in the earth.

If abandoned surface coal pits are not utilised as waste disposal sites, a grand opportunity is being missed to accomplish better environmental goals at greatly reduced costs to the government and the general public.

10.8 The U.S. Public Hearing Process Applied to Landfills

The widespread phenomenon of opposing proposed "bad neighbour" industry projects in a community by the local citizens, particularly waste disposal and treatment facilities and surface mines, is referred to as the NIMBY (Not in My Backyard) syndrome. As noted by Schoenbaum in his 1992 edition of Environmental Policy Law, "The NIMBY syndrome can frustrate siting a facility that is part of a larger federal or state strategy of environmental protection or pollution control. *** Few persons are willing to live in the vicinity of a waste treatment facility, and local residents generally band together to oppose the siting or expansion of a waste disposal site. Often such campaigns are waged in the name of preserving the environment. See, e.g., *Village of Wilsonville v. SCA Services, Inc.*, 426 N.E.2d 824 (1981)". (Schoenbaum, 1992, page 375).

10.8.1 *Nimby Syndrome Examples Defeating a Proposed Waste Site*

The following case is offered in evidence of the negative effect that the NIMBY syndrome can have in defeating the planning of larger federal environmental programmes. In *Village of Wilsonville v. SCA Services, Inc.*, 426 N.E.2d 824 (1981), the villagers brought a common law action in nuisance to require the removal of an approved site for a hazardous waste landfill.

The federal government imposes federal standards under the Resource Conservation and Recovery Act (RCRA) for hazardous waste treatment and disposal facilities, leaving the siting decision to the states. As Schoenbaum (op. cit.) notes, "This presents the important issue of the extent to which local opposition *** should affect the siting decision".

On hearing the case, the trial court was faced with divided expert testimony on the risk of future harm resulting from the planned location in Wilsonville, but granted the injunctive relief preventing the waste disposal site. On appeal, the Supreme Court of Illinois affirmed the granting of the injunction.

In opposition to a lower court's finding that the likelihood of substantial future harm was remote, the Supreme Court stated, "*** we think it is sufficiently clear that it is highly probable that the instant site will constitute a public nuisance if, through either an explosive interaction, migration, subsidence, or the 'bathtub' effect, the highly toxic chemical wastes deposited at the site escape and contaminate the air, water, or ground around the site. That such an event will occur was positively attested to by several expert witnesses. A court does not have to wait for it to happen before it can enjoin such a result". (389 N.E. at 837).

Another, more recent, re-occurring example of*Wilsonville*, employing the NIMBY syndrome tactics to defeat a landfill expansion in the guise of environmental concern, is found in *File et al v. D & L Landfill, Inc.*, 579 N.E.2d 1228 (Ill.App. 5 Dist. 1991), supra in Ch. 6. In *File,* the Concerned Citizens Group was less concerned with potential effects of pollution and groundwater contamination from the MSW site than with decreased values for their properties affected by the presence of the landfill.

Again, Schoenbaum makes the observation that "Local communities have also used local zoning and land-use requirements to exclude unwanted waste facilities". For support, see *County of Cook v. John Sexton Contractors, Inc.*, 389 N.E.2d 553 (1979). Public hearings on local zoning (land-use planning) are frequently misused with concerned citizen groups representing and using arguments for "environmental concern" to prevent "bad neighbour" industrial projects from entering their community, when in reality their "concern" is the lowering of their property values, not for the benefits of the larger common good.

As a further and more recent example, a lengthy litigation in Iowa, ending in 1996 after several years, illustrates the extent to which a county will go to prohibit and prevent reclamation of coal pits as landfills. The basis for the opposition was simple fear based on misinformation. and lack of correct technical information. Other Iowa counties were currently and successfully using and reclaiming mined-out surface coal mines by landfilling with municipal solid wastes.

In *Iowa Coal Mining v. Monroe County*, a coal mining lessee brought a suit against Monroe County, Iowa, after the County had enacted a zoning ordinance prohibiting the use of strip mining pits for landfilling. Two of Iowa Coal Mining Company's pit sites, Star 6 and Star 14, had been approved by the state for the landfill nonconforming use. The coal operator alleged a regulatory taking and for tortious interference by the County with his prospective contractual relationships.

Prior to the latest action and decision of 1996, the same parties had litigated over similar matters in 1993 [see *Iowa Coal Mining Co. v. Monroe County*, 494 N.W.2d 664 (Iowa 1993), referred to as *Iowa Coal I*]. In *Iowa Coal I*, the Iowa Supreme Court decided that as to Star 14, the zoning ordinance (1) was validly enacted as an exercise of the County's police power, and (2) did not effect a regulatory taking of the Star 14 site. As to Star 6, Iowa Coal's taking claim arising from the enactment of the ordinance was dismissed as premature.

Iowa Coal Mining Company originally strip mined three properties in Monroe County. In 1984 it became interested in landfilling them as a profitable method of surface mine reclamation and obtained a sanitary landfilling permit from the Iowa Department of Natural Resources (IDNR) for 10.3 acres of its 120 acre Star 6 site. In May 1988, Iowa Coal received a comparable permit for its Star 14 pit. Star 6 and 14 were located in the County's permissible A-2 agricultural zones.

The Court correctly observed that landfilling for a coal strip mine is relatively easy because both operations require essentially the same equipment except for a trash compactor. However, between 1984 and 1987, Iowa Coal expended much time and money preparing Star 6 and Star 14 for proper and safe, non-contaminating landfilling procedures, installing ground water monitoring wells, and negotiating contracts with waste disposal haulers.

Monroe County officials made it clear that they were opposed to combined strip mining and landfilling. On the day before Iowa Coal received its landfilling permit for Star 14, the County enacted Ordinance 6. Ordinance 6 was aimed at curtailing and ultimately extinguishing all nonconforming uses on A-1 and A-2 agricultural lands in the County. In a revised draft of Ordinance 6, landfilling was permitted, but not coal mining, as a conditional use on I-2 (heavy industrial) land. These changes meant that Iowa Coal could continue strip mining but could not combine the operation with a landfill. In the fall of 1988, the County denied Iowa Coal's application to rezone its property.

In the words of the Court, "The harder Iowa Coal tried to keep its business afloat, the harder the County tried to sink it". Continuing in the words of the court, "For example, the County had historically granted Iowa Coal permission to mine through a County road. On the pretext that such a practice would damage county roads, the County rescinded the permission in February 1988. In the fall of 1988, the County denied Iowa Coal's application to rezone its property. In addition — largely because of the County's actions — Iowa Coal lost landfilling business from sources inside and outside of Iowa". As a result of Ordinance 6, Iowa Coal brought the second action against the County claiming the ordinance deprived them of the only legitimate use of the property without paying just compensation. The district court ruled that the ordinance violated a state statute because it failed to develop an independent planning document before enacting Ordinance 6. The court awarded damages of $10,391,500 to Iowa Coal and $5,047,000 to the Company's owner for lost leased royalties.

On appeal in *Iowa I*, the Supreme Court held that the district court erred in its finding that an independent planning document was required before enacting a zoning ordinance; and, in invalidating Ordinance 6 on that ground. The Supreme Court reversed the takings awards and found that its sites were not ripe for a takings claim because Iowa Coal had not exhausted its administrative remedies with the County.

In May 1993, Iowa Coal filed the second suit, again alleging a regulatory taking claim for its Star 6 and Star 15 sites, believing the takings claim was now ripe for adjudication. Iowa Coal Mining also claimed damages for the County's tortious interference with Iowa Coal's contracts with waste disposal contractors.

In the second case, *Iowa Coal II*, (*Iowa Coal Mining Co., Inc. v. Monroe County, Iowa,* 555 N.W.2d 418 (Iowa 1996), much the same issues were brought in the new action by Iowa Coal Mining after the County passed a new but almost identical zoning ordinance, Ordinance 7, which allowed landfilling, but not in strip coal mines. The Court noted, "The long running feud *** took a heavy financial toll on the company. Between 1987 and 1990, Iowa Coal laid off most of its employees and substantially curtailed its operations".

In *Iowa Coal II*, the trial court entered judgement and damages for the coal company on the takings claim for the Star 6 site in the amount of $3,045,000 and $1,750,000. For the contractual tortious interference claim, Iowa Coal was awarded $850,000. The Star 15 site was granted no relief. The County appealed raising numerous issues of error.

On appeal, the Supreme Court held that:

(1) The County had no standing to challenge Iowa Coal's alleged noncompliance with its coal leases permitting it mine and landfill.

Because the County argued that Iowa Coal's leases had expired, Iowa Coal had no cause of action, no a capacity to sue. Iowa Coal's leases called for termination 90 days after cessation of coal mining and landfilling must also cease. However, Iowa Coal's lessor had not called for termination of the leases. Therefore, the County's challenge was rejected as the County lacked standing to challenge Iowa Coal's noncompliance with any of its lease terms because it was not a party to the leases.

(2) Iowa Coal held a valid state permit to accept certain wastes and was using the mine as a nonconforming use to continue. Iowa Coal failed to take advantage of this section of the ordinance. Exhaustion of remedies is necessary in a takings — without just compensation claim as no violation occurs until state proceedings have denied compensation or that the remedy is inadequate.

(3) Unless excluded by statute, every municipality is subject to liability for its torts. Here, the County was not excluded for interfering with a contractual relationship. The County's acts caused Iowa Coal's contract with Metro Waste depositing 125,000 tons of incinerated sewage sludge ash to fall through. Iowa Coal's action for tortious interference with a prospective contract was upheld.

The Court reversed Iowa Coal's takings claim as still not being ripe for adjudication and vacated the damage awards.

As previously argued in Ch. 3, ground water contamination is commonly conjectured and alleged as a future possibility in zoning and mine permitting hearings for "bad neighbour" applicants, even made in the post-environmental stringent regulation period. The allegations of potential water contamination are commonly made without basis, as well-illustrated in *Florida Rock Industries v. U.S.*, supra, where the federal government argued that quarrying in the local limestone would pose a risk of contaminating the sole aquifer supplying drinking water for the city of Miami, and Dade County, Florida. The government did not actually contend that limestone mining would contaminate the aquifer. The government's argument simply suggested groundwater pollution from Florida Rock's future quarrying. This speculative argument failed in view of the Court's noting of fact that none of the presently operating quarries in the immediate area in the same rock formation had polluted, nor were they presently polluting, the aquifer in question.

The power of intimation and suggestion that detrimental hydrological contamination "may result" from a newly planned surface quarry if approved, as used by the federal government in *Florida Rock*, failed under the scrutiny of the examining court. Unfortunately, this same "scare" tactic without foundation in fact has been too often successfully used in defeating quarry and landfill permitting at many public inquiry hearings. A weakness of the Public Inquiry Process (PIP) is that the hearings are not subject to a requirement to use the same factual legal scrutiny as the courts. The general public may enjoy defeating a permit for personal reasons on the basis of a whim, or at least, on unfounded and non-factual information, while the courts deal in fact and truth.

10.8.2 *Another Example of a Landfill Siting Defeat by NIMBY and Zoning Squabbling*

The following article was written in 1992 as a review of the frustrating, lengthy delay and costly, extended litigation that can be incurred by the public hearing and zoning process for siting of a MSW landfill. St. Peters, a city with a population of near 50,000, is located in St. Charles County, Missouri, with a population of approximately 225,000 people. Both city and county are in the St. Louis suburban area.

St. Peters, foreseeing a need for continued MSW landfill space in 1989, acquired an abandoned limestone quarry site of 29-acres two miles outside its city limits in an unzoned area of St. Charles County. It obtained the site from its landowners by threatened use of the power of eminent domain (condemnation) for the public good. In the ensuing litigation over zoning requirements with St. Charles County for landfill permitting, a local homeowners' association, Heatherbrook, joined in the jurisdictional squabble between St. Peters, St. Charles County and the state agency (DNR) responsible for waste disposal. St. Charles opposed St. Peters' claimed power of condemnation to obtain the landfill site. The homeowners' association opposed the landfill proposal for fear of depreciated land values for the nearby properties, a response typical of the NIMBY syndrome. The litigation has been the subject of several court trials starting in 1989 and has been heard by state Courts of Appeal three times, the last being in mid-1994.

Although the zoning and jurisdictional issues of *St. Peters* are pertinent to landfill siting, of greater interest and included are the state agency's environmental conditions imposed for a landfill.

A SUMMARY OF THE CURRENT LITIGATION FOR THE CITY OF ST. PETER'S PROPOSED QUARRY LANDFILL SITE LOCATED IN ST. CHARLES COUNTY, MISSOURI

by R. Lee Aston, J.D., LL.M., (1992) (unpublished)
Geological Engineering Department, University of Missouri-Rolla

Preface:
Siting approval for solid waste sanitary landfills, whether utilising an abandoned quarry, or not, frequently undergo the same objections by the public as do newly proposed aggregate quarrying sites. Proposed landfills and proposed quarrying operations are both victims of the NIMBY syndrome (Not in My Backyard) which result in highly emotionally-charged, antagonistic, arduous public and administrative hearings, only to be followed by long, drawn-out litigation. Such is the case for permitting of a mined-out crushed stone quarry as a solid waste landfill site currently caught up in litigation in the state courts of Missouri.

Background facts: In May 1989, the City of St. Peters, Missouri., (hereafter the City) acquired a mined-out, 29-acre, limestone quarry about two miles outside its city limits for the purpose of converting it to a municipal solid waste landfill and recycling facility. The intended site lies within an unincorporated part of St. Charles County. Eighteen months later, the City submitted to the Missouri Department of Natural Resources (DNR) its permit application pursuant to Sec. 260.205.2(3), RSMo. Included were copies of rezoning approval, a conditional use permit, and a refuse disposal area license. The City also submitted its argument to DNR that it was not legally required to have obtained those permits and authorisations from St. Charles County to develop and operate a solid waste disposal or processing facility.

Having conducted a preliminary review of the City's application, the Waste Management Program (WMP) of DNR [as of July 1, 1991, the Solid Waste Management Program] returned the City's application as "incomplete". WMP found under its Policy #405 that the City had failed to obtain a land-use permit and sanitary landfill operating permit from St. Charles County. The City appealed the decision in thirty days resulting in a suit against DNR. [*City of St. Peters v. Dept. of Natural Resources,* 797 S.W. 2d 514 (Mo. App., W.Dist., 1990)]. The Court ordered a rehearing of the matter for the City.

However, on rehearing, WMP found under Policy #405, the applicant may submit, as an alternative to the actually required permits, licenses and approvals, letters from local governments or agencies stating that unconditional approval for the required permit, license or approval will be issued on issuance of a DNR permit. In finality, the WMP accepted the City's argument that it was exempted from the sanitary landfill operating permit requirement. WMP was to conduct its technical review of the City's application and was ultimately satisfied that City had complied with all relevant local zoning regulations necessary to permit and operate a solid waste facility. A Joint Stipulation (J.S.) Order was signed by the hearing officer which allowed the technical review of the permitting process to proceed.

Recent litigation: On 21 August 1991, a local citizens' group, Heatherbrook Homeowners Association (Heatherbrook) et al, aggrieved by the decision of DNR to proceed with the permitting, filed a Petition for Review with the Circuit Court of Cole County, Jefferson City, Missouri, naming the Missouri DNR and the City of St. Peters as defendants in the suit. Heatherbrook sought an injunction to prevent DNR from issuing a permit to City to prevent "irreparable harm" to the site's nearby homeowners. Heatherbrook argued that there were two other law suits currently pending in the Circuit Court of St. Charles County relative to the issuance of the permit, viz., (1) *Heatherbrook v. Mo. DNR and City of St. Peters*; CV191 Cir. Ct. Cole Co. (1991), and, (2) *St. Charles v. St. Peters & Mo. DNR.,* No.CV190-7257CC, Heatherbrook contended that until the suits were settled further processing of the permit should be enjoined, and that:

(1) the Joint Stipulation Order of WMP and the City misrepresented the zoning controversy to the hearing officer who approved the permitting procedure for the City.
(2) the Joint Stipulation Order "was unsupported by competent and substantial evidence;
(3) the JS Order was not legally authorised;
(4) the JS Order was procedurally unlawful."

Heatherbrook also took issue on the City's claim that it was not legally required to obtain those permits and authorisations from St. Charles County to develop and operate a solid waste disposal or processing facility; and, that the City's position was based solely on a legal opinion by an attorney for the City; and that the opinion was an incorrect statement of Missouri law.

Hence, Heatherbrook sought:

(1) review of the Joint Stipulation Order by the Circuit Court of Cole County;
(2) that under the DNR rules, the City is subject to Ch. 260 and 10 CRS 80-2.020(2)(E);
(3) that the City is **not** exempt from St. Charles County's zoning requirements for location of a solid waste landfill in an unincorporated area; and
(4) the Joint Stipulated Order for permitting be set aside.

On the same date, Heatherbrook filed (August 21), the Circuit Court of Cole County ordered DNR to send up the record for review and commanded DNR to take no further action in the permitting process until ordered by the court.

Prior to the August 21 order, DNR had scheduled a public hearing for August 29 to be held in St. Charles County for the purpose of investigating the application of St. Peters for the solid waste disposal area operating permit. The Court's order negated the scheduled DNR public hearing on August 29.

In response to the Cole County Court's injunctive order, DNR filed its pleading that:

(1) modification, or setting aside, of the injunctive order should be made so that the public hearing could be had on 29 August as scheduled;
(2) the homeowners would not be prejudiced by the scheduled hearing;

(3) they would, in fact, have opportunity to express their views at the hearing.
(4) Heatherbrook's petition for injunctive relief had been heard *ex parte* (i.e., one sided);
(5) DNR had not been notified of the injunctive relief petition by Heatherbrook, or given an opportunity to oppose the relief granted to Heatherbrook (i.e., *bi-parte*);
(6) no investigation by the court was made as to the claim of "irreparable harm"; and,
(7) the court did not have jurisdiction in the matter since the decision of the DNR Director concerning the City's application for permit was "contested" and could be reviewed by the administrative hearing officer, i.e., not final.
(8) Heatherbrook did not have standing to file the action against DNR.

Determinative points of law concerning DNR's response:

(a) concerning DNR's arguments 4, 5, 6, above: only temporary injunctive relief may be granted *ex parte* where the "irreparable harm" is allegedly occurring, or is so imminent that there is not time to have a *bi-parte* hearing for granting of the injunction to prevent the occurring injury from doing more damage, or an "imminent harm" from occurring and causing injury to the applicant or his property. Obviously, here, the danger of harm from the investigatory permitting process by DNR was only speculatively *in futura* by Heatherbrook, certainly not imminent. There were to be more opportunities for Heatherbrook to join in any public opposition to the permitting process of the City's planned solid waste landfill. DNR claimed it had been denied "due process of law" by not being allowed to participate in the *ex parte* hearing.

(b) concerning DNR's argument (7), above: jurisdiction of the filed action by Heatherbrook against DNR to prevent issue of the permit to St. Peters was premature. Under administrative law procedures, until administrative remedies have been exhausted, final appeal for review of an agency decision cannot be made to a Circuit Court (Sections 260.235 and 536.150, RS Mo.).

(c) concerning DNR's argument (8), above: DNR argued that since Heatherbrook had not been a party to the contested case and had never attempted to intervene in the agency action, they lacked standing to file the action. DNR's argument against "standing" for Heatherbrook is subject to debate.

Heatherbrook was definitely an interested party in obtaining injunctive relief by virtue of its proximity to the proposed landfill and potential direct "irreparable harm" resulting from possible future granting of a permit by DNR. The fact that Heatherbrook's timing of filing against DNR was premature, holds greater weight. It is true that as a non-party to a formerly contested action between DNR and St. Peters, they lacked standing.

Total-action ban on permit's review process lifted:

The Cole County Circuit Court recognised the weight and strength of DNR's arguments, and for "good cause" modified its Order of August 21 completely halting DNR's investigation of St. Peters' permit application. A new Court Order of August 26 allowed DNR to hold the public hearing on 29 August as scheduled, and to complete its technical review of St. Peters' application, with the *proviso* that no affirmative permit decision be made by DNR until further order of the court.

Motions for Dismissal of Heatherbrook's complaints:

On 11 September 1991, DNR and City of St. Peters filed for dismissal of: (A) Heatherbrook's complaint; and (B) St. Peters moved for dismissal of the action of St. Charles County against St. Peters.

A. DNR's Dismissal Arguments for Heatherbrook's Suit:

1) **Lack of subject matter jurisdiction:**

As stated above, under determinative points of law, the Circuit Court of Cole County lacked subject matter jurisdiction to hear Heatherbrook's complaint by virtue of the fact that the owners' association failed to exhaust the administrative remedies for appeal under the Missouri code (see Sect. 260.235 RSMo.). Although Heatherbrook attempted to bring their action under Sect.536.150, that section expressly prohibited the action against a "contested" case, providing only for relief in "uncontested" cases. The issue between DNR and St. Peters over permitting was clearly a "contested" action in view of the former suit between those parties, *City of St. Peters v. Dept. of Natural Resources,* 797 S.W. 2d 514 (Mo. App. 1990) referred to above in Background Facts.

2) **Heatherbrook's Lack of Standing to file an Action:**

Heatherbrook lacked both standing and a vested right to seek preventive action of DNR for an administrative review of an application for a permit. Their attack on DNA's procedures and any subsequent findings were prematurely filed. Heatherbrook had failed to intervene as an interested party in the former contest between DNR and St. Peters; further, until after DNR makes its technical review, and only in the event it arrives at an affirmative decision awarding a permit to St. Peters, Heatherbrook has no vested interest to protect, and no right to file an appeal for review. Even then, Heatherbrook would have to file an administrative appeal to the Missouri Administrative Hearing Commission, not one for judicial review by a court. The issue was not "ripe" for judicial review.

3) **Lack of Demonstrated Irreparable Harm or Injury:**

Heatherbrook failed to show how they would be aggrieved, or "irreparably harmed", by DNR's decision to conduct a technical review of St. Peters' permit application. The alleged harm can only be speculative and in the future, in the event that St. Peters actually started a landfill operation. Heatherbrook's alleged fears of harm may still be precluded by the possibility that St. Peters' permit application may be denied by DNR. In the event the permit application is approved by DNR, Heatherbrook's remedies are not exhausted before a landfill operation could be started.

4) **Landfill Permit Application Review or Approval is Not Final:**

DNR's position was that the Joint Stipulation Order issued by the hearing officer was only interlocutory in nature, i.e., it was not final or ripe for purposes of review or contesting.

B. **St Peters' Dismissal Arguments for Heatherbrook's Suit:**

The City's arguments addressed the same issues posed by Heatherbrook as did the Missouri DNR's. St. Peters did cite one additional fact; that, although they claim they are not subject to the zoning of St. Charles County requiring a permit for the landfill site, they did acquire a conditional use permit from the County for operation of the landfill site.

On the same date, 11 September, St. Peters also filed its Motion for Dismissal of their suit with St. Charles County. This was one of the two pending cases Heatherbrook argued must be settled before DNR could proceed with the permitting investigation.

On 24 September 1991, the Circuit Court of Cole County suspended its Order of 21 August with reference to DNR's certification of the records for the *Heatherbrook* suit against St. Peters and DNR. Decision on St. Peters' application to use the limestone quarry as a sanitary disposal site awaits the outcome of the pending cases in St. Charles County.

Environmental Demands for a Conditional Use Permit for the Landfill

In St. Peters' case, the application was for 29 acres, located largely in a mined-out limestone quarry site adjacent to the Missouri River. St. Peters declared that the amount of material, of household refuse type, to be dumped daily was between 100 and 500 tons. After the collected materials had been processed through recycling, the residue was to be placed in the landfill. Recycling was to be done at the landfill site. Equipment to be acquired by St. Peters was to be loader, compactor, recycling, trucks, including water truck, and scales.

In granting the conditional use permit for the sanitary solid waste facility, thirty environmentally-oriented conditions were imposed on St. Peters by St. Charles County. The conditions are becoming fairly typical for waste landfills in or near large cities. A summary of the conditions imposed follows. Thirteen (13) were to be met before the solid waste landfill operation started. Seventeen (17) were required during operation.

1) Compliance with federal EPA and Missouri DNR requirements for installation and operation of a sanitary landfill;
2) Obtain all federal EPA and MDNR licenses and permits required;
3) No ingress or egress during operation until specified planned adjacent highway improvements had been completed;
4) Unless infeasible, construct the materials recovery facility (MRF) at an off-site location before accepting any solid wastes into the landfill;
5) Submit and obtain approval from St. Charles Co. Planning Dept. and the County Highway Dept.for an industrial site plan, including:

   (a) designated entrances / exits with any changes to existing highways to be paid for by St. Peters;
   (b) proof that all entrances / exits to property are at or above the 100-year-old flood plain level;
   (c) storm water controls for the property;
   (d) planning to prevent refuse trucks from depositing mud on public streets, by pavement of roads, or a wheelwash;
   (e) flood protection construction of levee to a 500-year frequency in compliance with Federal Management Agency and U.S. Army Corps of Engineers;
   (f) control of blown litter and trespassers by construction of an 8-foot high chain link fence, with lockable gates around entire property;
   (g) aesthetic controls by construction of a landscaped barrier visually screening site from adjacent road;
   (h) maintain a 200-foot buffer zone, with additional landscaping, between the landfill-pit and all adjacent residentially-zoned land;
   (i) indicate lighting; no lighting to reflect on to adjoining land.

6) obtain a land-use permit for planned improvements;
7) complete all planned and approved improvements;
8) extend and install water lines to the site; ensure other adequate fire protection services; place adequately spaced fire hydrants on the property;
9) construction of an enclosed system for leachate collection, which may include a treatment plant; spraying of leachate on the property is prohibited;
10) (a) Water quality monitoring: quarterly basis of water sampling from existing well(s) within 1,320 feet of property; or alternatively, place one monitoring well within adequate distance. St. Peters has option of providing an alternate source of water;
    (b) posting of routing signs for refuse haulers to access property;
11) Equipment requirements for daily operation, and backup machinery, required by DNR;
12) Dust control with on-site water truck;
13) Liner: If required by DNR, artificial; minimum thickness of 60 mm; spread at least 15-feet high along sides of sanitary landfill; if required by DNR, remaining sides to require a clay liner;

Operational Conditions:

14) obtain, maintain, construct, install and utilise all items for conditions 1–13;
15) inspection permission for all government entities between 9 a.m. and 4 p.m., Monday through Saturday;
16) receiving of waste limited to 6:30 a.m. to 6:30 p.m., Monday-Saturday; closed on Sunday;
17) no acceptance of yard waste (vegetation); any composting to be done off-site;
18) no transfer station to be on property;
19) equal access for dumping to all St. Charles County residents and refuse haulers, and same tipping fees shall apply to all, provided that all haulers meet the same stringent criteria as applied to waste from City of St. Peters;
20) solid waste only accepted by designated entrances; only from licensed haulers, except single axle trucks; no barge waste from Missouri River;
21) daily cover required for working face for bird and odour control;
22) daily litter control including approaches where abutting site;
23) regular basis testing for all monitoring wells, incl.# 10, as per DNR regulations; copies of results to St. Charles County; in event activities on site cause contamination of active wells within 2,640 feet of site, St. Peters will bear 100% of all costs of providing water to properties with contaminated wells;
24) reimburse St. Charles County for any reasonable costs incurred in monitoring landfill operations;

Conditions for Closure of Landfill Site:

25) no construction of completely enclosed structures on fill area;
26) ensure complete closure in compliance with all governmental regulations in effect;

Automatic Expiration of Conditional Use Permit upon occurrence of any of following:

27) a final and unappealable denial, revocation of forfeiture of permitting by federal E.P.A;

28) upon expiration of any state or federal permit for landfill or recycling operation; responsibility is on operator to obtain renewal of this conditional use permit before its expiration;
29) if City of St. Peters, or anyone on its behalf:
    (a) applies to any federal, state or local government for disposal of any hazardous or infectious waste at the property;
    (b) expands any uses on the property, including incineration, unless approved under a separate conditional use permit application;
    (c) relinquishes any portion of ownership, management or policy-making for any portion of the property, or any of its facilities to anyone other than city-employees;
    (d) violates any of the conditions of this conditional use permit, as determined by a final and unappealable decision;
30) If the County Commission,
    (a) determines that:
        (i) either the County Health Department or the Planning & Zoning Department notifies St. Peters of a violation of state statutes, or local health, or zoning order, or conditional use permit, and
        (ii) by written notice to City of St. Peters, and
        (iii) City of St. Peters has not rectified or corrected the violation within a reasonable time specified in notice; and
        (iv) said violations are a threat to health, safety and welfare of county residents; and
    (b) may enter an order revoking the conditional use permit when it has become final and unappealable.

St. Peters Case Update

During the prolonged litigation over the landfill siting, the City of St. Peters argued its right to the power of condemnation of the quarry property located two miles outside its city limits by relying on a Missouri statute which authorised 3d and 4th class cities to acquire property up to five miles beyond city limits (by condemnation or other ways) for incinerators, purification plants and sewage disposal plants. St. Peters had relied on the equitable maxim / rule of *ejusdem generis* (of the same kind, class, or nature), that is, that although a landfill site was not specifically named or enumerated in the statute, it was of the same nature as those named, viz., dealing with waste treatment.

After five years of litigation, 1989–1994, the last decision by the appeals court was that a 4th class city, such as St. Peters, does not have the authority granted to it by the state for the power of condemnation of property for a waste disposal site without its city limits. Therefore, the abandoned aggregate quarry (two miles outside its city limits) it had obtained for the landfill site, whether by condemnation or by threatened condemnation, could not be licenced for that purpose.

The court rejected St. Peters legal theory of *ejusdem generis.* The court countered with the equitable maxim / rule, *expressio unius est exclusio alterius* (the expression of one thing is the exclusion of another). Since the state had expressly named the purposes for which cities may acquire property up to five miles beyond their city limits, no other purposes, i.e., landfill sites, were not authorised.

Millions of dollars have been spent on the public hearings and litigation, ending in a defeat for the landfill permission. Although the landfill site had been defeated by a jurisdictional issue and turned on a hairsplitting point of law, the landowners and NIMBY had been victorious. In the interim years, St. Peters has disposed of its waste by contracting its haulage to the nearby combined quarry and landfill of Fred Weber Inc. at Maryland Heights, Missouri. (See Ch. 11 §8.2 — United States — An Example of a Successful Combined Quarry-Landfill Operation).

10.9 Conclusions and Comments

Treatment of resultant contamination of ground and ground waters from the misuse of former mining workings by the deposition of refuse without the necessary preparation to make the underlying ground safe from refuse deposition is not claimed to be the sole work and concern of the British people, or that they have singularly dealt with the problem. To be sure, examples may be found in other nations, but the suggestion is made that the British may be in the lead in dealing with such remedial measures. The rationality for Britain's earlier and successful dealing with refuse-filled derelict mining lands is suggested as being motivated by its population density along its racial aggressivness. Admittedly, there are several Asian nations with greater population densities. Thus, land being at a higher premium than in less populous countries, the practice of land conservation and reclamation has consequently a higher priority.

Thus, the obervation is made that it has been successfully demonstrated and accomplished in the U.K. that solid wastes in surface mine workings can be safely deposited even in the midst of a metropolitan area without injury to the environment or to the public and human well-being. The Bowmans Harbour, Poynter Street, Swan Farm and Newdale Hall-Telford examples given in populated areas illustrate that, even where the former mine site had been filled and contaminated by untreated and uncontained refuse and waste dumping, the mined land can be re-worked without harm to the local citizens, and the land conserved for the benefit of more public use.

It appears a trend is well-started in the U.K. for utilisation of surface mined lands to accommodate society's wastes and refuse. Further illustrations are entered in evidence in the subsequent chapter.

In Canada, in siting surface mines as landfills there appears to be a varied, an almost indifferent attitude, certainly not of one vital concern for waste disposal space urgency, whether utilisation of the surface mines is made or not. Blakeman, in his 1977 Land Use Studies to the Lands Directorate, Environment Canada, suggested a subsequent possible use for derelict quarries might be as landfills.

In 1993, Natural Resources Canada reported the number of major crushed stone and sand and gravel operations to be 250, nationwide. Based on the premise and correlation demonstrated between 'housing starts' and population centres, and the economic requirement of proximity of aggregate pits to the population centres and markets, the major Canadian producers are certain to be well-located as potential landfill sites for those population centres.

The inquiring survey in Missouri into the incipient use of abandoned surface coal mines as MSW landfills is indicative of the urgent need for waste sites in the present time and a critical need to solve the problem. The economics of waste disposal is forcing the investigation of unused mining voids as a lower cost way to solve the universal problem of "where to put the waste".

The quickest relief for easing the critical need for more landfill space would be by incinerating all MSW. As noted supra, incineration reduces the waste volume by 90%, and 65% by weight, which is a tremendous savings for landfill volumetric space. The 10% volume residue (35% by weight) would still require depositing in a landfill. This argument does not detract from the proposed recycling-use of surface mines as landfills. It would extend the lives of all landfills utilising surface mines, abandoned, currently mined, and future mined, as depositories and give society a surplus of future landfill sites to be considered. The urgency for landfill space would be eliminated in the future.

Furthermore, incineration has the very desirable attribute of producing energy. As noted by Hughes, supra, the energy content of landfill waste in the U.K. has been estimated to be worth £1 million per day (Hughes, 1992, p. 247).

Currently, a pertinent news item of interest in the December 1995 issue of New Civil Engineer, a U.K. publication, is summarised as follows:

MAKE LANDFILL LAST RESORT, SAYS
U.K. ENVIRONMENT SECRETARY

> "New targets for reducing the amount of rubbish sent to UK landfills and increasing the volume of waste that is recycled were set this week by the government. By 2005, the Department of the Environment wants to cut the amount of waste going to landfill sites by 14%, and wants to recovering value from 40% of municipal rubbish. ***
>
> "Today some 70% of the controlled waste in England and Wales is sent to landfill sites, but the government wants this cut to 60% over the next 10 years. Environment Secretary John Gummer intends *** to push on the process of cutting landfill use. A white paper was published which sets out the 'waste hierarchy', i.e., reduction being the first priority, followed by reuse. Only when neither of these are possible should recycling, composting or incineration be considered, and landfill should be the last resort.
>
> "Gummer said that local authorities needed to recognise the importance of incineration as an option, **but stressed that there was no place for 'nimbyism'. He urged councils to take planning decisions themselves on such projects rather than passing the buck to the**

> **DoE.** 'The local authority has a duty to say that an area needs an incineration facility where this is true, and they should not let someone else take the strain', he said". (emphasis added).

The point of greatest interest to this work is the Environment Secretary's advisory to avoid "nimbyism" by taking the decisions of planning and permission into their own "hands". Such a course comports with the recommended reduction of public hearing involvement and employing the command and control type of decisions for the public welfare. Permission for combined mining-landfill projects should be left to the discretion and decisions of regional / state or provincial boards in collaboration with environmental agencies. "Historically, one of the key features of our (the British) planning system has been the relative freedom given to decision makers to approve, refuse, defer or generally meddle with applications put before them". is an observation made by Oxford planning solicitor Scharf in Planning and Law. "The recent history of the planning system shows an evolution from an 'application-led' system in the 1970s, through an 'appeal-led' system in the 1980s into the 'plan-led' system of the 1990s". (June 1994, p. iv).

CHAPTER 11

LANDFILL TECHNOLOGY AND OPEN PIT FEASIBILITY

A Study of Pit-Liners, Current Lining Practice in Landfills — and New Techniques — To Make Worked-Out and Working Surface Mines Usable as Landfills and Prevent Water Pollution

11.1 Introduction

The main argument of this thesis is that reclamation regulation for mined-out, surface mines, or pits, should be re-examined and replaced by new legislation. Under current regulation in all three Anglo nations considered herein, reclamation of worked-out surface mines is not extensive enough to obtain the maximum re-use of the mining voids that were created. Current reclamation of non-coal surface pits is minimal. Regulation fails to utilise a residual natural resource that may be put to the far greater benefit for society. In a time of recycling used materials and resources, open-pit voids should likewise be recycled to extract maximum use. Current mine reclamation regulation is highly deficient and wasteful when critical national urgencies exist for waste depository and landfill sites. Complete restoration of mining voids by landfilling waste should be mandated by new environmental legislation mandating backfilling of mined land with refuse, solid wastes, or wastes that can be safely stored in the mined voids of: (1) abandoned mined lands (AML); (2) currently mined pits; and (3) future planned pits, and that they be environmentally assessed for simultaneous permission to mine and accept MSW, and / or other feasible wastes.

Current mode of reclamation for surface mines in the U.S. is to slope off the top bench on a one-to-three slope, allow the hole to fill with water, or to allow it to seek its natural level (which may, or may not, be some distance below the adjoining surface elevation), fence in the hole, landscape, and / or contour the surrounding surface, removal of all mine buildings and debris, and finally sow the surface with grasses and seedling trees. The water-filled hole may, or may not, become a recreational spot depending on size and the ownership, or simply may become a wildlife habitat watering hole if not too heavily frequented by man, or an aesthetic centrepiece to a new residential subdivision. As an example of the latter, a March 1995 "wanted" ad placed in *Pit & Quarry* magazine sought an abandoned quarry filled with water near a populated area in the USA or Canada, presumably to be the centrepiece of a planned residential area. Even more specifically, reference was made by the court in the *Florida Rock* case (see Ch. 5 §4.5.1) to the desirable landscaping and use of worked-out mines for residential areas and golf courses.

The premise of this thesis is that the majority of all non-fuel surface mines are naturally environmentally safer (or can be made safer) for the deposition of wastes than either (1) holes made in the earth's permeable overburden / dirt surface for landfills, or (2) landfills constructed on top of the earth's surface by mounding. In either case, permeable dirt is poor material in which to place or to use to cover wastes in the earth's surface. The use of impermeable clays, which, incidentally, is a surface-mined mineral, are commonly employed to prevent leaking of the excavation made in the dirt. The basic hypothesis is that a landfill made in a rock-lined pit has a better chance of containment of leachate from the infilled wastes than from a landfill in permeable dirt.

Thus, in advocating mandated use of surface mines for landfills by law, a review of the primary considerations for landfill follows, after which, examples are given as proof of the successful operation of surface mines as landfills in which containment measures have been taken to prevent pollution of ground and surface waters.

11.2 Problems of Solid Waste Landfilling

The primary problem that accompanies landfilling on any site is the danger of leachate escaping from the landfill and contaminating ground waters, and even surface waters. A related, secondary, but prominent problem is the gas generated from landfills, mainly in the form of methane and carbon dioxide, and in some instances, hydrogen.

Another current problem that confronts the consideration of surface mines as landfill sites are the firmly implanted environmental fears by landfills for water contamination that pervades the public mind of today. These fears are based on water contamination in recent years from landfills in general. Figures, true in 1991 as a result of investigation at that time, formerly served as legitimate basis for some of those fears.

In 1991, only 33% of the municipal landfills in the U.S. were provided with ground water monitoring; only 15% of them had liners; and, only 5% had leachate collecting systems. Furthermore, the report states that 35% of the sanitary landfills were in counties having geological faults; 14% were located on flood plains, and 6% were situated in wetlands. Obviously, those statistics must have applied to landfills that preceded RCRA's more stringent current monitoring and location controls for landfills, and were from a period when attenuated landfills were in abundance as opposed to current containment-type landfills. The fears of using surface mines and pits were, perhaps in some cases, justified, particularly if reliance on attenuation was the sole hope of prevention of ground water pollution. The degree of undeserved fear to which the use of surface mines had fallen is illustrated by a California law passed in 1988–89 "prohibiting the establishment of landfills on sites previously developed as aggregate mines". (Rock Products, Dec. 1995, p. 59).

It is readily seen that the public image of landfills being the cause of water contamination was partially justified. The public image of landfills lags behind landfill technology to the current detriment of landfill siting. Information for the public must be corrected and their knowledge made current.

11.2.1 *Highlights of Hydrology and Prevention of Ground Water Pollution in Landfills*

Elementary is the statement that when refuse or MSW is buried in a landfill, any contact with water produces a mineralised liquid called a leachate. The leachate, if not successfully and permanently totally contained within the landfill, may eventually seep into the ground water carrying with it the dissolved contaminants and cause water pollution problems.

The hydrogeology of a surface mine site should be known before a landfill within it can be properly designed. The controls of the movement of water through the area of the mining void could potentially affect the landfilled refuse if not properly contained within the site. Should there be a leak or seepage of the leachate from the contained waste within the landfill, the hydrogeology controls the migration and attenuation of the dissolved solids of the leachate as they are carried by the ground water. Three hydrogeologic factors considered in the design of landfills should be: (1) the position of the landfill site within the ground water flow system; (2) the position of the top of the zone of saturation, or water table, with respect to the deposited refuse; and (3) the texture and composition of the surrounding earth materials, which would affect their ability to transport and attenuate dissolved solids in any escaped leachate. Dissolved solids include all organic and inorganic components dissolved in the ground water. Whenever possible, it is desirable for the landfill base to be above the seasonally high GW table. At sites with a shallow ground water table, the landfill base may have to be constructed below the GW table. In that situation, total containment will be required. Consequently, because of the relative level of the GW table to the landfill's constructed base, either below or above, containment landfill construction will be either of two types, viz., total or partial containment.

The best solution to avoiding the potential leakage problem of leachate might be to prevent its production. The decision for making a landfill totally moisture-proof, i.e., totally contained, is guided principally by regulation requirements and economics of the site, except in an arid climate. In some areas where ground waters are not a great problem, total containment may not be an economically practical solution, nor may it be an entirely desirable one (see infra, for accelerated bio-degradation). Therefore, the next best solution is to minimise the amount of leachate produced.

The containment-type of landfill is designed to restrict leachate from seeping into the ground water or an aquifer, to minimise its degradation.

Containment-type landfills may be either partial or total, depending on their relationship to the GW table. If total containment is necessary, more than one liner will be required. A geomembrane laid over a clay liner offers a greater degree of permeability and assurance of ground water protection.

Maintenance of an optimal amount of moisture in a fill is necessary for the anaerobic process of biodegradation of the fill's refuse, as well as for its methane production and final stabilisation. The amount of run-in moisture during refuse infilling activities can be controlled and reduced by effectively controlling the runoff waters. Effective sealing of the mining void's bottom and walls with liners, followed by a tight capping on completion, will minimise leachate production.

As stated above, total isolation of the fill material from moisture may not be desirable. Moisture is necessary to the bio-degradation of the waste in order to stabilise the land. If the future restored surface is to have utility once more, the land must be stabilised. The sooner stabilisation is attained for the fill content, the sooner the benefits of surface use will be realised.

Earlier stabilisation may be accomplished by accelerating the waste bio-degradation through leachate re-circulation. It has been shown that controlled leachate re-circulation, including addition of nutrients to maintain optimum moisture and pH, can enhance anaerobic microbial activity, break down organics as evidenced by reduced toxic organic carbon (TOC)* and chemical oxygen demand (COD)**, convert refuse organics to methane and carbon dioxide, and precipitate heavy metals. Heating of re-circulated leachate to 86° F (30°C) also greatly accelerates stabilisation. Research at the U.S. Department of Energy's Argonne Laboratory indicates that circulating water through landfills could triple the speed of biological degradation, boost methane gas production, and make it possible to return the landfill to other uses in ten years or less (Bagchi, 1989).

* TOC is a test measuring the carbon as carbon dioxide; the inorganic carbon compounds present interfere with the test, hence, they must be removed before the analysis is made, or a correction applied (Salvato, 1982, p. 380).

** COD is usually measured in relation to certain industrial wastes. The chemical oxygen demand is the amount of oxygen expressed in ppm or mg / l consumed under specific conditions in the oxidation of organic and oxidisable inorganic material in the water. It does not oxidise some organic pollutants (pyradine, benzene, toluene) but does oxidise some inorganic compounds that are not measured, that is, affected by the biochemical oxygen (BOD)*** analysis (ibid).

(infra) BOD This characteristic of surface water, sewage, sewage effluents, polluted waters, industrial wastes, or other waste waters is the amount of dissolved oxygen in milligrams per liter (mg / l) required during the stabilisation of the decomposable organic matter by anaerobic bacterial action. Complete stabilisation requires more than 110 days at 20°C. Incubation for five days (carbonaceous demand satisfied) is generally used for domestic sewage, one pound of 5-day BOD is roughly equivalent to 1.5 pounds of ultimate BOD. If one pound of 5-day BOD is completely aerated, requiring 1.3 pounds of oxygen, 0.14 pounds of inert residue will remain (ibid).

11.3 Landfill Liners and Waste Water Treatment — In General

Liners, used in waste depositories designed to contain wastes and leachate within and to prevent pollution from leakage into ground waters, are currently made from various synthetic organic, polymer compounds. Resistance to corrosion, puncture, and mechanical strength are important features required for the liners. Liners are usually very large, often measured in acres. Joining of sheets to cover larger areas must be reliable in that there will be no separation during the installing process or under the weight and compacting of the deposited wastes. The strength of the material is obviously very important. At present, three types of synthetic liners dominate the market, viz., high-density polyethylene (HDPE), PVC and HYPALON (chlorosulfonated polyethylene). All three types are available with embedded woven fabrics providing high levels of resistance to puncture, tearing and bursting.

In the aggregate quarries investigated, clay is the common material used to line the quarry walls and floor to seal off the fill (see Figure 14, page 390). Where deemed necessary, additional methods for sealing the quarry walls may be used, e.g., synthetic geotextile curtains draped or hung on fill side of the quarry wall clay liners; guniting and grouting of the quarry's rock walls, where there are cracks, joints or open seams, may precede refuse filling and clay lining; admixtures, e.g., fibreglass may be applied to rock wall coatings. Another newcomer to the landfill scene that is creating great hope and expectation for

Clay Colliery Co. Ltd, Telford, U.K.

Figure 14 A Clay-Lined Surface Mine Repository Ready to Accept Waste

simultaneous treatment of leachate and leakage prevention is peat, or sphagnum. This natural material is discussed later.

11.3.1 *Liners — Natural and Synthetic Materials*

For smaller, local-type landfills, a knowledge of the contributing community's waste, or the wastes to be accepted will be helpful in determining the parameters of the liner(s) to be used. However, in larger, regional-type landfills where the distances for hauling wastes are seen to be greatly increasing in recent years, the knowledge of community waste is stressed less. Landfills will have to designate their parameters for the range of material they accept.

The types of materials used to construct landfill liners and final caps / covers are generally: clayey soils, clays, synthetic membranes, or amended soil and admixtures.

Clayey soil liners are compacted earth with a high percentage of clay. They may have its slight permeability further reduced by being treated with additives as bentonite, lime, asphalt, or cement. This type of liner material is not recommended for containment-type of landfills.

"Clay" is a loosely defined term for a naturally-occurring mineral and includes a large variety of minerals within the clay family. However, in practice, clay-like soil materials are sometimes referred to as clays. Geological and mineralogical definitions are given for the terms as follows:

> "**Clay mineral:** The clay minerals are finely crystalline, hydrous silicates with a two-layer type (e.g., kaolinite) or three-layer type (e.g., montmorillonite) ***. The most common clay minerals belong to the kaolinite, montmorillonite, attapulgite, and illite or hydomica groups.
>
> Mixed layer clay minerals are either randomly or regularly interstratified intergrowths of two or more clay minerals. (Howell / AGI Glossary, 1957)
>
> **Clay:** The term as used carries with it three implications: (1) a natural material with plastic properties, (2) an essential composition of particles of very fine size grades, and (3) an essential composition of crystalline fragments of minerals that are essentially hydrous aluminium silicates or occasionally hydrous magnesium silicates. The term implies nothing

regarding origin, but is based on properties, texture and composition, which are course, interrelated — for example, the plastic properties are the result of the constituent minerals and their small grain size (Grim, 1942).

2. Clays differ greatly mineralogically and chemically and consequently in their physical properties. Most of them contain impurities, but ordinarily their base is hydrous aluminium silicate (USGS).

3. A soil consisting of inorganic material, the grains of which have diameters smaller than .005 millimeters (U.S. Bur. Soils Classification).

4. Fine-grained soil that has a high plasticity index in relation to the liquid limit and consists mainly of particles less than 0.074 mm (passing No. 200 sieve) in diameter. (Waterways Expmt. Sta. Corps of Engrs. Tech. Memo. 3-357, 1953)". (op. cit.)

The landfill base below the GW table, even in a clayey bed, should still have a constructed liner placed on top because many clay beds have sand seams in them and minute vertical and angular fractures that are difficult to determine by subsoil permeability methods. See page of leachate can take place through the minute clay cracks. Differential settlement of the clay liner can also cause minute cracking in the clay allowing seepage through it. A deficiency of water in a well-sealed fill can cause the clay to dry denying it the moisture needed for maximum plasticity. Clay liners, used alone, are vulnerable to potentially harmful conditions resulting from the refuse degradation that may weaken the clay's impermeability. Hydrocarbons in the refuse can lower the impermeability of the clay. Acidic and caustic leachates can in some cases dissolve the soil binding agents in clay causing its impermeability to break down.

The addition of a synthetic membrane over the compacted clay liner, even on a naturally occurring clay sub-base, will ensure a clay liner against leachate degradation and give greater seepage protection against groundwater contamination.

Synthetic Membranes: Synthetic liners, made from polymers, are named after the major constituent polymer. Seven polymer-liners that were most commonly used in former years are: butyl rubber, chlorinated polyethylene (CPE), chlorosulfonated polyethylene (CSPE), ethylene-propylene rubber

(EPDM), low-density polyethylene (LDPE), high-density polyethylene (HDPE), and polyvinyl chloride (PVC). Each has disadvantages and advantages. The earlier HDPE and LDPE liners were preferred over others in many applications, but offered poor puncture resistance. Their puncture resistance has been improved and their popularity has continued.

With regard to puncturing of liners, great care must be exercised in laying them in the landfill and with ensuing passage over the membrane during fill and compaction to avoid tearing and puncturing. However, in 1994 a Japanese company developed a waste liner that sealed itself *in situ.* "The middle layer of the liner is made of a material that swells on contact with water. If a tear appears, the middle layer of the liner expands plugging the hole within minutes". (Mining Environmental Management, 9/94, p. 35).

Geogrids, geonets, and geotextiles are used as curtain drains and find use as substitute or reinforcement for landfill rock-drain beds or hung on the walls of quarry landfills to direct leachate downward within the fill to the bottom whilst protecting the walls from leachate seepage. One product boasts its new geotextile drainage composite to have "the capability of geosynthetic equivalency to a gravel drain with ten times the drainage capacity of conventional nets *** and capable of replacing several feet of stone or gravel". (ibid).

11.3.2 *Peat — The New Waste Treatment "Wonder" Material*

Increasing attention is being given to the use of peat-based technologies for waste treatment. The use of peat has great potential for lining in landfills, particularly for the MSW containment-type required herein, and holds great promise for making landfills truly, totally contained. In addition, peat has the prospect of neutralising leachate, or, at least, vastly reducing the contaminants within it to a point that it will not contaminate ground water surrounding the landfill if leakage were to occur.

According to a recently published paper on utilising peat in waste treatment, "It has been demonstrated that peat can successfully remove metals, nutrients, suspended solids, organic matter, oils and odours from domestic and industrial effluent.

"Peat is primarily composed of organic materials consisting of relatively undercomposed and decomposed plant remains. It accumulates in wetland areas, largely under moist, anaerobic conditions. *** Peat has a high adsorption capacity for transition metals and polar organic molecules.

"In natural peatlands, it was found that peat has a strong affinity for heavy-metal retention. Eger et al (1980) studied the drainage pattern of iron mining stockpile leachate through a white cedar bog. They found that the most flow occurred across the surface and through the top 0.30 cm of the bog. Contact with the peat resulted in more than a 30% removal of nickel and more than a 99% removal of copper. Eger and Lapakko (1989) further reported that the peat-bog treatment could have an extended lifetime of between 20 and several hundred years.

"Peatlands have also been used to remove nutrients from secondary sewage effluent. Tilton and Kadlec (1979) studied the removal of secondary effluent nutrients by peat when they applied it to a natural peat land in Michigan. They reported removal rates of 99% for NO^3-NO^2-N, 77% for NH^3-N, and 95% for total dissolved phosphorous. ***

"Frostman (1991) reported the use of a (constructed) peat / wetland system for treating iron mining stockpile seepage waters, which have low concentrations of metals. The moderate pH (5 to 6) of the waters are particularly conducive to metal removal by peat. The peat / wetland system was constructed so that seepage waters entered the system through an underlying limestone bed. The seepage waters then passed over and through a sphagnum moss peat bed. Initial reports on the system indicate that it is performing well in meeting the required nickel and copper water-quality standards.

"Peat / wetland systems have also been used for the passive treatment of coal-mine drainage. Passive wetland systems treat mine water by chemical and biological processes that decrease metal concentrations and neutralise acidity. Hedin et al (1994) describes three current types of passive technologies: (1) aerobic wetlands; (2) organic-substrate (peat) wetlands; and (3) anoxic-limestone drains.

Organic substrate wetlands promote anaerobic bacterial activity, which results in the precipitation of metal sulphides and in the generation of bicarbonate alkalinity. They report that this type of system removes acidity from mine water at rates of 3 to 9 g /m^2/ day".

For sewage-treatment systems, "researchers at the University of Maine conducted extensive laboratory and field studies on the ability of peat to treat septic-tank effluent (STE). *** full-size sphagnum mosspeat filter fields were constructed with 0.75 m thick layers. Either a gravity or a pressurised distribution system was placed within the layers. The bed widths were varied (i.e. 4.8 × 19 m. and 6.8 × 6.1 m) depending on the loading rates. The filter fields provided excellent fecal coliform removal (99%) with the 5-day BOD reduction exceeding 90% and the COD reduction exceeding 80%. At the University of Regina (Canada), Rana and Virarghaven 1987, Virarghaven 1993, studied the effects of filter depths and hydraulic-loading rates on the efficiency of peat filter beds. Even at high STE hydraulic-loading rates, the removal of BOD and total suspended solid (TSS) was excellent. Fecla coliform removal was good with effluent levels as low as 60 CFU / 100 mL".

Peat leach-mound systems were found to perform well under subarctic conditions. "Near Anchorage, Alaska, Riznyk et al. (1990, 1993) designed and monitored peat leach-mound fields that treated domestic septic tank waste water. During a two-year period, the quantity and quality of applied waste water was measured for BOD, COD, TSS, NO^3-N, TKN, total phosphorous, pH, fecal and total coliform bacteria, colour, turbidity, dissolved oxygen and temperature. The analyses indicated that the quality of the peat leachate is similar to waste water that has undergone tertiary treatment. In addition, sphagnum moss peat was tested to treat STE in New Brunswick, Canada (Daigle, 1993). The peat leach fields were 0.7 m thick, placed over 0.15 m of crushed rock. Distribution pipes were laid in a crushed stone layer on top of the peat. The pipes were then covered with an additional 0.20 to 0.25 m layer of peat. The STE and filter leachate were analysed for BOD, TKN, NH^4-N, NO^3-N, SS, pH and fecal coliform. The results from these three waste water treatment systems showed high efficiency, in the range of 84% to 99.99% for reduction of STE contaminants.

"In the early 1970s Farnham and Brown (1972) successfully treated secondary effluent through spray irrigation on constructed peat-sand filter beds. *** under maintained anaerobic conditions, fecal coliform was reduced by 95% and phosphorous concentrations were reduced from 7 to approximately 0.5 mg / L".

These encouraging results led to a further test conducted at the US Department of Agriculture Forest Service campground in Minnesota. "They reported that the peat-sand filter bed accomplished almost complete removal of fecal coliform bacteria and phosphorous. About 90% of the waste water nitrogen was removed during the second and third years of operation, but declined to about 50% by the fifth year because of oxidation of the peat and release of nitrogen from the peat itself".

With regard to metal removal, "The U.S. Bureau of Mines (Jeffers et al, 1991) developed small beads made of sphagnum peat moss and polysulphone for absorbing metal ions in low concentrations. They report that the beads effectively removed arsenic, cadmium, copper, lead, manganese and zinc from waste water. Metal concentrations in waste water were generally reduced to <0.1 mg / L, and treated effluent frequently met national drinking water standards". More than 100 different water samples have been tested, including acid mine drainage from mines and waste water from mineral processing operations, chemical plants and municipal water treatment facilities.

"Kadlec and Keoleian (1986) also reported on the strong affinity of peat to absorb metal ions. Humic acids were considered to be the primary metal complexing components in peat, especially in the well humidified sedge peats. *** An important finding was that metals have different selectivity for different peat types, and the design of peat-based metal removal systems must take this into account. Peat treated with sulphuric acid improves its ability to absorb copper".

Peat biofilters using peat moss have been highly successful in absorbing odourous gases. "Beerli (1989, 1991) reported two cases in Quebec, Canada, "*** one, removing malodourous nitrogen compounds released during treatment of hog manure by biological aerobic reactors on a hog farm; the other, at a waste water treatment plant at Magog from sludges in the filter press room where amines, ammonia, daimines, hydrogen sulphide, mercaptan, organic sulphides and skatole were emanating. The use of biofilters in Europe were reported by Coffey and Kavanaugh (1989) and Boehler (1991). Malodourous waste gases were being successfully treated at sewage treatment plants, vegetable and animal waste treatment plants. More than 100 industrial-scale peat biofilters were reported in operation in Europe in 1989".

The authors conclude that "There appears to be a high potential for using peat to treat waste water effluents from industrial and domestic sources". (Malterer, McCarthy & Adams, 1996).

11.3.3 *Future Treatment of Mine Waters*

It is thought that a proper understanding of the movement of water through mine voids, fractured material and collapsed rock material would help in devising an *in-situ* remedial treatment for contaminated mine drainage.

One such treatment method for the control of acid mine water quality uses bactericides. As reported at the International Mine Water Association's 1994 congress in Nottingham, U.K., "The *Thiobacillus ferro-oxidans* bacteria catalyse the oxidation of pyrite to acid which solubilises metals to contaminate mine water. Bactericides can effectively inhibit bacterial activity, curtailing acid production by up to 95% thereby reducing dissolution of metals in mine water, and reducing toxicity and treatment costs". (Mining Journal, 1994, p. 235).

11.3.4 *Daily Cover Material*

Most landfills in the U.S. use a daily cover at the working face of at least six inches in thickness as required by Subtitle D under RCRA regulation Although daily cover serves several essential purposes, it also uses up valuable landfill space. As an example, the total volume of daily cover material placed in a California landfill is reported to be 20% of the whole. Landfill void space can be better utilised for MSW rather than clean dirt. However, some other daily covers have been approved by regulatory agencies and are being used, e.g., a bituminous emulsion that can be sprayed on after compacting the daily fill material; or 4 mil thick Canvex plastic tarpoleans, over areas 100 feet × 75 feet which are removed on the next day's work. The use of thinner daily covers as mentioned, result in a 50% reduction of spoil material as daily cover and a substantial decrease in the loss of fill space. The rate of decomposition of the refuse should be increased without the obstructive daily cover lenses within the fill to slow down the percolation and circulation rates of moisture.

11.3.5 *Landfill Covers*

The purpose of a cover is to give the completed landfill a rain-proofed, attractive, well-contoured, natural-appearing dressing before returning it to beneficial surface use. To keep the land in that condition, the fill must be maintained in a stabilised condition, that is, future slumping of the surface must be prevented and avoided. As noted, above, final stabilisation of the fill might be hastened by actually encouraging infiltration and possibly adding nutrients such as domestic sewage to the infiltrating water. Once stabilisation of the fill has been realised, further decomposition by infiltrated water is to be avoided. The cover serves to minimise uncontrolled infiltration of rainwater into the landfill which will affect the decomposition, leaching, and settlement of the landfill.

Well-constructed landfill covers are placed in a series of layers. The lowest, or first layer, called the grading layer, gives the landfill a working surface to apply the subsequent layers over it. The grading layer is usually between 1 and 2 feet thick of a well-compacted coarser-grained material, such as sand, of relatively low permeability should be used. For better sealing, a synthetic membrane may be placed below the grading layer.

Next, the barrier layer is placed on top of the grading layer. The barrier layer is, as its name indicates, a barrier to prevent water penetration into the fill. If a relatively impermeable clay is used, the thickness should be about 2-feet thick. If a clay of higher impermeability is used, as bentonite, the barrier layer may be lesser in thickness, but not less than 1-foot. A thicker layer performs better in the event of differential settling of the waste. Again, a synthetic membrane may be placed on top of the grading layer in lieu of the clay. However, the usual practice is to use an impermeable clay to form the barrier layer.

The protective layer which overlays the barrier layer serves two main purposes, viz., to protect the barrier layer from freeze-thaw and desiccation cracking, and to provide a medium of sufficient thickness for root growth. The protective cover should be from 2 to 3.5 feet in thickness, depending on the geographic location and subjection to depth of freezing. Penetration by roots growth is to be avoided to prevent roots from making moisture entries into and through the barrier layer.

Lastly, the final cover is applied of an organic soil for facilitation of grass seed germination and growth. Necessary lime and fertilisers should be applied for several years to encourage a healthy, quick cover for prevention of soil erosion thereby giving stability to the cover layer and preventing moisture penetration.

11.3.6 *Landfill Gas and Collection as an Energy Source*

A related, secondary and important problem is the gas generated from landfills, mainly in the form of methane and carbon dioxide. The methane component is unstable and creates a hazard when escaping from landfill sites. 5 to 15% methane forms an explosive mixture with air. Left uncollected, landfill procedure to dispose of the gas is by venting and burning of it at the vent (called flaring). However, it was noted in 1986 by W.L. Hall at a conference in Solihull, England, that "a new industry has grown up in the U.K., the U.S. and Europe. *** to control, collect and possibly utilise the energy from landfill gas".

The larger volume MSW landfills, i.e., 30 feet or deeper, and large in area, e.g., 30-acres, are better producers of methane. The production ratio of methane to carbon dioxide varies for MSW, but is generally in the ratio of 3:2, depending always on the amount of organic material contained. Nitrogen, hydrogen and oxygen make up the minor contents of landfill gas. Hydrogen is usually present in larger amounts during the early stages of decomposition. Carbon dioxide increases the hardness and acidity of water; acidity increases the solution and leaching of non-organic soluble constituents in the refuse. This increase in acidity somewhat inhibits the generation of methane. An alkaline pH is more conducive to methane generation. The lives of methane producing landfills varies greatly with content and size, but can be expected to be in the 10 to 20 years range when of sufficient size. The production of methane can be encouraged and accelerated as discussed earlier, viz., by accelerating the waste bio-degradation through leachate re-circulation, heating of re-circulated leachate, or circulating water through landfills to boost methane gas production.

Landfill gas (LG), or methane, has potential as an alternative energy source. Recovering LG is an advancing technology. It offers economic advantages as a cheap source of energy to local governments and industries located in the area of a landfill operation. LG collection offers the added benefit of reducing potential environmental hazards from escaping gas.

Examples of profitable collection in connection with combined quarrying-landfill operations are offered in evidence at Ch. 11 §§8.1 -8.1, infra. See also U.K. examples of LG collecting for energy at Judkins and Allerton Park Landfills, et al, and at Fred Weber Inc., St. Louis, Missouri. Another LG-collecting site, though not a quarrying-landfill operation, demonstrates the economic potential in LG-collecting. A recovery facility operated near Fort Lauderdale, Florida, cost US$6 million to build in 1986. It is located next to a 100-acre landfill which produces 3.5 million c.f. of high quality methane gas per day, the equivalent of 250,000 barrels of oil. The gas is sold to Florida Natural Gas company for residential and commercial use (Seeney, 1988, page 11).

It should be noted that in the U.K., LG successfully collected at Judkins Quarry-Landfill (see §11.8.1.1, infra) and at Allerton Park (see §11.8.1.2, infra), requires planning permission for installation of a gas turbine engine for the generation of electricity whether the electricity is used on or off site. If gas collection and energy generation is pre-planned, planning permission is required before the landfilling operation begins.

11.4 Miscellaneous Advantages of Using Mineral Quarries and Pits

A favourable and arguable feature of utilising quarries and mining pits for landfilling is the available supply of cover material ordinarily found on the mining land. The overburden / dirt from the stripping process removed to get to the rock and minerals sought is stored in piles on the property. The available supplies of stored overburden will usually range from a stripped depth of anywhere between zero and 30 feet, or more for non-metallic mines. For strip-coal mines, the depths of removed overburden may be considerably greater, even surpassing over one hundred feet in depth. Consequently, piles of stripped overburden may run into millions of cubic metres that are available for landfill daily cover material. Having this material already present can mean a tremendous savings and advantage over an unprepared site.

A current problem, which may be curtailed by new regulations, is that topsoil, when present, may have been stockpiled separately and sold. Ordinarily, the sub-burden, that material immediately below the topsoil, is retained in site stockpiles, or used in bunds or berms. Consequently, none of the best vegetative coverage material from the original pit surface may be available at the time of closing the landfill. Reclamation regulations could prohibit sale of top soil from newly permitted surface mining projects.

By replacing the formerly stripped dirt, sub-burden and saprolitic rock in the landfill, it becomes unnecessary to import daily cover material from borrowed sites. The result is greater land conservation. The elimination of environmental disturbance of other properties for 'borrowed' cover material, then, becomes an additional benefit, as well as an economic benefit to the operation. This additional economic benefit, in turn, makes the operation of a mineral excavation for a landfill a less expensive operation and lowers the cost to the general public of waste disposal.

Another point in favour of the available supply of overburden of former quarries and pits is the advantage of having adequate cover material which reduces the problem of windblown litter. When in plentiful supply from quarry properties, its generous application can reduce portable / mobile and permanent fencing use and costs. Windblown litter can be a real environmental problem at landfills. It is a very visible criterion often used by the public in rating the operation of the landfill.

Still another possible advantage in utilising mined pits and quarries is the lack of concern for creepage of the earth's mantle which does not affect its stable rock walls. "Creepage" is the slow and minute movement of the mantle that takes place annually where there is any topographic slope. In surface mounded landfills placed on any topographic slope or inclination, the surface is subject to the minute movement of mantle "creep" which could, over a period of years, place a rupturing strain on polyethylene liners causing a tear for leakage of leachate. Thus, the stable rock walls of quarries, not being subject to creep, would not have the potential for rupturing liners placed within the pit.

Several more obvious advantages for using surface mines are those concerning traffic, noise, visibility and aesthetics for the proposed landfill site. Public objections raised to new landfill sites are generally for those issues. When using an established surface mine for a landfill, those four problematic concerns have already been resolved during the mine's former presence. There will be no increased, heavy traffic for landfilling as traffic volume and patterns already exist for mining and local residents have become accustomed to it.

Noise levels at a landfill are generally not as great as for mining. Noise screening has already been placed for the prior mining in the form of bunds / berms and tree growth. There should actually be a decrease in the level of noise from a landfill.

Visibility of a below-surface landfill will be less than for a landfill placed on the surface. Also, odours and the chances for windblown litter are reduced in a below-surface mine. Again, the noise screening placed for the prior mining is an accomplished feature that need not be constructed.

Landscaping and beautification of surface mine property entrances along public highways has been the order of the mining industry for over a decade. The aesthetics of the visible exterior of the landfill property have already been installed by the prior mining operation.

Anticipated concerns by the public over water contamination by an incoming landfill should be considerably diminished as the prior mining operation would have encountered them and been obliged to remediate the problem.

These numerous existing advantages of an already publicly accepted surface mine site can assuage the normal objections of local citizens to a new landfill site.

(i) **Elements of Modern Landfill Management** (Tomes, 1989, page 4)

1) Site preparation
2) Site operations
3) Environmental monitoring
4) Restoration and aftercare

The technology employed during the first three stages of landfill management, indicated above, are equally important to making the final stage successful. However, other than the prevention of leakage of contaminating leachate into the ground waters, and the control of landfill gas emissions which this work is concerned with, it is the final stage, the successful return of worked-out mined land and poor quality land to a more productive after-use that is the goal of this work. The details for those first three stages are not the purpose of this work, but are left to the landfill technologists. As pointed out by Tomes, "Adequate supplies of impermeable capping and restoration material are essential, but it is their application that is critical. Capping material will only restrict ingress of water and egress of gas, which is its purpose, if it is of the correct moisture content and is compacted in layers. The application of soils, conversely, must ensure compaction does not take place which would hinder agricultural development". (Tomes, op. cit.).

Thus, here again, one of the advantages of using a mining pit for landfilling is the generally "adequate" supply, if not abundant supply of restoration material found in the spoil piles of stripped overburden stored on the quarrying property. The stripped material from mining is being returned to its original place.

In keeping with the hypothesis herein, where ever lithologically, structurally and hydrologically possible, all open pits should employ landfilling with solid wastes as part of the complete land restoration process restoring the surface to its original elevation after the void of a mineral body has been removed. The contention herein is that virtually any surface mine can be made environmentally-safe for deposition of solid wastes by installation of properly-designed lining and leachate control systems. The only exception to landfilling a worked-out surface mine should be for the consideration of creating a small lake in the former pit for the purpose of land aesthetics, wildlife habitats and watering, or recreational uses.

(ii) **Remoteness for Landfill Sites is a Diminishing Problem:**

It is recognised that some argument may be made that parks, forests, wildlife areas, etc., are usually established in remote areas, and that any open-pit mines within those environmentally sensitive remote areas would be too remote from urban areas to use for landfilling operations during the reclamation process.

Towards refutation of an argument of remoteness of landfill sites from the waste-generating point or the collection point of accumulated wastes, long-distance transportation is becoming an ever increasing mode to waste disposal sites.

Trucking of wastes to disposal sites hundreds, and even thousands of miles, away from the generating, or collection source point, has become a present reality. Currently, railroads in the U.S. are promoting, with persuasive environmental arguments, their use for haulage of waste over long distances as opposed to truck haulage. An example in the U.S. is a public-private venture between Norfolk Southern Railway Company (NS) and Virginia's Roanoke Valley Resource Authority which features the "nation's first use of rail as the sole transportation link between a solid waste transfer station and a landfill". NS is publicising the fact that their "Waste Line Express" consisting of a 12-railcar train "can keep 96 garbage trucks off public highways leading to landfills". Added environmental benefits are stated in a comparison with truck haulage, "on a tonne-mile basis, railroads emit one-tenth the hydrocarbons and diesel particulates, and one-third the oxides of nitrogen and carbon; per gallon of fuel, a train can move a tonne of freight more than three times as far as a truck".

In the U.K., British Rail has developed a rail service by which regular shipments of waste are transported by the trainload at regular time intervals to distant sites for disposal.

New transportation developments indicate rapid changes forthcoming in the waste disposal industry. Remoteness of landfilling sites can be more desirable because of their lower visible profile.

11.5 United Kingdom

It was reported in 1972 "that England and Wales produced over 14,000,000 tons of domestic and trade waste. At 220 lbs. / cubic yard in the collection state, it would occupy 28,000,000 cu.yds. of space in its compressed tip state". (Skitt, 1972, page 1). With the predicted population increase in the U.K. of 13,000,000 by the year 2,000, John Skitt noted, "Together with enormous quantities of industrial wastes of almost infinite variety, the handling and disposal of these rapidly-increasing solid wastes constitute a problem of almost terrifying proportions for the societies of today and tomorrow". Skitt's further comment gives support to the thesis herein: "Excavations of minerals are said to exceed the rate of reclamation and the total of derelict lands is in excess of 90,000 acres. The annual rate of extraction (*from mining*) is five times the space required for house refuse. *** The relationship of this problem with refuse disposal might be solved mutually". (Skitt, 1972, page 7). (It should be noted that Skitt's figure of mined land includes that of coal mining).

In 1989, the volume of household (MSW), commercial and industrial waste for Britain was reported to be around 100 million tonnes. The 12th Report of the Royal Commission on Environmental Pollution, by Paul A. Tomes, stated the volume deposited in landfills for 1988 to be over 90 million tonnes, and that landfills were the "Best Practicable Environmental Option" for disposal of that waste. Tomes' 1989 report to the Institute of Quarrying stated that the mineral industry was creating "voids at a rate in excess of 200 million cubic metres per annum (Again, mined coal tonnage is included, and voids by overburden removal is not included). The restoration of quarries and pits involves considerable expenditure and landfill is arguably the most cost effective method of fulfilling that requirement subject to planning and technical suitability". (Tomes, 1989, p. 1).

In 1993, mineral extraction for"sand, gravel, hardstone and limestone" in Britain was reported to be 222 million tonnes, which equated to 120 million cubic metres of void space (1.85 tonnes/m^3). It should be noted that the volume of pit overburden removed for the mineral extraction is not included in the calculation of space voids, consequently, that figure would be somewhat larger (Tomes, personal communication, 1994).

Based on 1994 figures, Tomes further reports that "Britain's annual production of MSW (household waste) of 20 million tonnes requires 20 million cubic meters of space per annum of void space. (1 tonne/m^3 equates to 56.66 lbs/cu.ft. @ 35.3 cf./$m^{3;}$ with 27 c.f./cu.yd., the weight of a cubic yard would be 1,529.8 lbs. Compare this with John Skitt's figure of 220 lbs./cu.yd.) This means that void space created by mineral extraction in total exceeds household (MSW) production by a factor of 10. It should be noted that in Britain we produce 100 million tonnes of household (MSW), industrial and commercial waste per annum, the majority of which, is suitable for landfill, so we require an annual void space of approximately 100 million cubic metres. On the basis that at least 50 million cubic metres is unavailable for landfill *** Britain can be just about self-sufficient in landfill void made available from mineral extraction". (Tomes, personal communication, 1994).

The use of abandoned quarries and pits in Great Britain has been made for depositing materials as coal mining wastes, noxious sewage sludge, flue-gas de-suphurisation wastes, and pelletised incinerator residue for backfill.

11.6 United States

According to the USEPA (1994 report), the U.S. leads the world in waste production, generating some 200 million tonnes per year, and 3.6 lbs./day/capita. Landfilling has been the major method of waste disposal for MSW. In 1986, approximately 83% of MSW was being deposited in landfills. By 1990, it had been reduced to 73%. Between 1986 and 1990, incineration had doubled, from 7% to 14% of MSW, whilst recycling had increased from 10% to 13% during the same four-year period (Liptak, 1991, p. 24).

In 1986, EPA's goal had been to reduce landfilling to 55%, increase incineration to 20% and recycling to 25% by 1992. However, by 1994, incineration had only increased by 2% from 1990, to a total of 16% of MSW.

Two-thirds of that waste is placed in landfills. EPA's prediction is that much of the heavily populated East Coast will run out of approved landfill space in the current decade. EPA further estimates that 80% of the nation's landfills will close by the year 2000.

11.7 Canada

There has been virtually no adverse litigation to the siting of landfills. A personal communication with the New Brunswick Department of Environment emphatically states, "Mined out open pits do not meet the stringent siting guidelines for sanitary landfills in the Province and would not be acceptable under the regulations". And, the same prohibition is seemingly borne out in communication with the Saskatchewan's Municipal Waste Management which states that "mined-out open pits have not been used for landfills, *** and there were no plans to utilise mines for garbage disposal".

11.7.1 *A Canadian Solution for Safer Treatment of Uranium Tailings*

A safer procedure for the handling of uranium tailings has been found which would avoid serious contamination in the event of tanker spills during transportation.

An Ontario mining company refining uranium concentrates into uranium trioxide produces a by-product raffinate containing uranium, sulphuric acid and waste minerals. Formerly, the liquid raffinate was shipped by tankers to other uranium refining plants for extraction of the residue uranium whilst the remaining effluent was disposed of in managed tailings areas.

A new recycling plant has been built at Blind River, Ontario, which converts the raffinate liquid to a non-corrosive powder that is reportedly "virtually insoluble in water and easy to handle". The powder, which still contains small amounts of uranium, is stored in drums prior to shipment for further refining of the residual uranium. The raffinate has been made safer during public transportation and in the event of tanker spills while en route (Mining Journal, London, March 24, 1995, page 215).

11.8 Present-Day Examples of Safe Waste Depositing in Surface Mining Pits

The use of worked-out surface mines as depositories for solid waste is not unknown in the past, or in the present day. In older time, any topographic depressions might be used for filling with trash, garbage, solid or liquid wastes without regard to contaminating ground or surface waters. In more recent times, prior to rigid environmental regulations, it appears from the scant information available that where some concerns were applied, as prohibition of burning and requiring application of daily dirt cover, few other precautions, as clay liners, leakage prevention and monitoring was done at the time of filling.

However, in more recent years, successful examples, applying environmental precautions and safeguards against contamination of ground and surface waters, may be found at different locations throughout the three countries in this study. Currently-operated examples from each of the three Anglo Nations of the study are submitted in evidence that the use of non-fuel surface pits may be used for MSW landfills without fear of contaminating local ground and surface waters, and minimising harm to the local environment during the operation. Upon closure of the joint operations, the disturbed area has not only served society by providing essential minerals, then providing a depository for society's wastes, but has been finally restored to it former condition, and in some cases, restoration of the land has been an improvement over the original condition.

11.8.1 *United Kingdom*

Two examples are reported in detail, following, viz., (i) the Judkins Quarry and Landfill at Tuttlehill, Nuneaton, Warwickshire; and (ii) the prime example for consideration herein, the Allerton Park Quarry, a sand and gravel operation near Knaresborough, because it was planned as one project, i.e., both mining and landfilling for land reclamation. Additionally, infra, a listing of eleven other mineral extraction sites is given that have been infilled with wastes for quarry reclamation and complete restoration of the mined land to the original surface and to provide beneficial use in conserving land. Both examples of landfilling given are operated by Greenways Landfill, a division of ARC Ltd, which operates the landfills in conjunction with the ARC-operated aggregate pits. Greenways has the responsibility for the landfilling operations. In 1991–1992 Greenways Landfill disposed of more than 3.5 million tonnes of waste at its quarry-landfill sites. See Figure 15, page 409, for an idealised drawing of Greenways combined quarrying and landfilling operations.

It is interesting to note that Greenways has developed an "Environmental Initiative", a self-designed proclamation of dedication to quality in the management of waste disposal. In brief, Greenways states, "The Environmental Initiative is a concept of total quality of service, while ensuring the impact of our operations on the environment is kept to a minimum". Such environmentally-protective dedication bears out the argument made in the

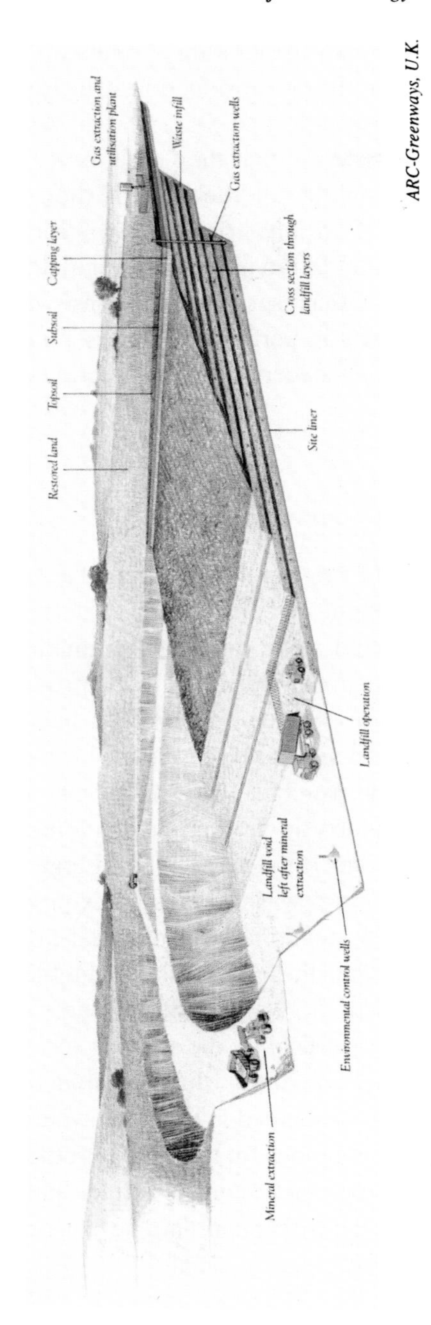

ARC-Greenways, U.K.

Figure 15 Idealised Drawing of Typical Combined Quarrying and Landfill Operation

defence of the "new breed", or new generation of operators in the solid waste and surface mining industries in the Introduction, Ch. 1, §1, "*** the Solid Waste Disposal Industry respond(s) to Defendants' Answer by way of Amendment to its Complaint saying that they are a **new generation,** being neither culpable, nor responsible for the acts of their antecedents and forebearers. Both, new-generation industries are responsible, conscientious, law-abiding citizens conforming to environmental regulatory controls. They are very aware and concerned with preservation and conservation of the earth's environment, and that their acts in performing its essential earth-disturbing job are carried out with care and concern for the protection of the public good and environment".

11.8.2 *Judkins Quarry and Greenways Landfill*

This quarry operation mines the pre-Cambrian Hartshill quartzite for crushing into aggregate for an assortment of construction material uses, e.g., road building, railway ballast, coated stone (asphalt / bitumen mix), ready-mix concrete and other concrete products. Its rate of production (1989) is approximately 450,000 tonnes per year. In 1989, it had a remaining life expectancy of eight to twelve more years (see Figure 16, page 411).

The quarrying operation was begun in 1845 and ownership was transferred to ARC Ltd. in 1973. The quarry is approximately 1-mile long (1.6 km), 270 feet deep (70 m.) and covers 195-acres (79 ha). Large areas of the tract are occupied by the pit, crushing, washing and screening plants, concrete products manufacture, product stockpiles, silt disposal lagoons, active and disused quarry waste tips (including the prominent "Mount Jud") and the landfill site operated in conjunction with the Warwickshire County Council.

After primary and secondary crushing, the stone is screened, washed and graded according to size and use. Washing, whilst cleaning the stone, removes fine, silt-size material which is suspended in the wash water. The waste wash water is conveyed to settlement lagoons for the fines to settle. The fine, settled sediment is periodically removed, stacked on the surface and dried for use as a refuse covering material in the landfill operation. The clarified lagoon water is subsequently discharged into the adjacent canal.

ARC-Greenways, U.K.

Figure 16 View of Judkins Quarry

In 1978, Greenways, working with the Warwickshire County Council came to an agreement that the Judkins Quarry would become a site for disposal of household and commercial waste. "In January 1981, the Judkins Quarry Working Party was established and comprised representatives from ARC Ltd., Warwickshire County Council and Nuneaton and Bedworth Borough Council". A frame of reference terms was formulated for the project. A Draft Brief was published in 1983 and submitted to the general public for their views and comments. The Brief was adopted in 1986 and served as guidance for the Judkins landfill development. By 1989, finding that the several planning permissions they held, with the oldest being over 30 years old, would no longer be adequate to regulate the activities and give comprehensive guidance on the eventual restoration and after-uses, ARC-Greenways, in conjunction with the Warwickshire County Council and the Nuneaton and Bedworth Borough Council, prepared a detailed planning brief. Its purpose was to provide a planning guidance for continued mineral extraction, landfill, but mainly for development and restoration of the Judkins site and its after-use. The final restoration concept was to develop a "Country Park" style of facility while preserving Mount Jud as a local landmark. (Mount Jud is not a natural-occurring feature. It is a man-made, conical spoil pile towering a few hundred feet above the surrounding land, which the community wants preserved.)

LG Collection as an Energy Source: A collection system of stone-lined ducts was installed to vent the gases in a controlled manner. Early accumulations of LG were flared until sufficient volume had built for a usable energy source. Encouraged by the 1983 Energy Act, which stimulated private investment in electricity generating, Greenways found sufficient quantity and quality of LG flowing from the landfill to power generators supplying 540 kilowatts of energy to the national grid. As the landfill progressed, gas production increased and power generation has increased to 1.5 megawatts.

Because of the large size and depth of the quarry opening, the estimated life of the landfill is forty years in serving the local area only (see Figure 17, page 413). Dried silt or clay from the quarrying operation and dredging from the settlement lagoons provide lining and cover material for the landfill. Ground water monitoring has accompanied all steps through the project.

ARC-Greenways, U.K.

Figure 17 View of Judkins Quarry — Landfill, U.K.

Upon closure of quarrying and partial surface plant area reclamation, land-use proposals will commence whilst landfilling will continue. Considered after-uses include an industrial office complex, small industries sites, recreational areas including a 9-hole golf course, public park and outing facilities, boating on the adjoining canal, footpaths. Collection of LG will continue whilst the landfilling continues, and for some years afterward. Greenways will continue to monitor the site for leakage from the landfill for years after it closure. After the landfill closure, a large area will become available for more and new surface uses (ARC Ltd., 1989).

11.8.3 Allerton Park Sand & Gravel Pit, near Knaresborough, North Yorkshire

This surface mine is of particular interest to this study because it was planned from almost its beginning as one operation, i.e., both mining and landfilling. This is the essence of the thesis, that new surface mines should incorporate into the planning and applied permission the process for landfilling with MSW as part of the reclamation of the mined land.

Allerton Park is seven miles North of Wetherby and four miles East of Knaresborough and adjoins the A1 carriageway. In 1987, ARC applied for permission on a 63 ha tract (155.67-acres) for extraction of sand and gravel, proposing to mine "about 4 million tonnes over a 15 to 20 year period, and to progressively restore the site to agricultural use by controlled landfill using domestic and non-toxic commercial and industrial wastes". Permission was first granted in 1988 for extraction by the North Yorkshire Council. In 1991, the site licence was issued and landfill operations began alongside continued mining. 41 ha (101.3-acres) of the site was designated for mining and landfill.

Project Statement: In its application statement, ARC pointed out the environmental acceptability of the location. It should be noted that the statements are those of the modern, environmentally-concerned and environmentally-responsible type of operators in the mining and waste disposal industries as put forth in the Opening Statement of the Introduction, Ch. I §1, i.e., "*** a new generation and 'breed' of operators with environmental awareness, ***".

Noting that "Minerals can only be worked where they occur naturally", ARC stated that "Much can be achieved at a modern mineral operation to minimise the potential nuisance at the site itself. Site screening; landscaping; tree planting; noise and dust suppression are accepted by responsible mineral operators as essential requirements for any mineral site".

With regard to the potential, omnipresent concern over added local traffic from the proposed project, ARC proposed building its own roadway on its property for direct access to a junction on the A-1 carriageway, adding, "The location of the proposed Allerton Park development has a great advantage because it will enjoy direct access to the A-1 carriageway, thus allowing the market area to be supplied using primary routes and avoid the local problems so often associated with quarry traffic".

Geology: With regard to site geology and protection of the ground water, a pre-proposal investigation revealed the Quaternary gravel and sand deposit was underlain by the Triassic Sherwood sandstone. "All boreholes penetrated some boulder clay, often to a depth of at least 20 m., so proving this area as the precise location of a large glacier which deposited its detrital load on shrinking northwards. Many of the boreholes also penetrated sand and gravel which has a wide thickness variation from 0.3 m. up to 15.4 m. This layer is the product of late-stage torrential outwash from perhaps a dam near to the glacier which first eroded through the boulder clay before depositing its relatively well-sorted load of sand and gravel. At its lowest point, proved so far, the base of this in-filled valley is at a depth of 26.7 m., corresponding to a level of about 30 m. above the Ordinance Datum.

"The boreholes also show that this sand and gravel layer was subsequently buried under a relatively large thickness of overburden, varying between 3.2 m. and 5.8 m., consisting in some places of the entire overburden sequence above (herein omitted) but in other, mostly thinner, locations, only top soil and boulder clay".

Ground water Protection: The Sherwood sandstone, below the sand and gravel deposit, "forms part of an aquifer and must be protected from pollution by leachate generated within the landfill. Preliminary meetings with the Yorkshire Water Authority were held to establish the protection measures which will be required.

"The boulder clay covering the mineral will be suitable for lining the base and walls of the cells and for capping upon completion. The base of each cell will be protected by 3 m. of compacted boulder clay in the areas where it is not already present. The side walls of the cells will be similarly protected where they come into direct contact with the mineral. Where extraction takes place below the water table, the level will be raised by compacted boulder clay to 1m. above the water table before infilling commences. Exploration and monitoring work will be carried out to establish the true level of the water table and obtain information on ground water flow through the gravel".

Landfill Method of Operation: "The landfill will take place in clay lined cells and progress from north to south. The cells will be sized so that waste can be brought to final levels and capped with clay before the field capacity of the waste is reached.

"The total volume of overburden on the site is in excess of 2 million m^3 which will provide sufficient material for the cell construction and capping. As the extraction progresses South, the mineral base falls so that the removal will take place below the water table. Before infilling commences in these areas, the void will be filled with compacted clay (1×10^{-9} m/sec) overburden to 1m. above the water table". Additionally, Greenways agreed to incorporate HDPE liners into future cells.

Urgent Landfill Relief for Local Area: In 1988, North Yorkshire County's MSW disposal at Rock Cottage Quarry at Ripon was filled and the Council was forced to find short-term replacement sites for its solid waste. The coming onstream of Allerton Park's landfill in 1991 was timely to provide long term facilities for the Harrogate / North Yorkshire area. The quantity of waste from the Harrogate area was reported to be 110,000 tonnes per year, which required 140,000 m^3 of void space. Allerton Park anticipated a total void of 4 million m^3 which would amply provide for that amount of annual waste input.

LG Energy Generation: An electricity generation scheme fueled by LG is planned as part of the Government's Non-Fossil Fuel Obligation (NFFO). It was anticipated that 0.5 megawatts of electricity would be supplied to the National Grid.

Land Restoration: Progressive restoration has been taking place during operations. As topsoil is taken up by stripping for new mining areas, it is attempted to replace it in newly completed areas without storing. Where this is not possible, topsoil is store in mounds and the surface grass seeded until it can be surface-replaced. When landfilling is completed, the area will be restored to forestry and agricultural use.

"Following final placement of the topsoil within each parcel of land, an Aftercare Programme is drawn up and submitted to the local Planning Authority for their approval. The results of such plans are reviewed every 5 years and supplemented where necessary".

(Quotations are taken from ARC's Allerton Park written statement in conjunction with their planning application, Estates Department, 1987.) (See Figure 18, page 417 for view of pit and landfilling operation.)

ARC-Greenways, U.K.

Figure 18 View of ARC-Greenway's Allerton Park Sand & Gravel Pit-Landfill, North Yorkshire, U.K.

11.8.4 *Miscellaneous U.K. Combined Open Pit — Landfill Operations*

A brief description of eleven other combined surface mine-landfill sites operated by ARC-Greenways are given below in tabulated form.

Table 3 Miscellaneous U.K. Combined Open Pit — Landfill Operations

Site name	County / Shire	mineral mined	date mining	date landfill started	gas collected started	area ha.	land use restoration
Stangate*	Kent	limestone / sandstone (ragstone)	1945–1975	1981	yes	33	agricultural
Offham	Kent	ragstone	1947	1981	yes	?	agricultural
Ongar	Essex	brick-clay	1895–1981	1985	yes	28.5	agricultural
Sutton Courtenay*	Oxford	sand / gravel	1930's	1977	no	360.	agricultural industrial amenity
Dix Pit *	Oxford	gravel	?	1985	no	32	pasture
Edwin Richards	West Midlands	dolerite	1850s	1991	yes	?	recreational
Burntstump	Nottingham	sandstone	?	1960s	yes	?	agricultural
Kaimes*	Kirknewton Scotland	whinstone (basalt)	18th century	1987	no	?	agricultural
Aldeby*	Norfolk-Suffolk	gravel / sandy marl	1953	1989	no	?	agricultural
Bradgate*	Leicester	diorite	1900	1980s	no	?	grassland
Llanddulas*	Clwyd Wales	limestone	late 1800s	1985–86	yes	?	agricultural woodland

(* indicates additional information below)

***Stangate Landfill** — serves some of the London boroughs, is licenced to accept domestic, industrial and commercial waste. The site uses the dilute and disperse principle and uses compacted ragstone-spoils over the base, which acts as an attenuation blanket. A combination of ragstone / hassock and HDPE lining is used for sidewall sealing. A tyre-shredding plant operates at Stangate to allow safe disposal of commercial vehicle tyres.

***Sutton Courtenay Landfill** — This landfill has a rail-siding and receives waste on to the site, thus reducing road traffic in the local area.

***Dix Pit Landfill** — Landfilling is designed and operated on the principle of total containment using *in situ* clay as an impermeable lining.

The site is of archaeological and scientific importance. It contained the Devil's Quoits, an ancient monument constructed by Neolithic man some 4,000 years ago. The monument was a stone henge in a great ditched enclosure in a circle of about 36 blocks of stone, each 6 to 7-feet tall and several tonnes in weight. Care was exercised in mining about the prehistorical monument. In the 1970s, mining revealed a clearer picture of the original form of the Quoits. During restoration of the site, Greenways worked with Oxford Archaeological Unit to restore the Quoits as a place of national historical importance over a three-year project.

Also, on the same site, Greenways, working with Earthwatch, has unearthed more than 200 bones and teeth of large prehistoric animals, as woolly mammoths, horse, bison and hyena. At least 25 mammoth tusks have been found, the largest almost 3 metres long and 300 lbs.

***Kaimes Landfill** — This site serves the city of Edinburgh and nearby communities. It is the only landfill site in Scotland that receives waste by rail; 120,000 tonnes of waste are received each year by rail in sealed containers.

***Aldeby Landfill** — This site is partially engineered with HDPE liners as wall seals.

***Bradgate Landfill** — Prior to depositing MSW, the walls and base of the quarry were lined with clay (Mercia mudstone) won from the quarry rim, overlaying the diorite.*

***Llanddulas Landfill** — Operated on a containment principle; base and side (wall) seals were formed with a comprehensive composite lining system consisting of clay and high density polyethylene (HDPE).

In addition to the above eleven quarry-landfill sites, Greenways operates 14 other quarry-landfill sites inclusive of Judkins and Allerton Park. These are scattered about the midlands and the Southeast, and one site each in Wales and Scotland. The 25 landfill sites and 8 household recycling centres handle about 2.7 million tonnes of waste per year. Landfill gas at six sites are reported to generate 9.6 megawatts of electricity. (Harding, 1996).

11.8.5 *U.S. Example of Combined Quarrying and Landfilling*

The Fred Weber Inc. Quarry and Landfill, Maryland Heights, (St. Louis Co.), Mo.

The following information on the Fred Weber Inc. operation was furnished by Mr. Marc Ramsey, Manager of Weber's stone plant and landfill.

The Fred Weber quarry began its aggregate mining operations in its present limestone pit in 1928. It currently produces 2,000,000 + tons / a. of crushed stone in various sizes. The landfill business was started in 1974. Three expansion permits have been granted since. The sanitary landfill area covers 54 acres (see Figure 19, page 421, for a panoramic view of the Fred Weber quarry-landfill operation).

With regard to zoning and public hearings for the combined operation, Mr. Ramsey feels that because of being a local industry of long standing and maintaining good community relations over the years, permitting with local approval has not been a problem. Due to their excellent past record, he states, "Our standing in the community and our operating record have been the key reasons for our succession of permit approvals. We anticipate this to be the case in the future for our additional requests".

R. Lee Aston, Univ. of Aston, England

Figure 19 Fred Weber Quarry, St. Louis County, Missouri, USA

The geological and technical features for the Weber landfill-quarry pit are as follows:

The quarried stone is the highly calcareous St. Louis and Salem limestone. Hydrogeological tests show that the limestone walls of the quarry have a low permeability. No other protective coatings on the quarry walls are necessary. The limestone formation is underlain by a 40-foot thick bed of very low permeable Warsaw shale which forms the floor of the quarry landfill. Floor depths in the quarry are as much as 200-feet below the surface elevations. The bottom of the landfill is about 120-feet below the normal water table.

The hydrologic key design of the landfill is based on an inward gradient ground water migration / flow. Any flow on water will be inward, i.e., there is no chance for leakage from the landfill.

Landfilling Materials: Accepted fill materials are MSW, construction and demolition wastes; mattresses and shredded rubber tire material is accepted; incinerated residue is not accepted. The landfill accepts between 800,000 and 900,000 cubic yards of waste annually; 60% is MSW, 30% is demolition waste,which includes metal (pipes, steel reinforcement bars), and 10% other.

Two other parts of the quarry have already been filled, covered with soil and revegetated.

Liners: no high density polyethylene (HDPE) liners are deemed necessary or used; only a compacted soil liner 12-feet wide is used along the sidewalls of the quarry. The soil liner is applied in 18-inch lifts, horizontally against the walls.

Monitoring Wells: Four wells are sampled quarterly for Subtitle D (CWA) parameters. Nineteen additional wells serve as piezometers to confirm that the hydraulic gradient is toward the quarry-landfill.

Leachate Collection and Disposal: Leachate is collected and pumped to a sewage treatment facility that serves the area. The leachate strength is weak and no special charges are incurred for disposal. The collections system is composed of a porous, 3/8-inch, pea gravel blanket, one-foot thick, sloped to one end of the pit. Leachate is collected at a low-end sump and removed via a lift station to the Metropolitan Sewer District trunkline.

Landfill Gas (LFG) Collection System: LFG is composed of 50% methane and 50% carbon dioxide. It is estimated by scientists and engineers that it takes at least a year, depending on the moisture content, for the trash to begin decomposing and producing gas. It takes one truckload, 20 cubic yards, of waste to produce 40,000,000 BTU's of methane gas. This is equivalent to seven barrels of oil.

Recovery of the gas is done by sinking six-inch PCV pipes 80 to 100 feet into the heart of the landfill. The pipes are connected to well head pipes at the surface level. The well heads are connected to a central point about a quarter-mile from the landfill where a 40-h.p. centrifugal blower draws the gas from the well heads at approximately 1,300 c.f. per minute. The gas is fed to the main plant where it is converted into energy.

Weber uses the landfill gas, as a substitute for natural gas, in a number of different applications in the plant: (1) seasonal heating of a commercial greenhouse operation with six unit heaters, each using 350,000 BTU's per hour; (2) asphalt plant hot oil heater, using 6 million BTU/hr. burner, continuous usage; (3) asphalt plant aggregate dryer using a co-mix of LFG with natural gas, using 20 million BTU/hr. LFG and seasonal usage; (4) steam boiler for asphalt ready-mix plant, 100 hp. boiler using 6+ million BTU/hr., seasonal usage; (5) plans are being made for heating supply of LFG to an adjacent high school for use in two 165 hp. boilers — to be brought on line in the fall of 1996.

Officials at Weber estimate that the company saves over $100,000 per year in fuel costs for their asphalt business. Also, Mr. Ramsey states that the adjoining greenhouse operation saves "about $2,000 to $3,000 per month in fuel costs from November to May where the temperature is maintained between 65 to 70 degrees". Weber has been selling plants from their greenhouse since 1983. Its crop includes 15,000 geraniums, 5,000 marigolds, a few hundred hanging baskets, et al.

11.8.6 *Miscellaneous Reported Examples of Quarries Used for Landfills*

The Lemons Gravel pit at Dexter, Missouri is currently used as a MSW landfill.
The Greenwood County, Kansas, landfill uses a former limestone quarry.
The Allen County, Kansas, landfill uses a former limestone quarry.
Eleven derelict surface coal mines in Kansas, Kentucky, Illinois and Indiana are given as examples of mine-landfill operations at Ch. 10 §7, supra.
A former sand and gravel pit at San Gabriel Valley, California, is used as a landfill.

11.8.7 *A Canadian Example of a Successful Combined Quarry-Landfill Operation*

The Walker Industries, Quarry and Landfill, Thorold, Ontario

The following information on the Walker Inds.' operation was furnished by Mr. Ron Plewman, P.E., and Operational Manager for the company.

Walker Brothers' crushed stone quarry had its beginning in 1887 and has been in continuous operation since. The quarry is located at Thorold, in the south-eastern part of Ontario Province near Niagara Falls, Canada, near the New York state border.

The quarry operates in the Gasport and Goat Island limestone member which is underlain by the Decew limestone and the Rochester shale. Stone production is about 800,000 tonnes per year. Landfilling began in November 1982 and has accepted a total of some 4,630,000 tonnes of waste since its beginning. It currently accepts between 300,000 to 600,000 tonnes of waste annually (see Figures 16 through 20, pages 411–426, for views and explanations of the aerial photographs and quarrying-landfill progression from 1986 to 1995.)

The area of active quarrying in 1986 has been the active landfill in the past few years and was filled to capacity and capped in 1995. A new cell was started as an active disposal area in June 1995 and has a life expectancy of 12 years. Future cells will be created as quarrying produces new void space.

The landfill design uses a compacted clay base and clay on the sidewalls. The "clay" is an impermeable type that is recovered from the fines of the mining operation. A liner is also placed on the sidewalls to direct leachate downward in the fill. This sideliner acts as a drainage medium and is of a geotechnic type, bonded on both sides to a geonet.

The site, although licenced for municipal waste, predominantly takes industrial and commercial waste. All waste must be pre-approved for disposal and is monitored on a continuous basis at the entrance. No recycling on the property is done before dumping of waste.

The leachate collection system is composed of 6-inch plastic pipe installed in a clean stone bed layed on top of the clay bed bottom liner. Leachate drains by gravity through the fill to the collecting system at the bottom and thence to a leachate sump pit with a pumping station. The leachate is pumped to the surface to holding ponds for testing to ensure discharge guidelines are met, treated on site if required, and then pumped through the company's private sewer lines that connect to the regional system and treatment plant. There is no collection of landfill-generated gas. LG is vented and flared.

A significant ground water monitoring system exists and was begun approximately 15 years before landfilling began. The ground water table level is reported to be 7 to 8 feet below the landfill's bottom clay liner. Daily waste layers are placed in 8-foot depths and a spent foundry sand is used as daily cover material. Clay is also used as the capping material.

As indicated in the two aerial photographs, Figures 20 and 23, taken nine years apart, in 1986 and 1995, and the accompanying explanatory sketches, Figures 21, 22 and 24, completed areas of fill have been restored to beneficial agricultural land uses for stock grazing, tree growth (sugar maple and reforestation), and vineyards (see Figures 25 and 26, page 431). Overburden soil from the quarry area was saved, enriched and spread in restored surface areas for growth of grasses, grapes and trees. The spring maple sugaring activity is now a tourist attraction drawing some 12,000 visitors each year (Plewman, 1996).

11.8.8 *Miscellaneous Canada*

A personal communication with the New Brunswick Department of Environment emphatically states, "Mined-out open pits do not meet the stringent siting guidelines for sanitary landfills in the Province and would not be

Walker Inds., Ontario, Canada

Figure 20 1986 Aerial view of Walker Inds., Quarry and Landfill, Thorold, Ontario, Canada

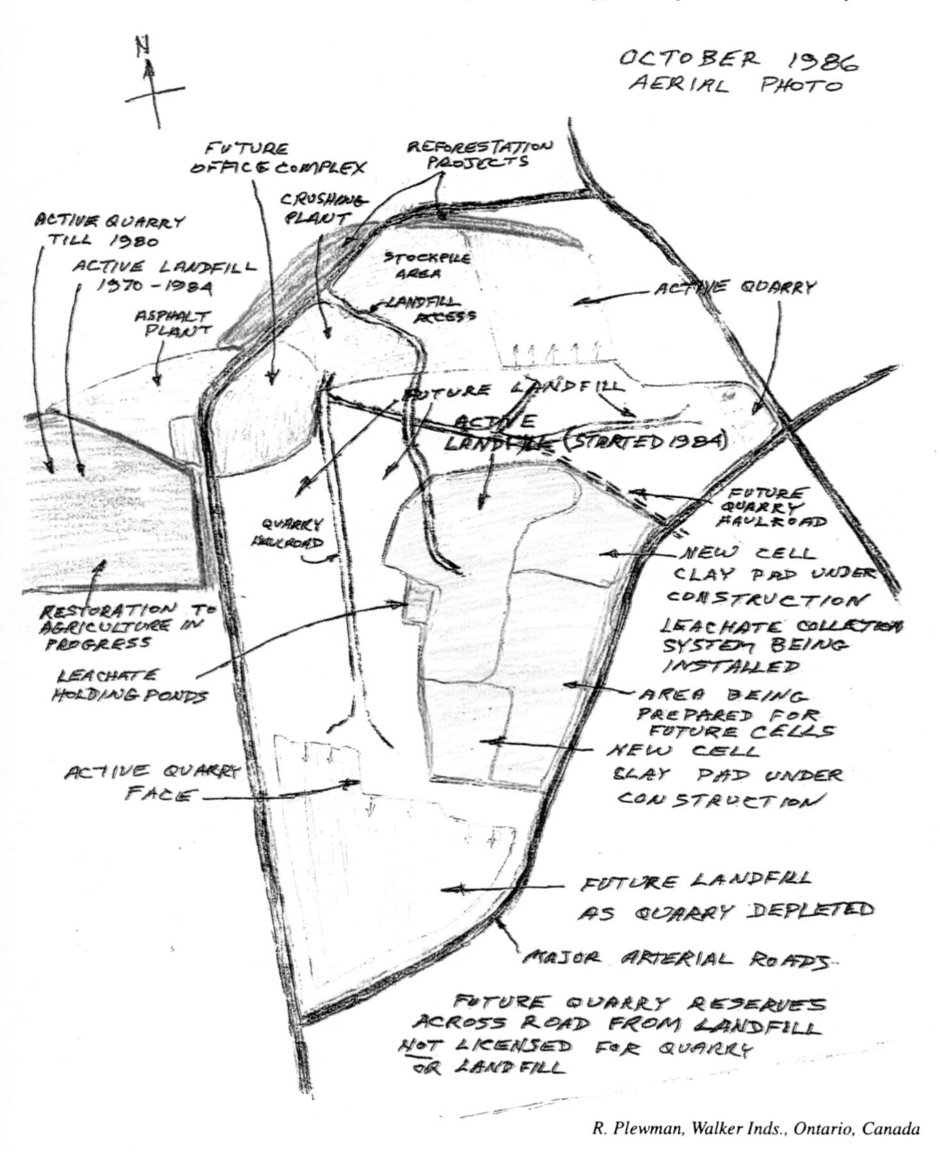

R. Plewman, Walker Inds., Ontario, Canada

Figure 21 Walker 1986 Explanatory Sketch of Figure 20 Quarry-Landfill

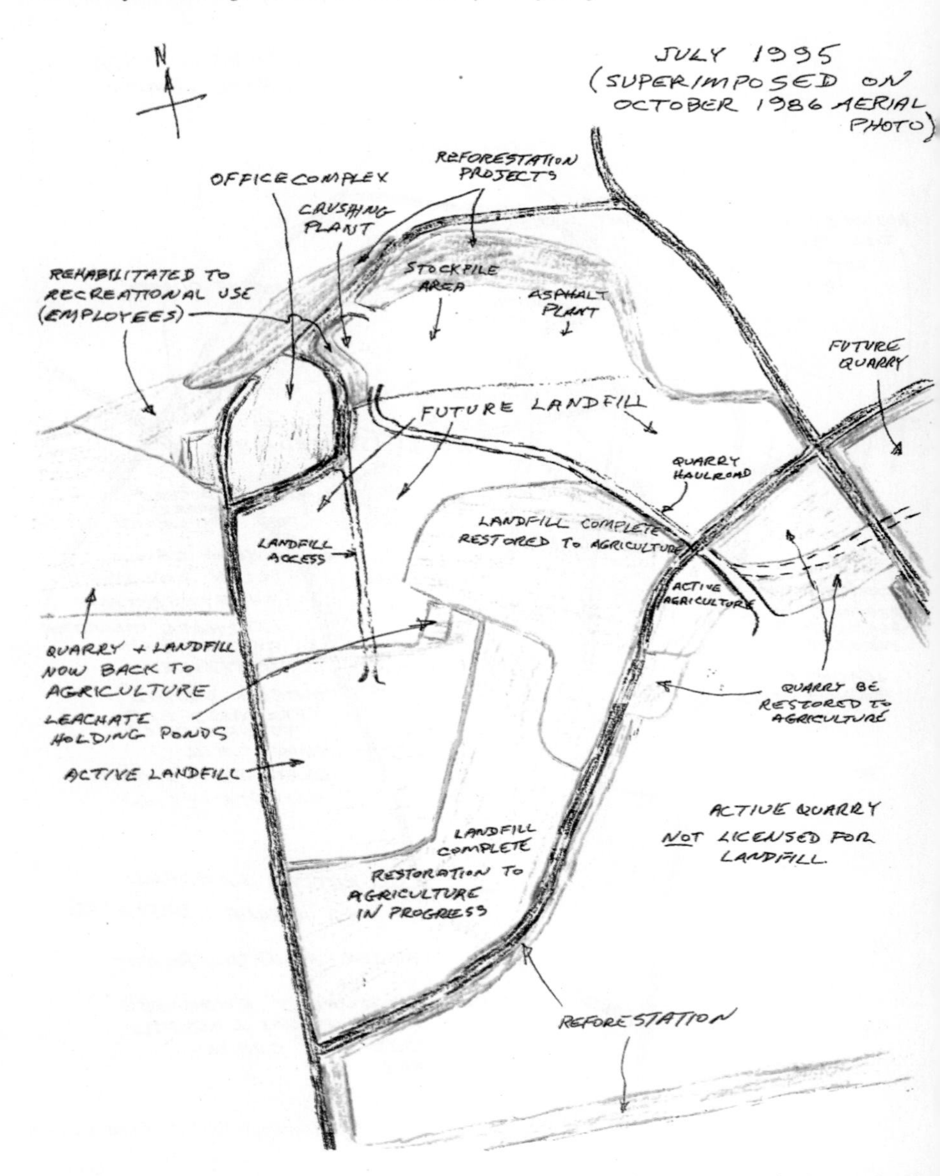

R. Plewman, Walker Inds., Ontario, Canada

Figure 22 July 1995, Explanatory Progress Sketch of Figure 20 — Walker Inds. Quarry & Landfill, Ontario, Canada

Walker Inds., Ontario, Canada

Figure 23 May 1995 Aerial Photo of Walker Inds. Quarry & Landfill, Ontario, Canada

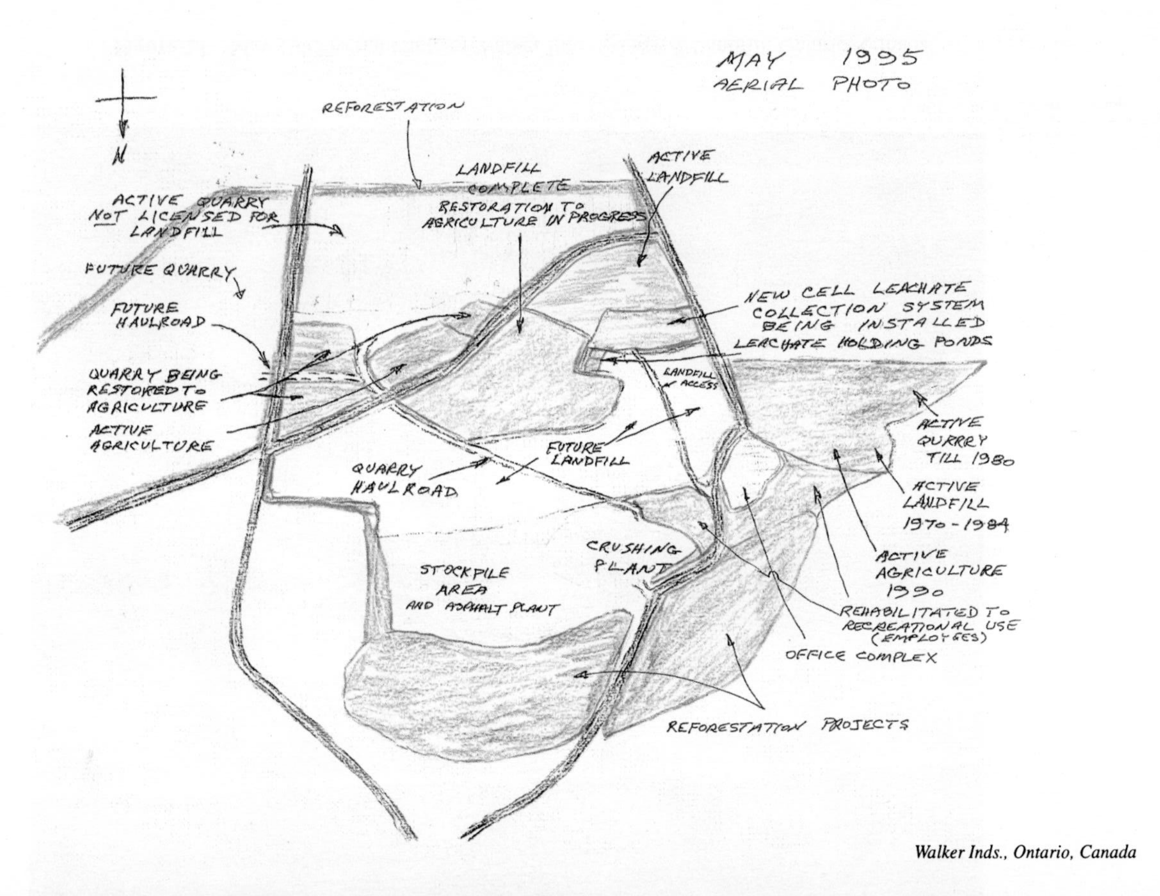

Walker Inds., Ontario, Canada

Figure 24 May 1995 Explanatory Sketch of Figure Walker May 1995 Expanatory Sketch of Figure 23

Figure 25 Walker Inds. Canada — Land Restoration After-Use — Agricultural Maple Sugar Production

R. Plewman, Walker Inds., Ontario, Canada

Figure 26 Walker Inds., Canada — Land Resoration After-Use — Grape Vineyard

acceptable under the regulations". And the same prohibition is seemingly borne out in communication with the Saskatchewan's Municipal Waste Management which states that "mined-out open pits have not been used for landfills, *** and there were no plans to utilise mines for garbage disposal". This was the only negative-sounding report from any of the provinces.

11.9 Summary: Conclusions and Comments

Regarding the use of peat in landfills, intense investigation for the use of peat in landfill base liners and in a leachate filter beds before entry to the leachate collecting tank is strongly recommended. With testing reports of such high potential for reducing contaminants in landfill leachate, the peat-treated effluent could be returned directly to surface waters without harm.

A problem for all landfills, whether without or within quarries, is the concern for the possible effects of differential settling. The amount of biodegradable matter present in the fill, the homogeneity of the waste materials, and the degree of compaction used, are all matters affecting the rate of settlement. Field loading tests of the fill may be appropriate. The degree or amount of liquids, particularly within containment-type fills, is of importance. Load-bearing structures on engineered containment landfills is inadvisable. With regard to quarry-landfills, numerous benches within the pit may be a concern causing differential settling, depending on the preceding factors of composition and compaction.

The problem of differential settling only becomes a major concern where load structures are intended to be placed on the completed surface. Where the intended surface is not load-bearing, as for agriculture or amenities, few problems of import would be anticipated. Unless the settlement rate appears to be uniform and stabilisation reached, the after-use of the landfilled surface should be limited to lightweight purposes, e.g., recreation, parks, parking lots and amenities with only lightweight structures.

The Omnipresent Critical Need for Waste Space: In spite of the current solution of waste deposition by landfilling a few of the surface mining operations in the U.K., until a far larger number of surface mines are designated as MSW depositories by law / regulation, as proposed herein, the waste space problem will remain unsolved. Surface mines easily create a sufficient amount of void space to accept all the waste generated in the U.K. Only a partial percentage of mining voids is being used. In spite of Greenways efficient handling of MSW in its 25 quarry-landfill operations, void reserves for waste disposal are reported to stand at a short 16 years of life at current input rates (Harding, 1996).

CHAPTER 12

FUTURE LEGISLATIVE PROBLEMS

12.1 Introduction

The approval and permitting, or licencing process for projects, both public and private, as reviewed in this work, seems to bog down most frequently in the early stages following proposal, i.e., during the public hearing stage, to be delayed and followed by prolonged appeals and litigation. This costly time period of tossing a project's application back and forth between proponents and opponents is an area needing much repair by legislation and regulation.

It is urged and advocated that the general public place more credence in the scientific and engineering studies available to various environmental authorities. In keeping with this, it is further advocated that the amount and extent of public hearings be minimised, in deference to scientific and engineering environmental studies. To avoid prolonged involvement and squabbling by the PIP which often results in defeat of vital projects and impedes environmental progress, an override decision power should reside in the government agency having permission control. Such override decision power may be used where a vital project's defeat by the public hearings was found to be based on public misinformation, by unfounded fears, or that an environmentally curable situation exists to overcome the opposition's reasoning.

At times, a discretionary power residing in a responsible governmental official is sufficient to eliminate an anticipated troublesome PIP. For support, the British procedure is noted supra, at Ch. 5 §4.8, Conclusions, viz., Public involvement in the planning and permission approval process for land development is limited. Hearings may be held at the discretion of the Secretary of State for the particular ministry, **but may limit those participating in the hearings. The Secretary of State has wide discretion in the hearings and**

may, or may not hold them. The ultimate decision for planning permission, whether contentious, or not, is placed in the discretion of the Secretary. (emphasis added).

Where discretionary power is not available, a power of decision override, or "command and control" type of environmental regulation should be extended to the licencing process for projects. After the public has voiced its reasons for opposing a project, an environmental study or assessment of the project, paying particular attention to the public's objections, may be made by the permitting board or council. Where there is substance found in the opponents' objections, the study may determine whether they can be treated by science and engineering to eliminate, or minimise to acceptable standards, the potential risk of the feared environmental harm. Where the opposition's objections are without genuine basis in fact, the override decision power may be exercised for approval of the project. The following chapter studies the present process in more detail.

12.2 *U.K. — Planning and Permitting of Future Surface Mining*

At the 1994 British Geological Survey's Minerals Extraction forum with the theme, The Economic and Environmental Balance, it was noted that the British mineral industry's main concern was with the growing difficulty in obtaining planning permission and the length of time it takes, particularly when the issues are contentious and a public inquiry proceeds. "Martin Kingston QC urged the minerals industry to take a more positive view of the public inquiry process (PIP) and to support the system, not undermine it. He saw no better alternative to the PIP for planning judgements and stressed the vital importance of testing evidence by cross examination. The main shortcoming of the PIP is that there is no requirement to provide the best available evidence. The appointment of an assessor early on, who could visit sites at the pre-inquiry stage would do much to expedite proceedings". (Mining Journal, London, April 22, 1994, page 286).

A primary flaw in Mr. Kingston's argument of urging "the minerals industry to take a more positive view of the public inquiry process (PIP) and to support the system, not undermine it" is that it is in conflict with reality and directed to only one side of the hearing process. His urging would be better directed toward all those who misuse it. Reference is made for this point by the article "Public

Misuse Of The Mineral Regulatory System", Ch. 8, §2.2. A questionable feature of his suggestion is the selection of an "assessor" to visit the sites at the pre-inquiry stage. Will the "assessor" be impartial in reporting on the site and the alleged environmental effects? The purpose of the environmental assessment report is to accomplish the same.

As previously argued, the public hearing and appeal process should be sharply limited to avoid delays, costly, vexatious and prolonged and repeated appeals and lengthy litigation. The weak point in the public hearing process, or PIP, is that allegations against projects can be freely made by the opponents without having to prove their assertions. Proposed projects may be slowed down, made unnecessarily difficult, or defeated by unsubstantiated claims of opponents. Even under the present PIP announcing the time and place, the public should be warned that frivolous claims cannot be entertained or heard by the board of inquiry unless substantiated by competent evidence. The preliminary inquiry should be for the sole purpose of opponents filing competent assertions. The inquiry council, or hearing board, may then study the evidence of competent claims filed. After determining those competent assertions that have carried a sufficient burden of proof against a proposed project, the hearing authority may then hold a public hearing on the merits of the opponents' submissions of proof. Those making claims that environmental injury, or harm to the public will result, should be required to carry the burden of proof as in any court procedure following a filed complaint for an action. This is necessary to avoid frivolous and unfounded claims simply to defeat a proposed project by slander.

After the opponents of a proposed project have shown cause for their assertions, and having sufficiently submitted evidence and proof that their allegations and claims are justiciable, thus having carried the burden of proof to the hearing authority, the public hearing process can proceed. Should a sharply disputed project between the proponents and opponents over environmental concerns continue, and where an EA report or study is not required for a local project, a mini-EA should be ordered by the proper hearing authorities to evaluate and to either prove or disprove the initially made allegations of environmentally-destructive assertions. The burden of proving the assertions made and the cost to carry them forward should be placed on the parties making them.

After the EA or study is made and reported to the authorities, their decision should be final, or at least, the number of appeals should be limited to no more than one, not continuing through numerous appeals to the highest court of the land.

In July 1994, the Scottish Office Development Department published a consultation paper entitled "Review of the Planning System in Scotland" inviting views on how the system might be improved. In its December 1995 review, the Scottish Office stated "The document generated a lively debate and 172 submissions were received making over 4,100 detailed comments about the planning system and the way it works. Details of the responses were published on 5 June 1995. *** It is clear that, while the majority of people indicated general support for the planning system, many share the Government's view that it is not working as well as it should. This manifests itself mostly in the operational aspects of the system, but also in the complexity of the legislation and the procedures it generates. Most of the problems identified are familiar and relate to delays in decision making, ineffectiveness and inconsistency. These shortcomings, in turn, tend to undermine the credibility of the system and its ability to safeguard the public interest. This has resulted in a wide range of suggestions for steps to be taken to speed up decisions and simplify the procedures while at the same time reinforcing the means for quality assuring both the planning process and the outcomes on the ground. *** There is a tension between a desire to streamline the overall process, on the one hand, and to build in safeguards in the interests of amenity and the local community on the other. The Government does not believe these objectives need necessarily be in conflict one with the other and that greater efficiency and effectiveness can be delivered through simpler, clearer and tauter procedures without abandoning the rights of interested parties to make their input in the decision-making process. The first priority, therefore, is to ensure that the basic tasks of processing applications, development plans and appeals are carried out as proficiently as possible.

"**Planning Permission Appeals** Considerable criticism has been directed at *** the delays encountered in reaching decisions on them. The upsurge in local plan inquiry work over the last 2 years, or so, and an abnormally large number of long planning inquiries resulted in a substantial backlog of appeal cases ***. The number of cases processed *** has more than doubled in the course of 1995. By November 1995, 70% of these cases had been dealt with inside 28 weeks". Performance targets for deciding public local inquiry planning appeal cases are 48 weeks, and 28 weeks for written submission cases, with 80% being decided in both types in the proscribed time limit.

Noteworthy are the comments made by a planning manager regularly involved in submitting plans for excavating minerals with full surface reclamation incorporated in all the plans. "CCCL used to pride itself that we had never had a refusal of planning permission and had never gone to Public Inquiry. This was because we did the 'spade work' before applying to ensure the best possible scheme the first time. *** Despite this, we have now had 1 or 2 refusals and, although we have not gone to Public Inquiry, we seem increasingly to be heading that way. This is because of —

a) increasingly strong environmental lobby;
b) 'Political' decisions contrary to officers recommendations to approve;
c) increasing number of 'greenfield' sites.

"**All** open-cast schemes attract some opposition — it varies from 2–3 letters of objection to full-blooded war (e.g., anti-open-cast residents' groups, petitions, public meetings, lobbying, media reports, production of videos and engagement of consultants by resident groups, etc.). On a competitor's site in Wales, objectors chained themselves to the bulldozers and occupied the offices.

"Appeal / public inquiries are very expensive (£50,000 to £100,000 plus) and are a last resort. The time scales are as follows —

Conventional planning application with no Environmental Assessment (EA) — local authority statutory time scale for determination is 8 weeks, although most take 4 to 8 months.

Planning applications with EA -16 weeks but more likely 6 to 12 months".

12.2.1 *A Case in Point: Defeat of a Multi-Appealed Mining Application in Scotland*

M.I. Great Britain Ltd. (M.I.) mines barytes from the Foss deposit near Alberfeldy in Scotland. Barytes (barite) is barium sulphate, a non-metallic mineral of high specific gravity used in drilling fluids by the international oil industry. The Foss operation produces between 50,000–60,000 t / y. of direct shipping grade ore. 97% of the Foss production goes directly into the North Sea oil industry. The Scottish operation meets about 25–30% of the drilling mud requirements for the North Sea oil industry, and because of its proximity to the North Sea area, it has a distinctive price advantage over imports from more distant mines in other parts of the world. Reserves of barytes at Foss are expected to last about four to five years more from 1995 at its current rate of production before playing out.

In looking ahead to carrying on its business, M.I. possesses a worldclass, unmined deposit of bartyes near Foss, in an adjacent location at Duntanlich. Proven reserves in the undeveloped deposit have shown it to be one of the largest known barytes deposits in the world. It could furnish as much as 200,000 t / y of direct shipping ore, sufficient to supply the entire needs of the North Sea market for the next 30 years. The deposit is better suited to underground mining, thereby having less impact and being less intrusive on the surface environment than surface mining. Production from the deposit would be even more competitive on the market than the Foss baryte. M.I. has estimated that its production could reduce the U.K.'s import bill by approximately £8 million per year. Additional benefits are those attributed to any industry, that is, supplying employment for a local workforce, and here, about 90% of the expenditures during the new operation would be spent in Scotland.

In July 1991, M.I. applied for planning permission for an underground mine at Duntanlich, but its application was refused by the Perth and Kinross District Council. M.I. appealed the refusal to the Secretary of State for Scotland in October 1992. A lengthy public inquiry lasting six months began in May 1993. An official "Reporter", appointed by the Secretary of State (similar to Mr. Kingston's proposed "Assessor", above) submitted his account of the proceedings three years later in June 1994. On the basis of the reporter's account, the appeal was refused. The issue and conflict has not ended even after prolonged public hearings.

M.I. has lodged an appeal of the Secretary of State's decision with Scotland's Court of Session. The charges made by M.I., *inter alia*, allege that "a number of the reporter's findings were not supported by the evidence at the public inquiry, and that the Secretary of State based his conclusions on a draft version of the National Planning Policy Guideline 4 for Mineral Working rather than on the final version which was materially different. In doing so, it is contended that he erred in law in holding that the economic benefits of the proposed mine were outweighed by the identified environmental uncertainties".

Although the environmental "uncertainties" were not elaborated, nor made clear in the report, they would clearly have to be major to surmount the socio-economic benefits to the local area and the U.K. as well.

The environmental disturbance at the mine site should certainly be minimal; there is no obtrusive surface mine working, nor surface plant concentrator, smelting or refining operations to mar the scenery or to pass gases into the air. The mine site is 9 km. from the nearest town, but is in a scenic area of the Scottish Highlands. The opening on the surface is minimal, being only a shaft for underground mining. There is to be no large plant for the surface other than for crushing, perhaps pulverising and loading for transport to Aberdeen. Aesthetic and surface intrusion could only be minimal. Those, in the present day, are ordinarily dealt with by plantings and screening from public view. It would appear that the socio-economic benefits clearly outweigh any biophysical environmental effects in this case. At worst, the environmental concern could be for surface subsidence, which could be dealt with as a condition precedent for underground mining permission by requiring pillars for surface support to be left in place in the mine (The Mining Journal, London, June 1995, p. 413).

The publisher makes an interesting conclusory comment: "M.I. may succeed in its appeal, but if it does not, the implications are ominous, not only for future mineral working and exploration, but for industrial enterprises in the Scottish Highlands generally. Preserving the region as a theme park heavily reliant on seasonal tourism to sustain the local economy may suit some but there are others who wish to work in the Highlands who may feel that a modest industrial base is a necessary ingredient of a well-balanced community". (ibid.)

12.3 U.S. Zoning / Planning Permission

Reclamation plans that contain fuller restoration of the surface for future amenity lands, are reportedly becoming of increasing importance in swaying local authorities in favourably obtaining quicker approval of new start surface mining operations and waste disposal sites. The days are passing in many communities when a new in-coming industry can sway approval by contributing to the local tax till, offer new jobs and pour money into the local economy. Applicants for permitting must convince the local citizenry as well as local authorities that any environmental disturbance will be close to nil, that no harm will result, and a reclamation plan for complete restoration of the land will be submitted, and that the new industry have financial responsibility to carry out the plan. Only a thoroughly convincing plan will be acceptable for gaining approval.

12.3.1 *U.S. — Withdrawal of Mineral Lands*

Over recent years, the legislative response to the public clamour for preservation and conservation of large areas of land in its natural state continues relentlessly to withdraw large areas from mining activity access. The latest example is the California Desert Protection Act which was signed into law by President Clinton on October 31, 1994. The Act transfers jurisdiction over nearly 3 million acres of desert land in California from the BLM to the National Park Service which effectively withdraws those lands from mining and mineral leasing. Additionally, the Act has placed another 3.6 million acres remaining under the BLM's authority in the wilderness category which severely limits access to exploration.

Some litigated examples of mineral land prohibitions follow:

(i) The prevention of further mine permitting in three national parks in Alaska was the issue in *Northern Alaska Environmental Center (Sierra Club, et al) v. Lujan (U.S.) (and Alaska Miners Assoc.),* 961 F.2d 886 (9th Cir.1992), before a federal court. In 1985 the environmentalists / Sierra Club challenged that the Park Service had not complied with the National Environmental Protection Act's requirement for an environmental impact study (EIS) before issuing a mining permit. The Park Service was then placed under an injunctive court order barring approval of mining permits until an EIS was prepared for each park. A gap of over five years occurred while no new mining permits were issued.

In 1990, the Park Service issued a record of decision (ROD) which recommended a plan for each park by which it proposed "that it purchase all existing patented and unpatented mining claims as funds became available. Mining claims that threaten the environment in the parks would receive priority for acquisition". In addition, the Park Service would seek a change in law whereby future patents of existing mining claims would convey the minerals only, making them subject to stricter requirements for reclamation of the environment to its original state. For those permit applications for claims that could not be acquired immediately, the Park Service would consider the cumulative impact each would have on the parks and require site-specific mitigation. The Park Service also sought lifting of the injunctive order.

The court granted the dissolution of the injunction and the Sierra Club objected and appealed. On review, the appellate court found that the EIS prepared by the Park Service adequately analysed the possible environmental impacts to make informed decisions for permitting mining claims.

Again, as in British Columbia, the costs to keep mining out of public lands appear to be of little concern to the environmental lobby. The mineral world continues to shrink with decisions for various governments' acquisition of mineral rights and prohibition of mining on untouchable lands.

(ii) In another 1992 case *Alliance v. Lujan (U.S.)*, 804 F.Supp.1292 (D. Mont. 1992) the federal district court of Montana decided in favour of wilderness groups that challenged the issuance of oil and gas leases in the Lewis and Clark National Forest as violative of the National Environmental Act (NEPA). The court found that cancellation of the leases was the proper remedy. The leasing agency had failed to give consideration to the "no leasing" alternative of the law while studying the impact statement.

The application of Section 522(e) prohibitions of the Surface Mining Control and Reclamation Act (SMCRA) continues to plague the coal mining industry. Section 522(e) provides that "subject to valid existing rights (VER), no surface coal mining operations shall be permitted after the enactment of SMCRA within" certain enumerated areas, specifically, on any lands within the boundaries of the National Park system, the National Wilderness Preservation system, and other environmentally sensitive areas; on any Federal lands within the boundaries of any national forest; on lands where the operation will adversely affect any publicly owned park or places included in the National Register of Historic Sites.

Although the Act grants surface mining rights in parks, forests and other restricted public lands, for owners of mineral rights with valid existing rights, the term VER has proven troublesome. Former President Bush's one year moratorium on the promulgation of a new regulation regarding VER and requiring that the 1986 policy provisions be used, will expire on October 24, 1993.

(iii) Valid Existing Rights (VER) was examined in a 1992 Pennsylvania decision, *Gardner v. Com., DER (Penn. Dept. of Environmental Resources)* 603 A.2d 279 (Pa.Cmnwlth.1992), where an owner held coal mining rights predating the enactment of SMCRA in land that had since been expropriated for a state park, but where the mining rights had not been taken by the state along with the surface. The owner claimed that the statutory amendment prohibiting all surface mining in public parks, except those subject to VER, deprived them of their right to surface mine coal they owned, and they were entitled to just compensation. The court found that the claim was not ripe since the claimants had not exhausted their administrative law remedies, i.e., they had not applied for a mining permit testing their claimed VER, and had not been denied the permit.

The conclusions that may be drawn that developed nations in placing their untouchable park, forest and wildlife lands beyond the reach of the mining industry for development of valuable mineral deposits contained in such land areas, are manifold. In creating hard and fast environmental regulations prohibiting mining, the national welfare of each nation suffers extensive detriment. The well-worn reasons are very much alive and still valid that a vibrant mining industry creates national wealth, employment and a valuable tax base. By comparison, beautification and conservation of the land, neither measurable in money, nor to be underestimated in value, has its proper place in the world. Both values may be had by not over-regulating ecological interests for the benefit of one to the detriment of the other. Both are essential to mankind and the environment.

Grievous error is made in the overall viewing of the modern mining industry as a selfish, destructive private interest. The mining industry creates national wealth and well-being for the people. Mining has made great strides in recent decades in becoming conservation-conscious and environmentally-minded in both spirit and practice. Numerous mining projects in the highly-developed and environmentally conscious nations have proven their ability to enter an ecologically sensitive area to extract the minerals, remove the valuable mineral deposits, restore the earth, and return the area to its natural environs, and frequently in better shape than it was found. Annual award-winning programs have been established for the best mined land reclamation project, for example, by the National Stone Association.

12.4 Canada — Withdrawal of Mineral Lands

Additionally, on the minus side of the ledger to the loss of the developed nations, the mining industry is moving to the undeveloped nations where the benefits of mineral development are far more appreciated in creating wealth and national well-being, and where there is less strict and unnecessary environmental emphasis. Advertisements by the developing nations offering attractive and enticing operating conditions appear abundantly in mining publications soliciting mining company investments to come to their countries. An examples is: the 1st Kazakhstan International Mining Exhibition, 4–7th October 1995, at Almaty, Kazakhstan. A full-page ad in the August 11, 1995 issue of Mining Journal, London, announces "Kazmin '95 is a must for any company wishing to participate in Kazakhastan's minerals industry. The conference will provide an excellent opportunity to establish ministerial and state contacts as well as dispensing information on local services, suppliers and 'doing business' in the country. *** The country's mining sector has undergone a major modernisation programme over the past few years and the Government is keen to attract international exploration and mining companies. As part of their commitment to foreign investors, the Government has introduced favourable tax conditions, stream-lined bureaucracy and removed all restrictions on the repatriation of profits after tax". A similar recent ad was placed by the government of Chad for an International Session for the promotion of Mining for November 1995. This solicitous attitude has placed Kazakhstan as the third most favourable country in the for mining away from the developed nations. This point is further supported by an article in The Mining Journal (London) of November 11, 1994:

CANADIANS ABROAD: CANADIAN MINING MONEY MOVES OVERSEAS

"Two years ago, a government working group in Canada identified some 420 separate mineral property holdings attributable to Canadian mining companies outside North America. The total number of companies involved was estimated to have been of the order of 200. Since then, the number of Canadian companies engaged in foreign activities is believed to have tripled. South and Central American countries have been the recipients of most of this activity, mainly Argentina, Bolivia, Costa Rica, Ecuador, French Guiana, Guyana, Peru and Venezuela. Elsewhere, similar growth levels have been achieved in Ghana and Tanzania.

"Most of the major mining companies in Canada have been engaged in offshore activities for a number of years and the recent dramatic increase reflects mainly the move into overseas activities by Canadian juniors. According to Toronto-based Gamah International Ltd, the amount of capital being invested abroad has soared.

"In 1993, it says, Vancouver Stock Exchange listed companies raised $C1,100 million in financing, up from $C390 million in 1992. Of the 1993 total, $C640 million was raised by mining companies and while a proportion of this equity has been allocated to diamond exploration in Canada, Gamah estimates that at least $C300 million has been allocated to the international spectrum of minerals exploration and acquisition. Supporting this, an analysis of Yorkton Securities' 1993 financings indicates it was a participant in $C280 million destined for international regions.

"Senior Canadian mining companies are investing similar amounts in construction of new mines in countries such as Chile, Tunisia and Turkey. When the full 1993 picture becomes clearer, Gamah surmises that it will show that Canadian miners are investing outside North America at an annual rate of some $C500 million. Mindful of this acceleration in offshore activity, Gamah has compiled a directory entitled 'Canadian Companies Active in International Mining' (CCIM). It highlights the current mineral exploration, mine construction and mining activities by 580 Canadian companies in more than 70 countries". (The Mining Journal, 1994e).

The loss of mineral production and expenditures by the mining industry in Canada is substantiated by released statistics from its Department of Natural Resources for the year 1993. The overall production value of metallic minerals decreased by 13.7%; value of non-metallic minerals decreased by 9.6%. Nevertheless, in spite of the explorational expenditures downturn in its own country, Canada has maintained its worldwide rank as the third largest mineral explorational spender after the U.S. and Australia.

Much of the land in every Canadian Province is open for mineral activities although certain lands are at times withdrawn for that category. For example, Professor Barry Barton reported "In 1986, 4.6% of the land in the Northwest Territories and 16.5% in the Yukon was not available for mineral activity because of parks, conservation areas, and native claim negotiations. *** In 1992, 18.2% of the land in British Columbia was inaccessible (parks, ecological reserves) or severely restricted (agricultural, Indian, mineral and placer reserves, class 1 watersheds, populated areas) (Barton, p. 169). Such vast acreage significantly affect and reduce the base for future supplies of mineral resources.

As noted in Ch. 5 §5.5.1, Canadian Mineral Land Takings By Environmental Regulation and Expropriation, access to mining in Strathcona Park in British Columbia has been prohibited even where pre-existing and valid mining rights to mineral claims were held under Crown and statutory grants, as in the given examples of recent cases of *Casamiro Resource Corp.* and *Cream Silver Mines Ltd* (supra).

Under Canada's 1990 Green Plan for a Healthy Environment, Canadian conservationists' proposals were endorsed to establish protected areas for each of the natural regions for a total land area comprising 12% of Canada. The percentage is significant as reduction of the land-base area of Canada by 12% is a very large area expressed in acres or hectares.

There are presently reported 39 Canadian National Parks and reserves which takes 2% of the nation's land surface. Under Canada's federal National Parks Act, all mineral exploration activity is prohibited, except under the Act and Regulations. There is, however, express authority to grant permits for the mining of sand, gravel and stone for construction purposes only within the park.

The conflict of park areas between mining interests and environmentalists has been in serious contention, but dealt with in several of the provinces. Illustrations are given.

12.4.1 *British Columbia*

About 1986, the provincial government appointed a Wilderness Advisory Committee in an attempt to resolve conflicts between mining and park considerations. In 1988, it adopted a formal parks policy with the Park Act and Mineral Tenure Act. Under these acts a downgrading from a Class B Park to the classification of a "recreation area" was allowed, which in turn allowed the mineral potential in an area being considered for a park to be evaluated by the government and private exploration. The designation of a recreation area is made by the Lieutenant Governor in Council acting on advice from the ministers for mining and for parks. The principal feature for resolving the conflict between mining and land conservation is featured in §19 of the Mineral Tenure Act. Briefly explained by Professor Barton, §19 works as follows:

> "Once designated (as a recreation area), the recreation area is open for staking and exploration. The next phase that §19 envisages is that the government will carry out an evaluation of the mineral potential of the area. The publication of the evaluation, along with a notice of intent to turn the area into a park, commences a ten-year period during which private explorationists are assured that the mineral claims they stake in the area will not be expropriated under the Park Act (B.C. §19(4) and B.C. Reg. 62/89, §17). If explorationists should turn up a major deposit within that period, the plans for a park would probably be shelved or cancelled. However, if at the end of the (ten) year period the area is indeed established as a park, the claims are liable to be expropriated under §19(5) which states: ... no compensation shall be payable to any person in respect of expenses incurred that relate to: (a) exploration and development, or (b) acquisition of the mineral title expropriated or any other mineral title". (Barton, pp. 178–179).

Professor Barton cogently notes that "British Columbia has therefore devised a system that responds to the mining industry's concern with land being "locked up" before its mineral potential has been ascertained and the concern that investment in a claim not be wasted by the sudden gazetting of a park". (Barton, p. 180).

However, in a more recent example in British Columbia, a province already checker-boarded with parks, the provincial government struck a devastating blow to deny mining of the Windy Craggy prospect, an undeveloped, but evaluated, world-class copper-gold deposit in a proposed provincial park. The proposed park would cover a million hectares in the Tatshenshini-Alsek region, northwest of Skagway, Alaska, and will be nominated as a World Heritage Site where mining will be totally prohibited. In its over-emphasis, cost appears to bear no object to keep mining out. The government later announced "talks" to compensate claim holders in the area. The fact that a world-class, valuable base metal deposit, already evaluated and planned for open-cast mining, will be lost with its accompanying benefits to society and industry, along with hundreds of new jobs and a valuable tax base for government, apparently means nothing. Strict mining environmental management controls were to have been employed for the Windy Craggy area as administered by British Columbia's EPA. In spite of what was purported to be compromising legislation between the Park Act and Mineral Tenure Act, the very large and valuable Windy Craggy mineral deposit located before the park was created, fell to environmental clamour to prohibit mining.

It is reported that British Columbia provided 12.8% of Canada's non-fuel mineral production in 1993. A report by Mr. Mike Smith of the respected accounting firm, Price Waterhouse, noted and corroborates the "drawing closed of the mining industry's purse strings" for mining explorational expenditures in British Columbia during 1993:

EXPLORATION: B.C. SPENDING AT RECORD LOW

> "A new report by accountants Price Waterhouse partly blames the New Democratic Government of British Columbia for the lowest level of primary exploration in the province last year since 1968. The report estimates that the mining industry spent only $C9 million on exploration in 1993, almost half the $C17 million recorded in 1992.***
>
> "Since 1991, only one new mine, Eskay Creek, has proceeded to the development stage in British Columbia. Moreover, the province's mining industry recorded a loss of $C14 million last year and has not made a profit since 1989.

> Mr. Mike Smith, author of the report, remarked that the current low of exploration investment in the province is "a significant concern" and is underpinned by high levels of taxation and the provincial government's sweeping environmental reforms, which have created "massive uncertainty" within the industry. In particular, Mr. Smith notes that the government has been slow to agree compensation to Geddes Resources following its decision last year to cancel the Windy Craggy copper project in favour of a provincial park". (*The Mining Journal*, 1994e).

12.4.2 *Ontario*

This province provides the largest share of non-fuel mineral production for Canada, accounting in 1993 for a third (33.6%).

This province has also suffered vacillation in its park planning-mining policy over several decades, waxing between permission and denial for mineral prospecting and exploitation. However, in 1983 a change in position was taken from total prohibition of mineral prospecting in parks. "The government announced 155 new parks, and *** permitted mineral exploration and development in 48 of those parks, or about 80% of them by area. A new government reviewed park policy in 1988 and reinstated the prohibition of mineral exploration and development as a non-conforming use in all parks in the wilderness and nature reserve categories. At present, exploration and development can only proceed on pre-existing claims, leases, patents, or licences of occupation, which are outstanding in 14 different parks. *** 53 new parks were announced along with the policy changes in 1988.

"In considering new parks, the Ministry of Natural Resources considers district land-use guidelines and different land use interests. In some cases, such as the Greenstone Belt of northeastern Ontario, the mineral potential (primarily for gold / Aston) has been considered to outweigh considerations that indicated park potential for the land". (Barton, pp. 176–177).

12.4.3 *Other Canadian Provinces*

Professor Barton notes that provincial "Governments have struggled to improve the way they handle mining and park issues. In Quebec the Minister of Energy and Natural Resources has the power to delimit lands 'for non-exclusive purposes of recreation, tourism or plant-life or wildlife conservation' and to impose special conditions and obligations on the mining rights in those lands or on their renewal (Que. §213.2). The prior approval of the minister is required to stake claims in these areas [Que. §32(5)].

It should be noted that Quebec ranked second of the provinces after Ontario in non-fuel mineral production, accounting for some C$2.78 billion, or 19.5% of Canada's total minerals in 1994.

"Similarly, Nova Scotia has introduced procedures to withdraw land from ordinary disposition and to reserve it for special licences and leases (N.S. §22).

"*** in the Saskatchewan Crown Minerals Act, (S.Stats. 1984-85-86 c.C-50.2 §10.1 added by Amendment 1992, c.25§5) where mineral dispositions may be cancelled after a negative environmental assessment, or after a Cabinet order for other environmental protection purposes. The Act provides that if there is a cancellation, compensation may be available under regulations". (Barton, pp. 175–7, 185). In 1993, Saskatchewan provided 9.6% of Canada's non-fuel mineral production.

In New Brunswick, Professor Barton cites the case of *New Brunswick (Minister of Natural Resources and Energy) v. Elmtree Resources Ltd.*, [1989] 101 N.B.R.(2d) 255 (Q.B.) as illustrating the friction and "*** frustration that builds between mineral explorationists and park officials, but in the context of a legal regime where the issue of a mining lease depended upon satisfying the government as to environmental concerns.

"The area that Elmtree staked was regarded by the government as a unique and sensitive area. Indeed, the Department of Natural Resources and Energy had identified it as a potential ecological reserve, although it had not yet formally established it as such. Consequently, Elmtree was refused a mining lease. Elmtree objected, and although it did not get the lease issued, it convinced the Mining Commissioner that the Department owed it a duty of care to inform it, at the time of recording the claim, of the identification of the land as a potential ecological reserve so it would not spend exploration money fruitlessly. The Mining Commissioner awarded over $5,000 compensation. On the appeal, the Court held that the Commissioner had no jurisdiction to award such compensation, so the question of whether any such duty of care exists was not dealt with". (Barton, pp. 186–187).

Additionally, Judge Stevenson, writing the *Elmtree* decision for the N.B. Court of Queen's Bench, stated that, "There is nothing in *** the Act that gives the (Mining) Commissioner specific authority to determine whether the Minister or his department has a duty to make a prospector or holder of a mining claim aware of proposed ecological reserves". Thus, it is apparent that mineral explorationist must second guess where the government will give future consideration for ecologically sensitive reserve lands that will prohibit mineral prospecting. Presently, under the *Elmtree* decision the Canadian government's mere consideration of ecological reserves land is tantamount to its withdrawal from mineral exploration and claim staking.

12.5 Minerals on Native Lands

A 1993 editorial in Mining Environmental Management, London, encouraged wisdom, temperance and moderation by both environmentalists and the mining groups in their pursuits. Hence, a realistic balance must be struck between continued development of natural resources and preservation of selected lands once occupied by ancient peoples.

A particularly variety of land withdrawal that confronts the present mining industry in developed countries is that containing highly restrictive laws for mining on aboriginal or native lands, particularly in Australia, New Zealand, Canada and the United States. As land areas are increasingly restored to aboriginal tribes, access to mineral resources on those lands become more difficult due to dual negotiations and approval, i.e., with the tribal councils and the administering governmental agencies. The United Kingdom, singularly, does not share the problem of aboriginal claims to lands.

12.5.1 *U.S. Aboriginal Land Claims*

Indians makes up less than 1% of the population in the U.S., but their role in ownership of mineral resources is considerably more impressive and important. 115,217 sq. mi. of Indian tribal lands are held in trust by the US government and contain nearly a third of national energy reserves of uranium and coal.

Current mineral development policy is governed by the Melcher Act (the Indian Mineral Development Act of 1982). Indian tribal councils may negotiate and make development agreements for mineral resources on tribal lands, subject to approval by the Secretary of the Interior (SI), which is to be reasonably given. Development can also occur on lands owned in fee simple by the tribes.

The Energy Policy Act of 1992 has provisions authorising the SI and Secretary of Energy to provide financial assistance to Indian tribes to promote energy resource development and renewable energy projects. In the manner of promoting self-governing programs, the Act authorises grants to certain tribes to establish tribal offices of surface mining regulation and reclamation for operations on tribal lands under their own adjudicatory system. Additional funds are available to assist tribal governments, along with technological assistance, for their development of environmental programs on Indian lands.

Most litigation involving resources on Indian lands is over oil and coal. In *Cheyenne-Arapaho Tribes of Oklahoma,* 1992, the aborigines prevailed in showing that the federal government had improperly approved an oil lease to the detriment of the tribes. In *Conoco v. Arkeketa,* No.92-C-014-B, 19 Indian L. Rep. (Am.Indian Law Training Program) 3085 (N.D. Okla.1992), the court procedurally upheld the Ponce Indian Tribe's oil and gas severance taxation on the oil company. The company had failed to exhaust the tribal hearing remedies before appealing to the district court.

Development has been detrimentally affected by allowing both Indian and state taxation on Indian land mining projects. The power of taxation on reservations was granted to the tribes as early as 1904, and upheld in *Merrion v. Jacarilla Apache Tribe.*, 455 U.S. 130 (1982). In 1991, the decision in a railroad case upheld the right of the Blackfeet Tribe to tax a railroad for its right-of-way across the reservation.

In *Northern Cheyenne Tribe v. Lujan (Sec. Interior),* 804 F. Supp. 1281 (D.Mont.1991), the aborigines successfully stopped the Secretary of Interior from proceeding with federal coal leases sold on land adjoining the reservation lands on the basis that the government had failed to comply with the National Environmental Policy Act and the Federal Coal Leasing Act 1976. The government neglected a special trust relationship with the Northern Cheyenne. An environmental impact statement found that coal development under the leases would have "significant" and at times "severe" social, economic and cultural impacts on the Northern Cheyenne Tribe.

In spite of frequent litigated disputes with aborigines, access to mineral development on their lands, in general, seems to more easily obtained than in the British Commonwealth nations (Aston, MEM, June 1993).

12.5.2 *Canadian Aboriginal Land Claims*

Aboriginal land claims are abundant, particularly in Canada's northern and western areas, the Northwest Territories. Negotiations with four aboriginal groups, the Inuvialuit, Inuit, Dene and Metis, have been carried on for several years. Large areas of land are involved and include the issue of mineral rights and ownership. In all cases, access to minerals on the lands are restricted. Canada may well accompany Australia with the term "shrinking Canada". Canadian examples of large land withdrawals and severe limiting of mineral activities follow.

The Inuvialuit people of the McKenzie Delta in Arctic regions reached a successful settlement for their extensive claim of aborigine lands in 1984, and in which an Environmental Impact Review Board was provided for. In 1990, a proposed drilling program by Gulf Canada was denied recommendation by the Board on the basis of Gulf's alleged inability to cope with a major blow-out.

The Denise Indians and the Metis people, hoping to attain the success of the Inuvialuit, claimed approximately 425,000 square miles of the Northwest Territories. An Agreement in Principle was made in 1988. The 'settlement area' agreed upon with the Canadian government covered 340,000 square miles in the NWT whereby claims to title by the Dene / Metis were extinguished for certain rights. To be granted was surface legal title to 71,000 sq. mi. and subsurface rights for 4,000 sq. mi. of the whole. Additional mineral rights and royalties are to flow to the aborigines from the entire 'settlement area'. A similar requirement for proposed mineral development on Dene / Metis lands, to that required in the Australian Aboriginal Land Act, calls for a pre-consultation with the aborigines for explanation of the project, expression of their views, and obtaining their general acceptance of the proposed mineral development.

On a lesser basis, in November 1990 Alberta made a settlement with the Metis people within the province for the transfer of fee simple title to 1,900 sq. mi. with rights (not title) for subsurface mineral resources. There are restrictive provisions for sale of the land and for access to explore, develop or work the minerals. The Alberta act provides for an Appeal Tribunal to hear disputes over access to existing mineral leases.

In 1990, the Supreme Court of Canada decided several cases which strengthened the aboriginal position with new powers and scope. In *R. v. Sioui*, [1990] 1S.C.R.1025, in a Quebec statute challenge, the Court considered the 1760 treaty with the Huron Indians treating aboriginal sovereignty as a basis for aboriginal claims. The Court maintained that aboriginal treaty rights should be given a "modern meaning … permitting their evolution over time".

A split-off aborigine group, the Gwich'in or Loucheux, habitating the McKenzie Delta in the NWT and the north slope of the Yukon, had participated with the Dene / Metis in the Agreement in Principle. After the larger group rejected the deadline set for the ratification of the Final Agreement with Canada, the Gwich'in negotiated their own Final Agreement (GFA). In exchange for their extinguishing all aboriginal claims, they received 2360 sq. mi., including subsurface minerals title, in the NWT, and 600 sq. mi. of surface-only title in the Yukon. They are to be free to make arrangements for mineral development on the collectively-held lands in the NWT, and are to receive royalties from all settlement lands and the entire McKenzie Valley. However, access to Gwich'in mineral resources is not as simple as it may sound. Under the GFA, an Environmental Impact Review Board (EIRB) was provided for, and no mineral development proposal may be approved until an initial assessment has been made by the Board for "significant adverse environmental impact" or for a "cause of significant public concern".

As in Shrinking Australia, access to Canada's mineral wealth on Crown lands is being whittled down by aboriginal claims and the national economy is seriously impaired when profitable mining projects may be rendered immobile by historical claims made centuries later (Aston, MEM, June 1993).

12.5.2.1 *Northwest territories aboriginal land claims update*

Land claims in the eastern far north have moved forward since 1990, thus giving a greater sense of security for mining companies to invest in prospective mineral deposits. In mid-1993, the Inuit people ratified their land claim in the High Northern and Eastern Arctic. The federal Canadian government agreed to pursue the creation of a new territory, called Nunavut, to encompass the traditional Inuit lands. The settlement gave the Inuit title to 350,000 km^2 of land to include subsurface rights for 36,000 km^2, plus financial and other participatory rights.

In the western Arctic, the Gwich'in people ratified their agreement with Canada which transferred 22,400 km^2 of land in the NWT and 1,550 km^2 in the Yukon Territory, with subsurface rights and a share of resources royalties included. The southern neighbours of the Gwich'in, the Sahtu Dene and Metis peoples followed suit, taking another 41,400 km^2 of land from the territories.

The Dogrib people to the South are still in the negotiating process for their land claims, while other aboriginals of the south Slave and Deh Cho districts have yet to file claims. On the more recent and brighter side for the mineral industry, the aboriginal peoples of Canada's territories are showing an inclination to support mineral exploration in order to raise their standards of living (Mining Journal, London, 1993, p. 3).

12.6 Proposed Mineral Evaluation of Public and Crown Lands before Preservation Purposes

As previously argued in Ch. 10, before government-owned lands are set aside in preservationist acts for parks, forests, wildlife, scenic and recreational areas, prior mineral studies should be made to locate or eliminate potential and valuable mineral bodies within their proposed boundaries. If economically exploitable mineral bodies are found within the planned boundaries, the mineral industry should be given the prior opportunity to mine the deposits of minerals, reclaim and restore the land to beneficial use before the lands become untouchable and set aside for preservationist purposes. This is a perfect example of the intended purpose of land use planning.

As part of future land-use planning, it should be recognised that cases may be presented where a potential mineral body may be located, but present-day mineral economics prevent it from being mined. In such cases, future planning should note the possibility of potential future mineral exploitation at a time when economics and mineral beneficiation technology would allow the mineral body's economically profitable development for the good of mankind.

12. 7 U.S. Mine Site Reclamation Legislation Proposed

Considerations for reclaiming derelict mine lands in the U.S. is under consideration by the U.S. Congress. Though the considerations have environmental merit, the weakest point in the discussions is the fact that it appears no thought or consideration has been given to putting those many surface mines to use as landfill sites to accomplish far better and fuller reclamation, to accomplish land conservation, and most importantly, to relieve the urgent need in the U.S. for MSW and hazardous waste disposal sites. All of those environmental goals appear to have been overlooked or by-passed in the proposed federal action to reclaim abandoned surface mines. A grand opportunity is being missed to accomplish better environmental goals at greatly reduced costs to the government and the general public. An article on the proposed U.S. mine reclamation legislation follows:

U.S. MINE SITE RECLAMATION

> "The cost of cleaning up more than 550,000 abandoned hardrock mine sites in 32 U.S. states could be between $32,000 million and $71,500 million according to a new report prepared by the Washington-based Mineral Policy Centre. The sites identified in the report include underground mine workings, open pits and waste heaps. The establishment of a national reclamation programme for abandoned hardrock mines is advocated to include a nationwide inventory of sites on public and private lands. A minimum of $400 million per year for the programme is recommended, to be met from fees and royalties imposed on the mining industry. The Mineral Policy Centre claims that it is not anti-mining but has called for wide-ranging reforms within the industry and has been a keen supporter of revisions of the 1872 Mining Law.
>
> "Not unexpectedly, its finding and recommendations have been strongly attacked by the mining industry's trade association, the Mineral

Resources Alliance, and also by Senator Larry Craig, a Republican from Idaho and author of a Senate-passed mining reform act. Senator Craig said the report admitted that the number of sites was unknown, hence the cost estimate for the clean-up could be much lower than the figure put forward.

"Of the 550,000 estimated abandoned sites, only 50 are on the Superfund hazardous waste list with an estimated clean-up cost of $12.5 million to $17.5 million. A further 500 sites with ground water contamination could cost $2.5 million to $7.5 million whilst another 14,000 sites with surface water contami-nation would cost between $14 million and $43 million. The report identifies some 230,000 sites where the landscape has been disturbed and 195,000 where little, if any, remedial measures would be required". (The Mining Journal, 1993b, p. 51).

12.8 Conclusions and Comments

Clarification is required for the initially, sounding and alarming "cleaning up of 550,000 abandoned mine sites" complained of by the Mineral Policy Center (MPC) in the preceding article (at §12.7). As close scrutiny of the article clarifies, the alarming figure can be immediately reduced by some 325,000 sites leaving a far lesser number to be concerned over. According to MPC's own published figures, about 195,000 sites are "reclaimed and / or **benign**" and require **"little, if any, remedial measures".** (emphasis added). Also, MPC states "231,900 sites have only landscape disturbance", thus needing only cosmetics to beautify the land (then why does MPC include these large numbers of benign sites to inflame the public?) Out of the 557,650 sites complained of by MPC, the truth of the matter is that the number of sites to be environmentally concerned about should be reduced by 76%, leaving only 24%, or 131,750 sites of potential environmental harm in any degree. Of the total, it should be noted and emphasised that only 50 of the sites (0.0089%) are on the U.S. government's Superfund clean-up list because of environmentally hazardous conditions.

MPC calls for a Hardrock Abandoned Mines Reclamation programme to be enacted. Utilisation of the abandoned hardrock mines as MSW depositories appears to be the solution, accomplishing not only the goal of MPC but relieving the urgent need for waste disposal space.

The point is made that the domestic mining industry is withdrawing from Canada and the U.S. (Even in the U.K., super-quarries for for future supplies of aggregates are advocated for location in remote places in other nations, e.g., Norway, and Spain, i.e., in someone else's "backyard".) The reason given by the mining industry is that mining has become uneconomical in their own country due to the high cost of overly stringent environmental regulation. Environmentalists argue that the present regulations must be maintained, without relaxing, and counter by calling for greater controls. A happy medium must be found between the two positions. If pollution is presently occurring from mining, either the regulations in place are not being enforced, or they are inadequate. If pollution from mining is not occurring, then present control is obviously sufficient without more control.

Section V

CLOSING ARGUMENTS FOR PROBLEM AND THESIS SOLUTION

CHAPTER 13

NEW DIRECTIONS FOR ENVIRONMENTAL LAW POLICY IN REGARD TO SURFACE MINED LAND RECLAMATION AND SOLID WASTE DISPOSAL

13.1 Introduction

A reasonable direction for improvement in environmental policy and law is sorely needed for solving and improving the two current, large environmental-creating problems, viz.,

(1) that of maintaining an unhampered, continuous supply of all minerals, particularly industrial and construction minerals obtained by surface mining with minimal injury to the earth's surface, and to be accompanied by the mined-out land's full and complete restoration for its re-use as a matter of land conservation and sustainability; and
(2) providing sufficient and ample space-volume for the disposal of municipal solid waste which is presently in critical shortage and in the foreseeable future.

At the same time, a related problem needs serious attention to make the flow smoother for the resolution of the first two problems, viz., that of minimising the defeating contentiousness of the public inquiry / hearing process (PIP).

13.2 Solution Considerations

A highly creditable conclusion was made by Paul Tomes in his 1989 technical paper at the Institute of Quarrying, (op. cit.), "Quarrying and landfill are two industries which are inextricably linked. We are fortunate in Britain that we are self-sufficient in stone, sand and gravel and can, therefore, be equally self-sufficient in landfill capacity, if we choose".

Although siting of mineral extraction and landfill operations is extremely contentious the general public, temperance must be sought and appealed to by education of the absolute necessity for such joint operations. The public requires construction materials for the construction of private and public works. It also sorely needs sites for disposal of its own refuse and solid wastes. However, when the public is faced with the nearby location of such a "bad-neighbour" development, the NIMBY syndrome surfaces in protest, whether the potential for real environmental harm exists or not, as basis for alleging future environmental injury. This fearful public attitude is corroborated in the 1992 U.K. Research Report, Environmental Effects of Surface Mineral Workings (op. cit. p. 14) which stated, "For a new mineral working or a major extension of an existing site, (public) apprehension may well exaggerate the anticipated problems". In fact, the U.K. report cites in its Introduction the basis for the report as "The need for the study arose from *** the increasing pressure on the provision of land for mineral extraction and the consequent difficulties that face operators in seeking planning approval". (op. cit., p. 1). The same report continues, "Such (public) apprehension can often be reduced, both before and during operations. It helps to ensure that those affected have contact with the operator and are given some understanding of what is involved in mineral operations and any compensatory measures that might be taken". (id., p. 14).

To illustrate that environmental lobbying has been effective in defeating mineral land development in the U.K., in November 1995, the Geological Survey of Northern Ireland, in promoting new mineral development, highlighted an announcement entitled "a window of opportunity". In it, the Survey called attention to lessening of a major concern to the mining industry by stating that, "The major concern that environmental lobbying is defeating economic consideration is diminishing". (Mining Journal, London, November 24, 1995, p. 388).

13.2.1 *Minerals Must Be Excavated Where They Occur*

The 1992 U.K. Research Report, Environmental Effects of Surface Mineral Workings concedes that "*** as mining can only take place where the minerals occur, there is a possible distinction that can be made in justifying compensation related to surface mineral workings but not for most other forms of development". (id. p. 116). A mining operation cannot be moved to some other location as could a manufacturing plant in order to accommodate the public's desire for its more remote and less intrusive location (i.e., NIMBY). It is true that there may be some small leeway in locating aggregate-producing materials since they are produced from massive occurrences of common rocks which frequently cover, or lie below, large surface areas, e.g., granite, gneisses, limestones and other igneous, metamorphic and sedimentary rocks. This is not as true for sand and gravel for there are more geological restrictions governing its mode of occurrence than for the other massive-occurring aggregate-producing materials. However, even these more commonly-occurring rocks have their economic limitations with respect to the depth of occurrence and overburden that has to be removed before mining can start.

13.2.2 *Consideration of Private Property Rights*

Are private property ownership rights being lost to over-regulation?

It has been noted earlier in the Introduction of Ch. 5, that the environmental regulations, particularly in the US, are of a "command and control" nature which encroach on previously-held areas of private ownership and self-control of individual property rights. The Endangered Species Act is a notorious prime example of taking away human rights for the sake of lesser species that has come to the for front of America politics. A prime example of loss of property rights was given in Ch. 5 §6, p. 6, where new bedrooms could not be added to the home of a young California couple because it was considered "harmful" to a rat habitat. This is a result of the "command and control" type of environmental regulation. In mineral law, for example, an owner of mineral-bearing property, whether an individual or a mining company, could previously mine any place on its property, or as many acres as it wished. After planning, zoning, permitting and licencing controls were in place, a landowner could not freely excavate or move about on his own property without the approval and licencing of the authorities. At times, even public notice hearings were involved before approval could be given for exercising excavation of the property owner's minerals on his own land. Clearly, the property rights of landowners have been severely encroached upon, limited, restricted and taken away from them. However, to obtain the desired results of environmental improvement for the good of the public, the new "command and control" system proscribing environmentally "harmful" acts and effects of mining requiring specific control measures must prevail to produce the governmentally and environmentally-desired benefits for the general welfare.

13.2.3 *Command and Control Environmental Regulation*

It is recommended that serious consideration be given to extending and employing the "command and control" type of regulation for the siting of future combined open-pit mining and solid waste landfill operations. Even limitations for public hearings on siting of the "bad-neighbour" developments may be necessary to the economy and welfare of the public. Too often, siting approval when subjected to the choice of the public at public hearings ends in rejections, or conditional approval is loaded with so many restrictive "conditions" and restrictions that the development is seriously hampered economically. In the U.S, rejection of sites have been made by the public in spite of preliminary approval of the site by environmental controlling agencies. Too often, public opinion succumbs to the unfounded fears of the NIMBY syndrome causing public disapproval. Waste disposal sites are frequently rejected in areas where they are sorely needed, rejected on unfounded fears of the NIMBY syndrome. (See *Village of Wilsonville v. SCA Services, Inc, File et al v. D & L Landfill, Inc.*, and *City of St. Peters v. Dept. of Natural Resources,* supra). If the "command and control" environmental regulation works for the Endangered Species Act, it can likewise work for the siting of mineral operations and landfills.

As in the Denver airport example given earlier, the critical need for construction materials was rejected by the NIMBY syndrome. Materials had to be found in a neighbouring state and transported at far higher costs. This is absurd when the very same construction minerals may be found near to the Denver construction site. In the end, the public pays and only the environment in Denver's backyard has been questionably served, not in Wyoming's. A similar case in the U.K., as evidence submitted of the NIMBY syndrome driving the costs of production higher, is the M.I. Great Britain Ltd. appeal in Scotland for denial of permission to mine the barytes deposit at Duntanlich (supra at Ch. 12 §2.1). Without production from that unmined world-class barytes deposit, the cost of North Sea oil production will rise; along with it, the price of fuel oil in Britain will also rise.

13.3 The Spreading of the NIMBY Syndrome as an International Problem

Hitherto, environmental concerns in the undeveloped nations have ranged from nil to minor. As argued in Ch. 1, the Introduction, at §1.1: "The underdeveloped and the developing nations are striving *** to develop their mineral resources. These Third World nations have openly solicited foreign mining industry to invest in the development of their wealth of minerals without the costly and stringent environmental and mine permitting regulations of the developed nations".

As evidence of an already changing attitude in a developing nation, inhabitants of India are already adopting the NIMBY syndrome accompanied by unfounded fears according to the following report from The Mining Journal, London, of June 30, 1995.

OPPOSITION STALLS INDIAN BAUXITE MINES

> "An increasingly strident campaign is being waged against the development of a major bauxite mining project and an alumina refining project in the Indian state of Orissa, sparking much discomfiture amongst the state administration and the project developers. A similar incident in 1990 forced Bharat Aluminium to abandon plans for mining bauxite at Gandhamardan in Orissa after having spent nearly $US1 million because of local opposition.

> "According to reports in the London *Financial Times,* the local inhabitants of the Rayagada and Kalahandi districts of Orissa have been told by the newly-formed Anchali Suraksha Parishad, which is spearheading the lobby, that the proposed complexes would 'disturb the environment, displace the tribals and cause a cultural shock to the local people'. The developers, struggling to counter what they claim is a campaign of disinformation, maintain that these fears are largely unfounded and that the projects will 'generally comply' with accepted international environmental requirements". (op. cit., p. 479).

Ironically, the same article goes on to say that a competitive joint venture in the same geographic area and dealing with the same governmental body, "have been able to convince the Orissa government that they will offer adequate compensation for any resettlement and will protect the tribal culture. India's Industry Minister vowed that the projects will be implemented, stating, "In fact, we want more such projects in the state". (op. cit., p. 479).

13.3.1 *A Continuing Landfill Siting Problem — The Public Hearing Process or the NIMBY Syndrome Takes Over*

As stated in the Introduction, Ch. 1 §1, a third problem for this work is to propose a less contentious, less controversial and less combatant way for both private and public projects to obtain approval, be permitted and licenced, than through the extended PIP / public inquiry process. More public trust and confidence should be placed in the scientific management by the in-place regulatory agencies already charged with the protection of the public from environmental harm. Similarly stated by Browning Ferris Industries' brochure on Solid Waste Disposal, "The key reason for many disposal problems is the lack of public support for waste disposal facilities of any kind. While many citizens and public officials recognise the need for sanitary landfills, they may not be aware that a well-designed and professionally operated facility can actually protect the environment".

In a preponderance of projects suffering rejection at the hands of the NIMBY syndrome, the defeat is illogical. The stigma of the "bad-neighbour" industries still prevails. It shows a lack of public trust, or a lack of information by the public, or blatantly ignores the engineering profession's modern technological advances for environmental protection against pollution and contamination. All aspects of centuries-old industrial repugnance are now treated for environmental protection, e.g., dust, noise, water, visual intrusion, wildlife harm, et al. In view of present environmental technology, there is little basis for defeat of siting projects when socio-economic considerations clearly predominate.

Landfill site selection has become both overly complex and time-consuming. Browning-Ferris continues, "Today, before a permit application is tendered, the landfill operator must ensure that zoning and land-use restrictions allow for the construction of a sanitary landfill. These curbs are often major obstacles in successfully siting a sanitary landfill facility.

"Regulatory bodies also demand detailed assurance of technical and fiscal responsibility. These agencies have regulations that are meticulously written to guarantee the public that any future landfills will be operated in an environmentally sound manner. For example, before any site receives state operating permits, that site will have undergone rigorous analyses involving the study of the site's geology, hydrology, soil conditions and land-use compatibility. In addition, a detailed engineering plan is prepared and submitted for approval by regulatory authorities. There are also financial requirements that must meet minimum criteria established by state and federal agencies".

With those and other assurances required by the USEPA, the NIMBY syndrome has no basis or place in the PIP.

13.3.2 *An Offered Solution to Combat the NIMBY Syndrome*

To obtain the desired approval for the public good, the "command and control" type of regulations should be considered and introduced in proscribing environmentally beneficial planning regulations in place of the PIP to produce the governmentally and socio-economically desired benefits for the general public's welfare. A balance must be found between the bio-ecological and socio-economical considerations of the present and future.

Many state waste management laws already exist which provide a mechanism for state override of local authority which may block the siting of hazardous waste facilities, e.g., N.C. Gen. Stat. §143B–216.10 et seq. The same override power for landfill siting can be enacted by legislative bodies.

The power of override by the U.S. federal government in other areas of environmental legislation has been reserved by it, e.g., under the federal Insecticide, Fungicide and Rodenticide Act (FIFRA), the EPA Administrator may suspend the authority of the State until such time that the State can and will exercise adequate controls. Similar provisions exist for other federal environmental acts where the states may have primacy for implementation of the statute, but the federal authorities retain supreme authority.

It should be emphasised that under the best practice model law herein proposed, the mandating of abandoned and active non-coal surface mines to be used as MSW landfills in their reclamation process will automatically eliminate the public inquiry process for their approval and permitting as landfills. To circumvent the PIP for future surface mine-landfills, a simple legislative act that exempts their future siting from the PIP would resolve future problems.

13.4 Deciding the Procedure for Permission of Projects

Between the British and U.S. philosophies for application of environmental regulation and exacting compliance, the choice appears to be between the one extreme of British temperatism employing persuasion and a large degree of discretion by authorities, including voluntary acts by the developer, or that of the U.S., for the "command and control" procedure, with strict adherence and strict liability under severe penalties of the law, inclusive of jail sentences for environmental "crimes". This choice is mirrored by Hughes' posed question, "*** is environmental control best effected by bodies possessing great discretionary powers and preferring to follow time-honoured paths of persuasion and cooperation, or should it be achieved under strict mandatory regulations administered by publicly accountable bodies"? (Hughes, 1986, p. 73). As Hughes notes, the latter is "not commonly the British way".

Hughes further argues that "Those who create standards for environmental protection must discover an appropriate compromise between rigid legalism and uncontrolled discretion, bearing in mind **the need to achieve economically**

efficient and cost effective results. What is more, they must frame their standards in such a way that when enforcement action should be taken the circumstances are such that the enforcer feels not only legally, but also socially and morally justified in acting". (Hughes, 1986, p. 74; emphasis added). In the end, it is the electorate that will determine the moral degree of punishment to be exacted for the violation of environmental regulations, and whether violations are to be viewed as criminal, or not.

Whether enforcement of the environmental regulation by imprisonment is morally correct, and to what extent sanctions or punishment is meted out for violation, is of concern for all proposals and works in achieving the end results and product of an improved and cleaner environment.

13.5 The Main Issue for Joint Quarrying and Landfilling Approval

However, the main issue of this study, work and proposal is the taking advantage of two already in-place and permittable developments of universally continuous-need, by combining the need for the greater efficiency and improvement of existing environmental works thereby yielding greater efficiency, economic benefits, and improved land reclamation and conservation.

The theory herein proposed is that non-fuel surface mines or pits, upon completion or being "worked-out, should, by legislative enactment, automatically become part of the planning and the regulatory process mandating their reclaimed use for controlled tipping of local and municipal solid wastes as part of the reclamation process in returning the derelict land to public usefulness.

Query: Why excavate other holes in the ground or pile and cover waste mounded on the surface just to deposit wastes when earth voids (holes) are already provided to society by the mining industry? Or, put another way, why make landfill sites on the surface in more permeable soil when more impermeable pits in rock are available and safer? On evidence of research, those excavations / pits made for construction materials, i.e., stone aggregates in particular, are either naturally environmentally suitable for landfill sites, or can readily be made environmentally safe and acceptable as landfill sites.

There are two main, essential characteristics of aggregate (stone) pits that make them desirable and amenable to landfill sites for controlled tipping, viz., (1) the mineralogy of the mined material, i.e., rocks [as granite, gneiss, diabase (trap rock), et al] are generally chemically neutral, and do not contain toxic metals which may be released, infiltrate and contaminate the ground waters from pit workings; and, (2) the location of crushed stone (aggregate) pits are generally found near urban areas making them accessible to the population concentrations as landfills where most waste is generated.

Under (1), above, crushed stone pits, being generally chemically inert, do not have the undesirable characteristic of coal pits to provide acid water drainage. For those pits that are within or below the water table, the pit walls are generally solid rock. Where there are prominent "cracks" (faults, joints) in the walls of the pit, they may be gunited thereby closing them to leakage, or sealed with vertical HDPE liners or impermeable clay lining. Both processes for sealing have been successfully done in landfill sites.

Under (2), above, a desirable cycle of ready-made openings for landfill sites in urban areas where most waste is generated is currently in place, and regularly re-occurs over each generation or two, and should continue to re-occur for a long time to come. In the cycle, urban areas continue to grow and expand, many into megalopolises, or the joining of metropolises. The demand is great in these areas for construction materials, principally rock for aggregate, and limestones for cement, clay for bricks, etc. All are low-priced and low-profit materials that generally cannot tolerate a high transportation cost to get them to the construction site / market. Therefore, economics has dictated that the source sites for stone aggregates, in particular, must be, and are, close to the urban areas where the construction growth is concentrated. As population growth continues to occur and expand in the metropolises, the older stone pits are encroached upon by dwellings, et al, and are forced to close and move further away from the population to areas beyond the new expanding edges of the population / growth centres because of their undesirable "bad neighbour" characteristics (noise, dust, traffic). The old, abandoned pit sites should be reclaimed and returned to beneficial surface use. The governmental reclamation trend has been to allow pits to fill with water-making lakes and landscaping the surrounding surface. Ensuing development around the former pits then becomes desirable residential areas, golf courses, parks and other recreational uses. The opportunity to utilise the land for critically-needed landfill sites has been lost and cannot be recovered without zoning difficulties. New residential areas view landfills as "bad neighbours", and do not tolerate the abandoned pit as a landfill site unless designated as such before extensive encroachment.

Support to that part of the cycle referred to as pit location in proximity to population centres, supra, is corroborated in Blakeman's report to Environment Canada, "Although mineral aggregates and industrial minerals are essential to the national economy as are metallic minerals and fuels, *** due to their high bulk / low value characteristics which affect transportation costs, the centres of their production are normally much less geographically remote from populated areas". (Blakeman, 1977, p. 5).

Other, non-fuel open pits, generally located more remotely from urban areas, should not be excluded for consideration as landfill sites, perhaps even as hazardous and toxic waste burial sites.

13.5.1 *A Reclamation Alternative for Sand and Gravel Pits*

Rubblefills are a viable alternative for reclaiming these mining pits. In areas where environmental regulations prohibit refuse or MSW depositing in sand and gravel pits, building construction waste materials are often acceptable. The universal lay prohibition for depositing MSW in gravel pits is based on the non-technical dogma that "all sand and gravel are located in aquifers and refuse will contaminate the water source". Without combating the lay argument, there is a need for depository sites for construction and demolition waste. Using sand and gravel pits for such waste has merit. By not placing demolished construction waste in permitted MWS sites, more critical space is made available for MSW.

As an example, the 125-acre sand and gravel pit in the San Gabriel Valley, California, was started in the 1940s and mining was finished in 1993. Local regulations prohibit depositing MSW, but construction materials waste is acceptable. The pit became a rubblefill in 1983. Currently, it receives between 250 and 300 truckloads per day at a tipping fee of $50 per load. Construction waste is generally inert, thus, generates little leachate, and the lay-fear of water pollution is assuaged.

13.6 The Best Practicable Environmental Option

Tomes concluded his 1989 technical paper, saying, "Due to environmental pressures now being exerted on the landfill industry, higher standards and a more professional approach are required. Even though the result of this will mean higher landfill costs, this method (i.e., landfilling quarries) of waste disposal will still be the 'Best Practicable Environmental Option'. Restoration of mineral workings will continue to be the favoured option for landfill sites, and at the same time, landfill will be one of the most cost effective methods of restoring pits and quarries". (op. cit., p. 20).

A noted contemporary British civil engineer, John Skitt, County Waste Disposal Engineer, Staffordshire County Council, in his 1979 Waste Disposal Management paper on the subject of land conservation, wrote, "Excavations of minerals are said to exceed the rate of reclamation and the total of derelict land in the U.K. is in excess of 90,000 acres. The annual rate of extraction is five times the space which would be required for the disposal of house refuse". His apparent insight is further emphasised in a following paragraph, "The relationship of this problem with the disposal of refuse will probably be obvious; **the two might be solved mutually**". (emphasis added). *** Strange as it may seem the mutual solution of the two problems concerned does not always appear to have been visualised by the planning authorities; ***".

Similarly, in a paper written and given by Paul A. Tomes, FIQ, M. Inst. W.M., Company Landfill Manager-ARC Aggregates (at the time), 2d October 1989, and presented to the Institute of Quarrying, Annual Conference Symposium at Bristol, he made the statement in his Introduction, "Every year in Britain we produce around 100 million tonnes of household, commercial and industrial waste". Further on he states. "The mineral extraction industry creates voids at a rate in excess of 200 million cubic metres per annum".

13.7 Final Argument For Thesis Proposal

Thus, with successful examples of surface mines used as landfills already in existence, there is no great novelty in the proposed procedure. However, the novelty of the proposition argued herein, is that the use of quarries as depositories for waste should be intentional by regulation and planned rather than the occurrence of landfilling a surface mining pit being one of happenstance, after-thought, or an *ad hoc* basis, or in desperation as by a community with an already overflowing landfill forced by "extreme" urgency and a last resort to find a place to deposit their MSW someplace just to get rid of it. Witness: the siting approval for the Allerton Park (U.K.) quarry (supra at Ch. 11 §8.1.2, as a landfill was strongly influenced by the critical need to relieve the North Yorkshire County of its landfill requirements. The Harrogate area's MSW disposal facility at Rock Cottage Quarry, Wormald Green, had expired and filled in April 1987, just prior to the time of ARC's June 1987 application. The Council had to temporarily expand the Rock Cottage Quarry site until 1991 when area MSW disposal could begin at Allerton Park. Allerton Park quarry held promise of meeting the Council's waste needs for the next 25 years or more. Allerton Park landfill is currently receiving about 60,000 tonnes / annum of MSW.

The procedure of dual functions, i.e., quarrying and landfilling, should be done by regulation with the benefit of specific pre-planning and permission for a dual operations on approved sites from the very beginning, particularly where the ground water conditions present no great hydrology problem to the filling of the mining void. Where hydrology might present a problem for waste filling in the former attenuation-type fill, a total-containment type landfill will enable the mining void to be used. Consequently, all mining surface mining voids can be utilised as depositories of MSW without potential injury of contamination to the water resources. Only those mining void sites that were located in potentially geologic zones of disturbance, i.e., subject to earthquake disturbance, should possibly be eliminated as landfill sites.

The three successful operations of crushed stone quarries, offered in evidence in Ch. 11 §8, viz., Judkins, Weber and Walker quarries, subsequently given permission to deposit MSW as filling after years of mining operation and before closure of the mine, prove the successful employment, utility, practicality, and plausibility of depositing MSW in rock quarries. Inadvertent, or unintentional as it may have been, the procedures have set examples that should, henceforth, be intentional as mandated by environmental regulation, planned from inception of permitted new surface mine as part of its full reclamation.

Consequently, landfilling derelict mined land with waste is not only beneficial in relieving the critical shortage of depositories for MSW, but supplies the greater environmental need to replace the mined-out voids from which the minerals were removed. By replacing the void with waste filler, the original surfaces can be restored to the land and the land can once more serve society with a beneficial surface use, e.g., a park, golf course, recreational site, parking lot, or a host of other lightload building and residential surface uses.

Lastly, land use is not only conserved by the restoration of the surface to a beneficial use, but also conserved by using the same ground for two essential uses, mining materials for public consumption and depositing the public's solid wastes, as opposed to using two separate tracts of lands for the two separate uses, i.e., mining and waste disposal.

The Mineral Policy Centre, Washington, D.C., calls for a Hardrock Abandoned Mines Reclamation program to be enacted in the U.S. (see Ch. 12 §§6 and 7). Utilisation of the abandoned hardrock mines as MSW depositories appears to be the best solution for reclamation, accomplishing not only the goal of the MPC, but at the same relieving the urgent national need for waste disposal space. Instead of requiring multi-billions of dollars for reclamation as MPC claims, the utilisation for MSW as herein proposed would be a self-paying, even profitable, way for full reclamation of the hardrock sites. The new way for the public to view a surface mining operation is not as the "raping of the earth by mining" but the creation of an essential new waste disposal site to deposit society's waste in, with resultant reclaimed land at an increased value.

13.8 Summary of Supportive Evidence

The use of surface mines for MSW landfills is justified by the evidence presented in this thesis research and is summarised under the following headings: (1) Need and Justification; (2) Environmental Security; (3) Mined Land Restoration; (4) After-use of Conserved Land.

13.8.1 *Need and Justification*

Governmental-source information and statistics indicate that there is a current critical need for MSW landfill space, being greater in the U.K. and the U.S.

Canada does not suffer the overall urgent need except in its more densely populated areas. That critical need will extend into the foreseeable future with continued population growth.

The nature and logic of the repetitive "urban quarry-landfill cycle" justifies the use of construction mineral surface mines / pits for MSW landfill sites. The urban quarry-landfill cycle is summarised as follows:

1. past to present industrial mineral economics has dictated that construction mineral (mainly aggregate) quarries must be located on the perimeter or near the markets of growing populations centres;
2. the urban population centres enlarge, encroach on and engulf the perimeter quarry sites;
3. to avoid nuisance claims, the quarries must relocate further away from the expanding perimeter residential areas of the growing population centres, leaving, at best, partially unrestored open voids and unusable land in the growing area;
4. the new growth centres that forced the quarries away generate new and added volumes of waste with resulting demand for more landfill space to an already critical MSW load;
5. starting with 1, above, the cycle repeats itself, again and again over new generations of population as the megalopises increase and expand in size.

The use of surface mines / quarries for MSW landfills is justified —

i) to resolve the present and future critical space requirement for waste depositories;
ii) to furnish fill material for the voids created by mining of the construction minerals since there is insufficient volume of earth material left to restore the mining void to its original surface;
iii) MSW landfilling accomplishes both (i) and (ii).

13.8.2 *Environmental Security*

Environmental security of MSW landfill sites must ensure the integrity of the quality of groundwater and surface water resources. Landfill technology has progressed to a superior point of even a decade ago with safer and stronger synthetic membranes for containment. Attenuation (leaking the leachate into the subsurface) of landfill leachates can no longer be tolerated. Landfills and their attending leachates must be placed in total-containment landfills with the leachate being withdrawn by pumping, treated and disposed of through the municipal sewer systems, or sewn on the earth as fertiliser. Security for ground waters at landfill sites is already in place under regulatory-required monitoring.

The discovery of peat as an omnipotent absorbent and purifier of waste liquids, and construcetd wetlands, makes their proposed usage for treatment of landfill leachate collection systems practical for ensuring the integrity of ground water quality in the immediate areas of landfill sites.

Landfill gas has been found to be a usable and profitable source of energy and can be collected at a landfill site. Where the type of waste deposited does not generate sufficient volume of gas to warrant investment of a collecting it, it may be vented, flared and monitored for safety against explosions and preventing air pollution.

13.8.3 *Mined Land Restoration*

Derelict and fully unreclaimed mined lands can be fully restored to society for beneficial re-use by landfilling with MSW, while simultaneously resolving the problem of where to place the waste.

Construction mineral and stone aggregate pits, except for the shallow overburden which was removed and remains stored on the surface of the property, excavates and uses all of the rock mined for production of construction products. Unlike coal pits, little material is left to replace in the mined void. For example, a surface stripping coal mine may remove 60 (or more) feet to reach an 8-foot coal seam. After mining the 8-foot thick bed of coal, for reclamation, the coal miner has 60 feet of spoil to replace in the 68-foot deep pit, or 88% of original backfill material. By comparative example, the aggregate miner may remove 8 feet of overburden to mine 175 feet of rock. The stone quarrier has only 5% of original material to replace, thus, still leaving a major void. Therefore, the aggregate miner usually leaves the stored overburden on the surface. It would be wasted effort and money to place 8 feet of dirt in the bottom of a 175-foot deep hole.

Therein lies another positive argument for utilising stone quarries for landfill sites. The un-replaced, stored overburden furnishes a ready supply of daily cover material for a landfill thus saving further earth disturbance for cover material as required at non-quarry landfill sites. Surface coal mines have abundant daily cover material for landfilling. Clay beds are attendant to coal seams and provide impervious clays for sealing of a landfill.

MSW landfilling furnishes otherwise unavailable material for infilling the mining void. The spoil piles of overburden may be used for both daily cover of wastes, and for the final cover in complete restoration of the surface.

Additionally, land conservation is obviously well-served by utilising a useless derelict mining void for landfilling. It further reduces the use of land for surface landfilling sites alone. The same land is used twice; first for mining and second for landfilling. The amount of land saved should be roughly double that of the present scant, non-use of surface mining voids for landfilling. Presently, derelict and abandoned mined lands are virtually useless, and at times become misused for illegal dumping of refuse, thereby potentially contributing to pollution of water resources.

Landfilling accomplishes full restoration of the mined land for new surface uses to the benefit of society. Monitoring against air and water pollution must be maintained for a safe period of time after landfill closure, usually mandated for 20 to 30 years after closure of the landfill.

13.8.4 *After-Use of Conserved Land*

The restored and conserved land will generally have greatly increased in value by virtue of its location in a newly increased populous area and become an asset to the property owner (new residences inevitably seem to flock around quarry operations — known as "coming to the nuisance"). Generally, reclaimed surface mine land by landfilling re-uses are of a light-load nature, unless otherwise planned for. Well-known restored land surface uses are for amenities, recreation facilities, golf courses, parks, water recreation where ponds and small lakes are creatively left or provided, nature conservatories, wildlife habitats, parking lots, shopping centres, and light residential buildings. Planned after-use lightload residential housing and lightload business structures placed on containment landfills need careful attention in per-planning. Generally, engineered containment landfills should not have heavy-structure buildings placed on them as they would likely destroy the integrity of the impervious cap. Additionally, any piling driven into or through the landfill into the underlying strata would inevitably compromise the containment. Highly acid liquids within the fill could also affect the stability of the structures, attack and deteriorate the construction materials.

By using the near-to-population-centres derelict and active stone quarry sites, the end result amenity and residential uses are fortuitously and conveniently located to the new and expanding population areas.

13.9 Closing Arguments

1) Should legislation as proposed herein be enacted, mandating the reclamation of abandoned, presently mined, and future permitted surface mines by infilling with MSW, the PIP will be clearly eliminated for the first two categories since it will be required by law. However, siting of future combined surface mines and landfills will still be problematic if subjected to PIP unless some provision is made to deal with it as well. A simple legislative act that exempts future siting of new mine-landfills from the PIP would resolve future problems.

2) As opposed to utilising surface mined pits for landfilling, consider one of the leading alternative considerations for relieving the critical and urgent need for landfill space and sites, that is, to continue and increase MSW disposal by land raising / mounding. Land raising for MSW is to pile the waste on the earth's surface and cover it, there by raising the natural elevation of the surface. (Landfilling of earth depressions, e.g., canyons, ravines, etc., are not considered land raising. Furthermore, the use of earth depressions, as ravines, interferes with local drainage patterns and is not recommended.)

Such an alternative can hardly be considered as "land conservation" as undeveloped property must be used for the site. Material for daily / intermediate cover, and for capping the site would have to be taken from some other undeveloped land for the landraising waste disposal site, thus, further detracting from land conservation and only creating more disturbed land. Additionally, the use of derelict mined land site is being wasted, compounding the injury to land conservation.

3) With the prospect for mines being in more remote locations, at least for aggregate and construction mining minerals, as for coastal super-quarries and sea-bed mining, a less combative and shorter process of public hearings may follow with a higher percentage of approval. However, environmental contentions and opposition will still be present for the coastal locations of the super-quarries, regardless of their remoteness. Witness the current environmental opposition to the super-quarry permission at South Harris Island, Scotland (see Ch. 9 §5, supra).

Thus, it is submitted that the case for land conservation, preservation and sustainability of land by utilisation of surface mines as MSW landfills in their complete reclamation process, thereby restoring the derelict land to beneficial surface use, has been proven.

It remains only for the Anglo nations to adopt the best practice model law herein to make it an accomplished fact for the improvement of the environment by surface mining and muncipal solid waste industries. The generalised formats for proposed legislation in the three Anglo nations follows in the Appendix.

Appendices

APPENDIX A

A PROPOSED BEST PRACTICE MODEL LAW FOR LAND CONSERVATION AND RECLAMATION OF SURFACE MINED LAND BY SOLID WASTE IN-FILLING

The author concludes with a proposed national / federal law making reclamation conditionally mandatory for all abandoned, currently operated, and future-opened aggregate and surface mines by waste-filling, thereby relieving the national urgency for waste disposal sites, and greatly conserving land by restoring its original surface to valuable after-uses. The author contends that all surface mines, if not already safe for ground waters, can be made environmentally safe by total containment of MSW under present technology.

Explanation of rudiments of the proposed legislation:

Mandatory conditional reclamation are the key words to the proposed regulation.

"Mandatory" means that all surface mine properties are required to be registered and seriously considered and environmentally assessed for possible reclamation of the mined void by backfilling with municipal solid wastes (MSW) or other wastes.

"Conditional" implies that reclamation with MSW is only mandatory if the environmental assessment of earth conditions of the surface mine site are environmentally safe and approved by EPA, or some corresponding equal federal agency, for acceptance of MSW, or other acceptable wastes, under total containment conditions. If the site is in a geographical area needing an MSW disposal site, or will need one within a specified near-future period, then it must be reclaimed with acceptable wastes.

"Mandatory" does not mean that every property owner or lessee of an abandoned surface mine, or a currently operating surface mine, or every future surface mine operator opening a new surface mine, must be the party to reclaim the pit by backfilling with MSW.

If the surface mine is found to be acceptable as a potential MSW depository site by EPA, the surface mine property owner (for older, abandoned surface mines), or the current surface mine operator who is responsible for the mine's reclamation under mine regulations, may lease to, or contract with, a waste disposal operator to landfill the site for its complete reclamation. The surface mine operator, small or large, is not being "forced" to go into the landfilling business in order to accomplish full land reclamation.

When a surface mine operator applies for mine closure, if his surface mine has been approved as an eligible MSW site, it would be placed on a list for reclamation by filling with MSW, or other acceptable wastes. The responsible national, or federal agency for surface mine reclamation with wastes would hold reclamation in abeyance until a waste disposal operator is found to complete the reclamation by operating it as a disposal site. The surface mine operator may then contract with the disposal operator, be it private or municipal or local county government, to complete reclamation of the surface mine site by landfilling. The mine operator would be relieved of the responsibility for mine reclamation for an eligible disposal site, at least during the abeyance period.

A suggested "abeyance" period of five or ten years would appear reasonable. If no waste disposal operator is found in the abeyance or holding period, the regulating agency must determine whether an extension is feasible. Feasibility of extension would be based on the time-prospects of finding a waste disposal operator for the site.

For simultaneous operations, i.e., landfilling while mining, the landfilling operation may be contracted with a disposal operator who will work with the mine operator. For example, in the U.K., ARC-Greenways operates 25 quarry-landfill, combined operational sites. ARC is the aggregates quarrier, and Greenways is the landfill operator. They coordinate their operations. (However, in the ARC-Greenways example, both are owned by the same corporate entity, Hanson Industries.)

THE LANDFILLING BUSINESS IS PROFITABLE

Serious consideration should be given by the Parliaments in the U.K. and Canada and the U.S. Congress to enacting Aston's proposed Surface Mine-Landfill Reclamation law. The cost of reclamation of the abandoned surface mines properties could be accomplished at far less cost to the government, hence, the taxpayers, as contrasted with other environmental surface mine reclamation proposals. In many cases there would be no cost at all to clean up abandoned surface sites, and in many cases reclamation could be done at a profit, by simply utilising them as landfills. All future surface mines should automatically be considered by law as potential landfills and used for such when determined to be safe waste disposal sites. Additionally, the nation's critical need for landfill space would be totally relieved. **The reclaimed land's increased value would inure to the owner's benefit.**

It is also proposed that where future natural resources are located in properly zoned industrial development areas, mining be automatically permitted without the public hearing processes, and be subject only to state mine licensing controls. Permitting denials for necessary mineral resources, the subsequent reclamation and landfilling projects by the usual process and subject to the NIMBY syndrome would therefore be eliminated by the advocated mandatory mine permitting and reclamation procedure of landfilling with waste. The Surface Mine and Landfill Reclamation Act would be of the "command and control" nature similar to the United States' Endangered Species Act, but, hopefully, with less extremism than the ESA.

A1 Introduction

The model environmental and land conservation law for reclamation of non-fuel surface mines by back-filling with municipal solid wastes, or other acceptable wastes, is proposed in the following general formats, whether enacted by the British Parliament, the U.S. Congress, or the Canadian Parliament.

The arguments for the law are summarised in the Findings and Purposes of the U.S. Congress under the U.S. format, A-3 §3, infra. The introductory formats for the U.K. and Canada are similarly proposed in following §2 and §4, respectively.

A2 The United Kingdom

ELIZABETH II

DERELICT MINED LANDS RESTORATION AND WASTE CONSERVATION ACT 199x
199x CHAPTER XX

To establish regulatory reform, legal standards, a responsible agency and procedures for carrying out the complete reclamation of all non-fuel surface mines, including abandoned, active, and future mines, making mandatory that such reclamation of the mining voids created by surface mining shall be fully restored to the land's original surface elevations and contours; that the reclamation operation shall be by a landfill procedure utilising municipal solid waste as backfilling material, while simultaneously employing absolute, environmentally-safe technology measures for the protection of water resources quality with the ultimate and final purpose to restore the surface of the mined land to beneficial uses of the general public, thereby conserving land for the sustainability of present and future generations.

Be it enacted by the Queen's most Excellent majesty, by and with the advice and consent of the Lords Spiritual and Temporal, and Commons, in this present Parliament assembled, and by the authority of the same as follows: -

ARRANGEMENT OF SECTIONS

PART I

THE SURFACE MINED LAND RESTORATION AGENCY AND THE SCOTTISH SURFACE MINED LAND RESTORATION AGENCY

CHAPTER I

THE SURFACE MINED LAND RESTORATION AGENCY

Establishment of the Agency

[The detailed goals for establishment of a responsible governmental agency to carry out and fulfill the goals of the Act are given in the following section, A-3.3 — United States — under the Findings and Purposes by the U.S. Congress.]

A3 The United States

The name of the proposed Act shall be —

Surface Mined Lands Restoration, Landfill Relief, and Conservation Act of 199x

or, alternatively

Landfill Relief and Mined Surface Lands Restoration and Conservation Act of 1996

AN ACT

To establish regulatory reform, legal standards, a responsible agency and procedures for carrying out the complete reclamation of all non-fuel surface mines, including abandoned, active, and future mines, making mandatory that such reclamation of the mining voids created by surface mining shall be fully restored to the land's original surface elevations and contours; that the reclamation operation shall be by a landfill procedure utilising municipal solid waste as backfilling material, while simultaneously employing absolute, environmentally-safe technology measures for the protection of water resources quality with the ultimate and final purpose to restore the surface of the mined land to beneficial uses of the general public, thereby conserving land for the sustainability of present and future generations.

BE IT ENACTED by the Senate and House of Representatives of the United States in Congress assembled,

SECTION 1. SHORT TITLE AND TABLE OF CONTENTS.

(a) SHORT TITLE: This Act may be cited as the "Landfill Relief and Mined Surface Lands Restoration and Conservation Act of 1996".
(b) TABLE OF CONTENTS: The Table of Contents is as follows:

TITLE II — NATIONAL RELIEF FOR URGENTLY NEEDED LANDFILL DEPOSITORIES

Sec. 201. Need and Justification
Sec. 202. Environmental Security for Landfiling

TITLE III — COMBINED ENVIRONMENTAL CONTROLS

Sec. 301. Mandated Combined Operational Planning and Permission; After-use planning
Sec. 302. Mandated, Automatic Landfill Licensing for Abandoned and Current Surface Mines
Sec. 303. Mandated, Automatic Licensing for Future Surface Mines as Landfills
Sec. 304. Regulation of Combined Works by State Environmental Agencies

TITLE IV — LAND CONSERVATION

Sec. 401. Restored Land Environmental Protection
Sec. 402. Restored Land After-use Options (parks and recreation, agriculture, wildlife habitat, residential)
Sec. 403. After-use Controls

Sec. 2. FINDINGS AND PURPOSES.

(a) FINDINGS: The Congress finds that —

(1) there are a great number of abandoned surface mines from former years left unreclaimed in the U.S.;
(2) there is virtually no current federal or state regulation for full reclamation of hardrock surface mining; present reclamation of non-coal quarries is near impossible due to the nature of the mining. Hardrock surface mining requires the removal of a very high percentage of the minerals during open-pit mining, thereby leaving virtually no earth material left to be backfilled in the mined void.

Thus, current hardrock surface mine reclamation only requires sloping of the top bench of the quarry, removal of all debris, equipment and buildings from the surface, contouring and planting of vegetation of the disturbed, but intact surface, allowing the mined pit to fill with water. Full restoration of the land to surface use once more is infeasible and very unlikely.

(3) similarly, there can be no expectation that future hardrock surface mines can be restored to beneficial surface uses under the imposed physical and economic conditions;

(4) non-coal surface mined land is virtually useless as left with only water-related amenity-use possible;

(5) full restoration of the surface of open-pit mined land is economically infeasible for lack of fill material;

(6) a simultaneous, critical national need exists for municipal solid waste landfill space. One study reports that all but one state is running out of suitable locations for landfills.

(7) the two problems can be solved as one by landfilling the surface mining voids, thereby solving the national urgent need for MSW depositories, while there is more than ample mining voids available from abandoned sites to accommodate the nation's volume of MSW, plus new mining void space made each year;

(8) land filling of mine voids with MSW is entirely feasible, logical, appropriate and environmentally safe with current containment technology without endangering ground water sources;

(9) MSW landfills have the distinct possibility of supplying a fuel-saving source of energy by capturing the landfill gas that is generated within the fill;

(10) aggregate quarries generally have large volumes of space within them which would provide landfill space for many years at each location; metropolitan areas generally have several aggregate quarries nearby to solve the critical MSW space for many years in the future for each high density waste generating area;

(11) mandating by law the complete restoration of surface mines by landfilling of all abandoned and active mines will eliminate the public inquiry process and its inherent "nimbyism", consequently, the law will expedite the restoration of surface mined land to beneficial after-uses;

(12) licensing of future surface mines as landfills will be compulsory and accomplished in a one-step approval and permitting process which is not to be denied licensing by "nimbyism" when located in a proper zoning area;
(13) there is a critical shortage of landfill space in the U.S. which must be met by this twofold process;
(14) that other methods of waste reduction are not meeting sufficient reduction in volumes of waste necessary to relieve the shortage of space for disposal;
(15) that burial of waste is the most satisfactory method of disposal; and there is always residue for burial from other waste reduction methods;
(16) landfilling at sites other than in mining voids requires additional earth disturbance at other locations to obtain daily and final cover material, impermeable bottom lining clays and unnecessarily leads to duplicity of earth-disturbed land, while filling of mining voids would reduce new earth-disturbed locations by at least 50% thereby greatly conserving land uses;
(17) since many surface mines, particularly those of the aggregates and construction minerals industry, have removed nearly all of the void volume space in mineral extraction, there is little material left to replace in the mined-out pit. Filling and full restoration of the surface is impossible without obtaining earth materials from new locations, thus compounding earth disturbance. Surface mines do have adequate material remaining to supply daily covers, lining and capping earth materials;
(18) due to the past and present economic requirements of the aggregate industry requiring proximity to the construction sites of growing metropolitan areas, and additionally, since the same growing metropolitan and urban areas are the volume generators of MSW, the near-city aggregate quarries are logical, appropriate, convenient and well-located sites for landfilling of the metropolitan generated MSW; finally, that the surface mined voids and the need for landfilling space are mutually made for each other;
(19) the process of landfilling surface mined voids enables formerly derelict mined lands to be salvaged and recycled, to have their surfaces restored, and the land returned to beneficial after-uses for private and public good. This process for total restoration of used land is pure land conservation.

(b) PURPOSES: Based on the environmental regime already established in the United States under the National Environmental Protection Act 1969, and the many other resultant environmental acts that have flowed from NEPA's creation, the purposes of this Act are to further promote environmental security, the restoration of derelict and potentially contaminating surface mined lands in keeping with the conservation of the land —

(1) by establishing the legal process for requiring derelict surface mined lands to be restored to further beneficial surface uses in pursuit of land conservation, and
(2) to concurrently relieve the national critical urgency for MSW depositories, and
(3) to establish the legal process for requiring the use of non-fuel surface mines as MSW landfill sites, and
(4) to reduce the national dependency of foreign gas and oil supplies by utilising landfill-generated gas as a domestic energy source; and
(5) to establish the primacy of state authorities to implement the successful attainment of the goals of this Act.

Then, the general form of statutory enactment to be followed by the U.S. House of Representatives.

A4 Canada

The format for the proposed law to be enacted would be as follows:

THE HOUSE OF COMMONS OF CANADA	CHAMBRE DES COMMUNES DU CANADA
BILL NO.	PROJET DE LOI NO.
An Act to establish complete reclamation of non-coal surface mines, to provide for municipa solid waste depositories and conserve land use.	Loi de mise en oevre, etc.
AS PASSED BY THE HOUSE OF COMMONS (DATE)	ADOPTE PAR LA CHAMBRE DES COMMUNES
Preamble WHEREAS the Government of Canada seeks to achieve sustainable development by conserving and enhancing environmental quality and by encouraging and promoting economic development that conserves and enhances environmental quality; ETC.	Preambule ATTENDU: que le gouvernement federal vise au developpement durable par des actions de conservation et d'amelioration de la qualite de 'enironnement ainsi que de promotion d'une croisaance economique de nature a contributuer a la realisation a la de ces fins;
NOW, THEREFORE, Her majesty, by and with the advice and consent of the Senate and House of Commons Canada, enacts as follows:	Sa Majeste, sur l'avis et avec le consentement du Senat et de la Chambre des communes du Canada, edicte:
SHORT TITLE	TITRE ABREGE
Short title 1. This Act may be cited as the Canadian *Surface Mine Restoration, Waste Relief and Conservation Act 1996*	etc.

A5 Closing Statement to the Honourable Members of H.M Parliaments in the U.K. and Canada, and to the Honourable Members of the U.S. Congress

Seldom does a better opportunity present itself to accomplish multiple goals for the betterment of the environment and for the general public's health and welfare than this present one for enactment by government legislative bodies as proposed herein.

By one legislative act, derelict surface mined lands of the past will be fully reclaimed; complete restoration of current and future surface mined lands will be provided for returning the land surfaces to the beneficial use of society; potential water contamination from the sites will be greatly reduced, if not eliminated; land use will be conserved and sustained for future generations; present and future relief given for the critical space deficiency for disposal of municipal solid and other wastes, whilst providing a new source of energy and conservation of natural fuels by the landfill gas generated; and the contentiousness, delays and high costs of the public hearing process and frequent ensuing litigation greatly reduced.

It remains only for the legislative bodies of the Anglo nations of the United Kingdom, Canada and the United States to refine and adopt this best practice model law as proposed herein to make it an accomplished fact for the improvement of the environment by surface mining and muncipal solid waste industries. We submit the issues have been proven and rest our case.

APPENDIX B

AN OVERVIEW OF THE U.S. GENERAL MINING LAW, 1872

MINING MAGAZINE — INDUSTRY VIEWS
October 1994, pp. 217–218

There has been considerable debate in the U.S. about the need, or otherwise, to amend the 1872 Mining Law. Mr. R. Lee Aston, a member of the International Bar Association, who is an attorney, mining engineer, geologist and adjunct professor of mining and environmental law at the University of Missouri-Rolla, U.S., provides some insight into the legal position:

"After the metallic mineral discoveries of the Far West in the early 1800s, Western miners wanted to continue with as little federal control as possible in order to preserve their acquired mineral rights and insure that free mineral land acquisition would continue. Even then, U.S. opinions differed, with Eastern groups feeling that there should be some financial return to the federal Government from the public lands, either through royalties, taxes, or by sale of the mineral lands. That sectional division persists to some degree at present.

"Federal control over mining on public lands was initiated with the Mining Laws of 1866 and 1870, with the former applying to lode deposits and the latter to placers. The Mining Law of 1872 largely replaced the two former laws and made additional provisions for orderly staking, registering and working requirements of claims for proving the location of a 'valuable mineral deposit'. The fundamental basis of the mineral location system of the 1872 Mining Law is the right of self-initiation. Unless mineral entry has been restricted or withdrawn from mineral entry in the public domain, a mineral prospector may enter the public domain at will to search for minerals, attempt to prove a valuable occurrence of a metallic mineral deposit by hard work and expenditure of funds to prove its existence. If successful in proving the existence of a 'valuable deposit of locatable minerals', his find may be protected by applying for ownership rights through a patenting process for title to the land. The application fees and charges for obtaining title are very small. This right is restricted to U.S. citizens, and aliens are not allowed the right to locate, excepting those that have declared their intent to become U.S. citizens.

"Four types of mineral locations can be filed under the 1872 Law, viz., lode, placer, mill sites, (up to five acres in addition to the mineral claim) and tunnel sites. The required acts for locating a claim are: (1) discovery of a (prospective) valuable mineral deposit; (2) posting notice at the location on the ground; (3) marking the claim on the ground; (4) discovery or development work; (5) filing notice or certificate of location with the county recorder's office, and with the Bureau of Land Management (BLM) state office. The maximum size for a single lode or placer claim is 20 acres. A maximum of 160-acres for placer claims may be filed by a group of 8 persons.

"Early Congressional debates in the 1860s centred on the argument that it was essential to the health and growth of the Nation to continue and expand mineral development in an unfettered manner. To tax the minerals produced would be analogous to taxing farm produce, neither of which would stimulate growth of the Nation. To stimulate the much needed mineral industry for national growth and wealth, the miners' prospecting efforts had to be encouraged. Present-day opponents of the 1872 Law say that day of need is long-gone. Today's proponents argue the necessity to perpetuate the Nation's wealth and economic well-being, saying the same stimulation for the mineral industry is still needed. Furthermore, they argue that prospectors gamble against great odds for riches and that opponents fail to realize that the prospector's odds for 'winning' a valuable mineral deposit are far less than at the Las Vegas gaming tables. If the rewards for the risk of searching for and finding valuable mineral deposits are taken away from the U.S. mining industry, the U.S. will find its mining industry increasingly drifting away, even more heavily than at present, toward Central and South America and to other more mining-friendly countries. Reference is made to the recent trend by the Canadian mining industry away from its homeland as a lesson to be learned from.

"Opponents say the law is archaic, 'a relic from America's pioneer days' that has outlived its purpose and no longer serves a modern mining industry.

Criticism is made of the number of Mining Law cases that been litigated because the law is outmoded, citing 200 that have been before the U.S. Supreme Court in over 100 years, and 'thousands of cases' in the lower federal and state courts; that the mining law has not kept up with the advances of times; the patenting process has been greatly abused to obtain lands for non-mining purposes where no mining has taken place, resulting in so-called "sandscams". In these occurrences, common mineral claims have led to patenting of land for real estate-recreational purposes that have re-sold for many times in value. Criticism is made that the 1872 Law fails to provide a fair return to the government for minerals owned by the public; that government charges are 'give-away' amounts; that public domain mining land can be 'bought' (patented) for ridiculously low prices, e.g. $2.50 an acre for placer claims, and $5 per acre for lode claims; that miners can mine for years without paying any royalties to the government; that the fees to maintain a claim are too low; that claim work / development requirements of $100/a. are too little to obtain exclusive rights and ownership of the land. The charge is also made that the old Law contains no environmental protection for public lands from mining.

"Proponents counter that the government is not to profit from the sale of public lands, only to administer for development.

"Proponents contend the law is not archaic but has been kept vibrant with the advances of the mining industry and environmental concerns through amendments and developing case law. Several mining acts have been enacted over the past decades which have amended and augmented the 1872 Law. Various additions have addressed many of the law's deficiencies charged by opponents. For example, there are three different federal policy systems governing the exploration, development and production of minerals on unappropriated public lands, viz., (1) hardrock (generally, metallic) claims under the General Mining Law of 1872; (2) mineral leasing under The Mineral Leasing Act of 1920; and (3) by the minerals disposal system under the Materials Act of 1947. Specific minerals were removed from coverage of the 1872 Law and placed under the Leasing Act. The Multiple Mineral Development Act of 1954 reserves all leasing act minerals to the U.S. prior to issuance of a patent. The Multiple Surface Use Act of 1955 codified previous court decisions by statutorily precluding uses of mining claims, prior to issuance of patent, for any other purposes than mining or incident to those purposes. This Act eliminated the locator's exclusive right of possession of the surface for unpatented claims. The U.S. retained the right to manage and dispose of the vegetative surface resources and to use the surface for any purposes as long as not interfering with the work of the mineral claimant.

"The Materials Act makes available common minerals as sand, gravel, pumice, volcanic cinder, stone and clay at a market price usually determined by competitive bidding (contract sale). Uncommon varieties are locatable under the Act of 1955. Non-metallic minerals are acquired by mining claim or by a sale contract which is based on their chemical composition, physical properties and prospective use. Common quartz varieties as jasper, obsidian, opal, etc., are not locatable unless they exhibit uncommon qualities. Semiprecious minerals and precious stones are locatable. Also, geothermal steam and associated geothermal resources on BLM lands are leasable under the Geothermal Steam Act of 1970. Under the Mineral Leasing Act and the Materials Act, agency permits are required for explorational activity. The leasing method requires annual rentals until production, with royalties paid thereafter. The responsible agency has complete discretion to accept or reject offers.

"Proponents admit it is true that the nomenclature for lode and placer deposits should be expanded to include the currently exploited low-mineral value porphyry and disseminated type of deposits.

"As to the number of cases litigated over the 1872 Law, the number is far less than for many other federal statutes, particularly the National Environmental Protection Act (NEPA) 1969. For the same reason of excessive litigation, proponents ask 'would the environmentalists favour dismissing NEPA?' The plethora of environmental statutes and regulations following NEPA has probably generated more litigation in one year than the 1872 Mining Law has in a century.

"With regard to the 'give-away' prices for public domain patented mineral lands, proponents note that of all lands patented under various government programmes for public lands, as for grazing, homesteading, etc., only 3% have been for mineral lands. It prompts the question, if such a great give-away, then why hasn't more been taken? The poor odds for a successful mineral discovery is the answer.

"The Senate and House have each passed its own version of a mining law reform bill. The next step is for the conference committee to establish a compromise bill that will win or lose approval in both houses of Congress.

"The Senate's bill, approved in May 1993, provides a 2% net royalty, allows patenting of mining claims for fair market value of the surface, leaves current reclamation procedures in place, and retains the existing system for filing and maintaining claims.

"The Representative's bill, approved in November 1993, provides an 8% gross royalty, does away with patent (ownership) rights, imposes federal reclamation standards, establishes a new system for claim location, provides environmental safeguards to withdraw lands unsuitable for mining.

"A compromise bill for reformation is expected to reach the respective floors of each house during the current legislative session.

"In spite of opponents' claims that the 1872 is archaic, on-going revisions have been made periodically and made again in 1993 by Congress, the Department of Interior and other agencies concerned with mining on public lands. Case law, resulting from litigation of the 'relic' law has continuously updated and clarified the law for modern applications This has been true throughout the old law's existence.

"In late 1992, Congress passed the Department of Interior 1993 Appropriations Act which requires most holders of unpatented mining claims to pay a rental fee of $100 per claim per year in lieu of the assessments work of the 1872 Mining Law. First and second year payments were due before August 31, 1993. This provision expires on September 30, 1994, but was later extended. Exemptions were made for small claim holders, i.e., with less than 10 unpatented claim under a valid notice or plan of operation, and less than 10 acres of unreclaimed surface disturbance from such mining activity or exploration work. Exemption requires that the claimant must be either producing between $1,500 to $800,000 in gross revenues, or performing exploration work that discloses, exposes or makes known possible valuable mineralisation. Qualified exemptions may choose to pay the annual rental fee or do the required assessment work and meet the filing requirements of the Mining Law and the Federal Land Policy and Management Act (FLPMA).

"In 1992, the BLM issued rules governing and restricting the use and occupancy of unpatented mining and millsite claims. The Interior Board of

Land Appeals (IBLA) issued decisions from reviews of public land mining plans that will effect wilderness study areas; the abandonment of unpatented mining claims for failure to file timely evidence of assessment work; trespass for road building in a wilderness study area; the rehabilitation of wilderness study areas affected by unauthorised mining operations; surface protection bond requirements for a mining claimant under the Stock-Raising Homestead (Taylor Grazing) Act, 1916; and numerous other environmentally sensitive issues.

"In 1993, IBLA's decisions continued to 'tighten-up' on public domain regulations already in place, for example, the sufficiency of evidence required to establish a 'discovery'. Although environmentalists accuse the government of laxity of regulation enforcement, the disproving of mining claims is not new and is illustrated by an older IBLA case from 1990: In *U.S. v. Michael R. Ware*, the prospector's claim samples were suspected as being high-graded and challenged by BLM as being a valid "discovery". At hearing, the BLM showed that mining costs would be 15 to 20 times higher than costs from miner's sample values; claim would be uneconomical and not a valid claim. On appeal, it was held that: (1) BLM had proven a case of non-discovery; (2) claimant's assay evidence alone was inadequate to refute BLM's assays; and (3) it was claimant's responsibility to refute the BLM's evidence.

"And, in an older example, a 1985 case, *U.S. v. Jos. Fahey Co.,* concerned unpatented gold claims, the 'Open Fields' exception to requirement of search warrant for governmental sampling unpatented claims was developed. Fahey was convicted of 14 counts of mail and wire fraud in connection with four unpatented, non-producing mining claims near Death Valley, in Nevada. Fahey objected to the government's taking of samples from the site. Sampling was found permissible under the 'open fields' exception to the warrant requirement for search and seizure.

The federal court upheld the government's position that legal title to the unpatented claims was held by the U.S. Fahey, as holder of an unpatented claim had a non-exclusive possessory interest in the land for purposes of mining only; the U.S. retained the right to manage and dispose of the other resources on the land, with right of entry, and the general public's use for recreational or other purposes as long as they did not interfere with mining.

"Opponents have made much of 'wrong-doing' and environmental harm over the 'sandscam' of the high-purity sand deposit land patented in the Oregon Dunes National Recreation Area in 1989. In the case of *Duval,* the BLM issued a patent for Duval's claim of 1,000 acres in the sand dune area. Duval had valid existing rights to the deposits since he held the claims 13 years before the Dunes Recreational Area was created. After lengthy litigation over the claim, Duval prevailed because he was able to establish that the sand was a 'valuable and uncommon' variety not to be mined for sand, *per se*, but for its constituent minerals, viz., the quartz and feldspar, where there was a prominent lack of impurities and grain size was uniform.

"Proponents are wary of the opponents' environmental safeguards to withdraw lands 'unsuitable' for mining. Large acreages of the public domain are currently being withdrawn from mineral entry in the name of wildlife studies, endangered species, primitive and recreational areas, and other environmental reasons. Such terminology included in a revised Mining Law could continue to whittle down the available acreage for mining exploration to a mortality level for the industry.

Opponents' charge that the 1872 Law affords no environmental protection for public domain mining is erroneous. Federal and state mining and environmental laws require operator's bonds, plans of operation, plans to prevent undue degradation and reclamation of the site. Public domain mining sites are subject to obtaining mining permits, the Clean Water Act, and regulations by the BLM and / or the National Forest Service.

APPENDIX C

LIST OF REFERENCES

Agricola, G. (1556), *De Re Metallica*, translation from Latin by H.C. & L.H. Hoover, 1950, Dover Publications, Inc., New York.

American Jurisprudence (Am Jur) (1966), 2nd Edition, Vol. **26**, *Eminent Domain,* I §1–9, In General, Gulick, G.S. and Kimbrough, R.T., Eds., The Lawyers Cooperative Publishing Company, Rochester, N.Y.

American Jurisprudence (Am Jur) (1971), 2nd Edition, Vol. **54**, *Mines and Minerals*, Irwin J. Schiffres, J.D., Section Editor, The Lawyers Cooperative Publishing Company, Rochester, N.Y.

ARC, Ltd. (1989), Judkins Quarry Planning Brief, published jointly by ARC Ltd, Nuneaton & Bedworth Borough Council and Warwickshire County Council, U.K.

Aston, R.L. (1990), 'A Quarryman's Reply *Wolfeboro* Decision)', *Legal Briefs*, Pit & Quarry, March, pp. 38–39, Hunter Publications, Chicago, Illinois.

Aston, R.L. (1992a), 'A Current Study of Enviromining Laws for Industrial Minerals and Rocks', MSc Thesis, University of Missouri-Rolla, USA.

Aston, R.L. (1992b), 'A Summary of the Current Litigation for the City of St. Peter's Proposed Quarry Landfill Site Located in St. Charles County, Missouri', (unpublished), U. Mo.-Rolla, USA.

Aston, R.L. (1993a), 'Enviromining Law', *Mining Environmental Management,* Vol. **1,** No. 1, March, pp. 16–17, The Mining Journal Ltd., London.

Aston, R.L. (1993b), 'Land Title: Recent Court Cases', *Mining Environmental Management,* Vol. **1,** No. 3, September, pp. 8–9, The Mining Journal Ltd., London.

Aston, R.L. (1994), 'The U.S. General Mining Law 1872 — An Overview', pp. 217–218, The Mining Magazine, October, London.

Aston, R.L. (1995a) 'Nimby Syndrome Continues for Ohio S & G Operator ', *Legal Briefs,* Aston, R.L., Legal Editor, Pit & Quarry, April, Advanstar Publications, Chicago, Ill.

Aston, R.L. (1995b), 'Assays from the Legal Vein (TM) — Replacement Water Supply Mandated by SMCRA', *Engineering & Mining Journal*, February, p. NA–16LL, Chicago, Ill.

Aston, R.L. (1995c), 'Assays from the Legal Vein (TM)-Texas Surface-Destruction Test Re-Surfaces', *Engineering & Mining Journal*, February, p. NA–16LL, Chicago, Ill.

Bagchi, A. (1989), *Design, Construction, and Monitoring of Sanitary Landfill,* John Wiley & Sons, New York.

Bainbridge, W. (1900), *The Law Relating to Mines and Minerals,* 5th Edition, Brown A., Ed., Butterworth & Co., London, U.K.

Bartlett, R. (1984), 'The Right to Mine and Extent of Ministerial Discretion', Chapter 3, Bartlett, R., Ed., *Mining Law in Canada*, Continuing Legal Education, Law Society of Saskatchewan, Canada.

Barton, B.J. (1993), *Canadian Law of Mining*, Canadian Institute of Resources Law, Calgary, Alberta, Canada.

Beerli, M. (1989), 'Use of Biofilters in Odour and Volitile Organic Solvent Control', Third International Symposium on Peat/Peatland Characteristics and Uses Proceedings, May 16–20, Spigarelli, S.A., Ed., Bemidji State University, Bemidji, MN, pp. 335–348.

Beerli, M. (1991), 'Odour and VOC Control by Use of Peat Biofillers' Symposium '89 Peat and Peatlands Diversification and Innovation, Vol. II, New Products Proceedings, August 6–10, Overend, R.P. and Jeglum, J.K., Eds., Canadian Society for Peat and Peatlands, Quebec City, Quebec, Canada, pp. 136–146.

Black, H.C. (1968), *Black's Law Dictionary*, 4th Edition, West Publishing Co., St. Paul, USA.

Blackman, S. (1996) pers comm, Canadian Institute of Resources Law, The University of Calgary, Canada.

Blakeman, W.B. (1977), *Statutory and Environmental Protection Aspects of Mineral Aggregate and Industrial Mineral Production*, March 1977, W.B. Blakeman & Associates, Ottawa, Ontario, Canada.

Blakeman, W. (1996), pers comm, Senior Policy Advisor, Policy and Programme Development Section, Ontario Ministry of Environment and Energy, Toronto, Canada.

BGS (1995), British Geological Survey, 'Minerals in the National Economy', United Kingdom Production of Minerals 1989–1995, Keyworth, Nottingham, U.K.

Boehler, A. (1991), 'The Uses of Biofilters in Odour Control', Symposium '89, Peat and Peatlands Diversification and Innovation, Vol. II, New Products Proceedings,

August 6–10, Overend, R.P. and Jeglum, J.K., Eds., Canadian Society for Peat and Peatlands, Quebec City, Quebec, Canada, pp. 119–121.

Browning-Ferris Inds. (1991), *Solid Waste Disposal*, Brochure # 9/91 GCP 10M, Houston, Texas, USA.

Charlier, M. (1995), Staff Reporter, Article in March 1, 1995 Issue, Wall Street Journal, N.Y., N.Y.

Coffey, P. and Kavanagh, R. (1989), 'Specialised Peat-based Waste Treatment Systems', Third International Syposium on peat/Peatland Characteristics and Uses Proceedings, May 16–20, Spigarelli, S.A., Ed., Bemidji State University, Bemidji, MN, pp. 324–334.

Crommelin, M. (1974), 'Mineral Exploration in Australia and Western Canada', 9 UBCLR. 38, p. 42, School of Law, University British Columbia, Canada.

Daigle, J.-Y. (1993), 'Peat Moss for Wastewater Treatment', Open File Report 93–3, New Brunswick Department of Natural Resources and Energy, Mineral Resources, Bathurst, New Brunswick, Canada.

Dodds-Smith, C.A., Payne, C.A. and Gusek, J.J. (1995), 'Reedbeds at Wheal Jane', *Mining Environmental Management*, Rosin, N., Ed., Vol. **3**, No. 3, pp. 22–24, The Mining Journal Ltd., London.

Dominie, K. (1992), *A Report on Provincial Waste Disposal Sites and Waste Management*, Department of Environment and Lands, Government of Newfoundland and Labrador, June 1992.

Drake, R. (1996), pers comm, Editor, Rock Products, Cleveland, Ohio.

Eger, P. Lapakko, K. and Otterson, P. (1980), 'Trace Metal Uptake by Peat: Interaction of White Cedar Bog and Mining Stockpile Leachate', Sixth International Peat Congress Proceedings, Duluth, MN, August 17–23, Grubich, D.N., Ed., U.S. National Committee of International Peat Society, Eveleth, MN, pp. 542–547.

Eger, P. and Lapakko, K. (1989), 'Use of Wetlands to Remove Nickel and Copper from Mine Drainage', Constructed Wetlands for Wastewater Treament; Hammer, D.A., Ed., Lewis Publishers, Inc. Chelesa, MI, pp. 780–787.

Engineering & Mining Journal (1992), 'Wisconsin Permits First Mines Since World War II-Almost', February, McLean Publishing Co., Chicago, Illinois.

Engineering & Mining Journal (1994) 'Georgius Agricola, 500th Anniversary', July, pp. 38–39, WW, Chicago, Illinois.

Evans, P. B. (1994), The Clean Water Act Handbook, SONREEL, American Bar Assn., Chicago, Illinois.

Farnham, R.S. and Brown, J.L. (1972), 'Advanced Wastewater Treatment Using Organic and Inorganic Materials', Fourth International Peat Congress Proceedings,

Otaniemi, Finland, June 25–30, Otaniemi, Finland, Finnish National Committe of International Peat Society, Helsinki, Finland, Vol. **IV**, pp. 271–286.

Gates, P.W. (1968), *History of Public Land Law Development*, U.S. Government Printing Office, Washington, D.C.

Garrett, T.L. (1994), Book Publications Chair, Committee Reports, Environmental Group, '*Water Quality*', Managing Editor, Mansfield, M.E., Natural Resources, Energy, and Environmental Law, 1993, Year in Review, American Bar Assn. & The National Energy Law & Policy Institute, University of Tulsa, College of Law, USA.

Grad, F.P. (1978), *Environmental Law*, Vol. **2**, Matthew Bender, New York.

Halsbury's Laws of England (1980), 4th Edition, Vol. **31**, Mines, Minerals and Quarries, pp. 256–259; Butterworths, London.

Harding, D. (1996), Public Affairs Manager, Greenways, U.K., pers comm.

Hatheway, A.W. (1993), *Chronological History of Industrial and Hazardous Waste Managment,* unpublished, (copyright 1993), University of Missouri-Rolla, USA.

Hatheway, A.W. (1995), *Perspective No. 26, Blasting Damage; Dynamite and Groundwater,* EG News, Vol. **38**, No. 3, Fall 1995, pp. 37–39, Sudbury, MA.

Hedin, R.S., Nairn, R.W. and Kleinmann, R.L.P. (1994), 'Passive Treatment of Coal Mine Drainage', U.S. Bureau of Mines, IC 9389, 35 pp.

HMSO (1991), Research Report, *Environmental Effects of Surface Mineral Workings,* p. 13, Department of the Environment, HMSO, London, U.K.

HMSO (1994a), *Landfilling Wastes,* Waste Management Paper No. 26, 6th Impression. Department of the Environment, HMSO, London, U.K.

Hodgson, E.C. (1966), 'Digest of Mineral Laws of Canada', Mineral Report 13, Mineral Resources Division, Department of Energy, Mines and Resources, Ottawa, Canada.

Howell, J.V. (1957), *Glossary of Geology*, American Geological Institute, **J.V. Howell**, Coordinating Chairman, National Academy of Sciences — Natural Research Council, Washington, D.C.

Hughes, D. (1986), *Environmental Law*, Extraction and Infill; Mineral Extraction, Ch. 8, Butterworths, London.

Hughes, D. (1992), Environmental Law, 2nd Edition, Butterworths, London.

Institution of Civil Engineers (1991), *Pollution and its Containment*, Thomas Telford, London.

James, C.A. (1898), *Mining Royalties,* Longmans, Green and Co., London.

Johnson, W. (1992), *Clay Colliery Company Ltd. Bowmans Harbour Wolverhampton, Environmental Statement,* Vol. **1**, Non-Technical Summary by Johnson Poole & Bloomer, February 1992, Telford, U.K.

Kadlec, R.H. and Keoleian, G.A. (1986), 'Metal Ion Exchange on Peat' Peat and Water, Fuchsman, C.H., Ed., Elsevier Applied Science Publishers Ltd., Essex, England, pp. 61–93.

Kennett, S.A. (1992), 'Issues and Options for Intergovernmental Cooperation in Environmental Impact Assessment ', Canadian Institute of Resources Law, No. 39, Summer 1992, University of Calgary, Canada.

Kennett, S.A. (1995), 'The Environmental Management Framework Agreement', Canadian Institute of Resources Law, No. 52, Fall 1995, University of Calgary, Canada.

Kent, H.G. (1996), Planning Manager, Clay Colliery Company Ltd., Telford, England, U.K., pers comm, September 1993 and January 1996.

Kent, J. (1896), *Commentaries on America Law*, Vol. **II**, Part VI, 14th Edition, Little, Brown and Company, Boston.

Kusler, F. and Meyers, J. (1990), 'Litigation: Is the Claims Court All Wet?', *National Wetlands Newsletter,* Nov./Dec. 1990, Washington, D.C.

Lanciani, Rudolfo (1890) (republished 1980), *The Destruction of Ancient Rome*, Arno Press, A New York Times Company, New York.

Lindley, C.H. (1914), *A Treatise on the American Law Relating to Mines and Mineral Lands* aka *Lindley on Mines*, 3rd Edition, Bancroft-Whitney, San Francisco, USA.

Liptak, B.G. (1991), *Municipal Waste Disposal in the 1990s*, Chilton Book Co., Radnor, Pennsylvania.

Mac Swinney, R.F. (1897), *The Law of Mines, Quarries, and Minerals,* 2nd Edition, Sweet and Maxwell, Ltd., London.

Malterer, T, McCarthy, B. and Adams, R. (1996), Use of Peat in Waste Treatment, *Mining Engineering*, Vol. **48**, No. 1, January 1996, pp. 53–55, Littleton, Colorado.

Mining Environmental Management (1993), 'Enviromine-Mining's Ancient Past', Rosin, N., Ed., Vol. **1**, No. 1, p. 24, The Mining Journal Ltd., London.

Mining Environmental Management (1994), 'Technology', Rosin, N., Ed., Vol. **2**, No. 3, p. 35, The Mining Journal Ltd., London.

Mining Environmental Management (1995a), 'Potential Use of Metal Hyperaccumulators', Chaney, R., Brown, S., Yin-Ming, L, Angle, J.S, Homer, F and Green, C., Vol. **3**, No. 3, Sept., pp. 9–11, *The Mining Journa*l, London.

Mining Environmental Management (1995b), 'Farming for Metals?', Nicks, L.J. and Chambers, M.F., Vol. **3**, No. 3, Sept., pp. 15–18, *The Mining Journal*, London.

Mining Environmental Management (1995c), 'Reedbeds at Wheal Jane', Dodds-Smith, M.E., Payne, C.A. and Gusek, J.J., Vol. **3**, No. 3, Sept., pp. 22–24, *The Mining Journal*, London.

Mining Environmental Management (1995d), 'Biotreatment of mine drainage', Bender, J. and Phillips, P., Vol. **3**, No. **3**, Sept., pp. 25–27, *The Mining Journal*, London.

Mining Environmental Management (1995e), 'Enviromine', Rosin, N., Ed., Vol. **3**, No. **3**, Sept., p. 28, *The Mining Journal*, London.

Mining Environmental Management (1995f), 'Minerva', Rosin, N., Ed., Vol. **3,** No. 4, Dec., p. 3, The Mining Journal Ltd., London.

Minor, J.C. (1986), 'Environmental Law: The European Dimension', Ch. 4, *Environmental Law,* Hughes, D., Butterworths, London.

Natural Resources Canada (1995), Mineral and Metals Sector; Market Trends for Industrial Minerals, publication # 1004; Importance of Mining and Minerals to the Canadian Economy, pub. # 1005; Canada's 1994 Mineral Production (News Release), pub. # 1006; Mineral Production of Canada, 1992, 1993 and 1994, and Average 1990–1994, pub. # 1901; Cement 1994 Review and Outlook, pub. # 5105; Mineral Aggregates 1994 Review and Outlook, pub. # 5117; Stone 1994 Review and Outlook, pub. # 5131.

Orlik, M. (1992), 'Legal Obligations in Waste Disposal', Waste Management Conference, Civil Engineering, Aston University, Birmingham, U.K., 5 November 1992.

Outerbridge, C. (Ed.) (1986), *American Law of Mining,* 2nd Edition, Rocky Mountain Mineral Law Foundation (RMMLF), Matthew Bender, N.Y.

Owens, M.D. and Elifrits, C.D. (1995), 'Survey of the Use of Abandoned Surface Coal — Mined Land for State-of-The-Art Solid Waste Disposal Facilities', University of Missouri-Rolla, USA.

Pickering, J.R. (1957), *The Mines and Quarries Act, 1954,* Butterworth & Co., London.

Plewman, R. (1996), pers comm, Manager, Walker Inds., Ltd., Thorold, Ontario, Canada.

Priestley, J.J. (1968), 'Civilisation, Water and Wastes', *Chemistry and Industry,* (March 23, 1968), pp. 353–363.

Ramsey, Marc (1995), pers comm, Manager, Fred Weber, Inc., Maryland Heights, Missouri.

Rana, S. and Virarghavan, T. (1987), 'Use of Peat in Septic Tank Effluent Treatment-Column Studies', *Water Pollution Research Journal of Canada*, Vol. **22**, pp. 491–504.

Riznyk, R.Z., Reid, Jr., L.C., Rockwell, Jr., J. and Reid, S.L. (1990), 'Peat Leach Mound Treatment of On-site Domestic Septic Effluent in Cold Region Environments', *International Conference on Peat Production and Use Proceedings,*

Jyvaskyla, Finland, June 11–15, 1990, The Association of Finnish Peat Industries, Jyska, Finland, pp. 383–392.

Robertson, J.D. and K.D. Ferguson (1995), 'Predicting Acid Rock Drainage', *Mining Environmental Management,* Rosin, N., Ed., Vol. **3**, No. 4, pp. 4–8, The Mining Journal Ltd., London.

Rock, C.A., Brooks, J.L., Bradeen, S.A. and Struchtemeyer, R.A. (1993), 'Use of Peat for On-Site Wastewater Treatment: 1 — Laboratory Evaluation', *Journal of Environmental Quality*, Vol. **13**, pp. 518–523.

Rogers, A. (1876), *The Law of Mines, Minerals, & Quarries in Great Britain and Ireland,* 2nd Edition, Stevens & Sons, London.

Salvato, J.A. (1992), *Environmental Engineering and Sanitation,* 4th Edition, John Wiley & Sons, Inc., New York, New York.

Schoenbaum, T.J. (1985), *Environmental Policy Law*, 1985 Edition, The Foundation Press, Westbury, New York.

Schoenbaum, T.J., Rosenberg, R.H. (1991), *Environmental Policy Law*, 2nd Edition, The Foundation Press, Westbury, New York.

Scottish Office Development Dept. (1995), 'Review of the Planning System in Scotland: The Way Ahead', December 1995, ahd01228.115.

Seeney, K. (1988), 'Landfill Trash: Energy From Buried Resource', *Missouri Resource Review,* Mo. Dept. of Natural Resources, Vol. **5**, No. 2, Jefferson City, Mo.

Shinn, C.H. (1884), *Mining Camps, A Study in American Frontier Government*, Vol. **12**, 2nd Series, Johns Hopkins University, Baltimore.

Silliphant, Dave (1994), pers comm, Director, Operations Branch, Dept. of the Environment, Fredericton, New Brunswick, Canada.

Skitt, J. (1979), *Disposal of Refuse and Other Waste*, Charles Knight & Co., Ltd., London.

SME (1995), 'In One Day', Society for Mining Metallurgy, and Exploration, Inc., Foundation for Public Information and Education, Inc., Littleton, Colorado.

Scott, W.D., Meine, F.J, Chouinard, U.S. Eds., and Wallace,W.S., Canadian Ed., (1959), *The American Peoples Encyclospedia*, Spencer Press, Inc., Chicago.

The Mining Journal (1993a), 'Northwest Territories, Country Supplement', Vol. **321**, No. 8253, p. 2, London.

The Mining Journal (1993b), 'U.S. Mine Site Reclamation', p. 51, London.

The Mining Journal (1994a), 'Bangladesh approves BHP's Coal Plans', p. 155, London.

The Mining Journal (1994b), 'British Geological Survey's Minerals Extraction Forum — The Economic and Environmental Balance', p. 286, London.

The Mining Journal (1994c), 'Exploration: B.C. Spending At Record Low", p. 20, London.

The Mining Journal (1994d), 'Mine Water Congress', p. 235, London.

The Mining Journal (1994e), 'Canadians Abroad: Canadian Mining Money Moves Overseas', p., London.

The Mining Journal (1995a), 'Focus and Comment — Living with Minerals', by Confederation of British Industry (CBI), p. 89, London.

The Mining Journal (1995b), 'U.S. — Canadian Mining Company Echo Bay Moving Abroad', p., London.

The Mining Journal (1995c), 'Environment — The Blind River Expansion', p. 215, London.

The Mining Journal (1995d), 'Mining Reform Cost Jobs', p. 258, London.

The Mining Journal (1995e), 'Focus and Comment — Caravan vs. Trucks', p. 413 London.

The Mining Journal (1995f), 'Opposition Stalls Indian Bauxite Mines', p. 479, London.

The Mining Journal (1995g), '1st Kazakhstan International Mining Exhibition' p. 104, London.

The Mining Journal (1995h), ' A Window of Opportunity — Northern Ireland', p. 388, London.

Tilton, D.L., Kadlec, R.H. (1979), "The Utilisation of a Fresh-Water Wetland for Nutrient Removal from Secondary Treated Wastewater Effluent', *Journal of Environmental Quality,* Vol. **8**, No. 3, pp. 328–334.

Tkachuk, D.M. (1993), *Alberta's Wetlands: Legal Incentives and Obstacles to Their Conservation,* Canadian Institute of Resources Law, University of Calgary, Canada.

Tomes, P.A. (1989), 'Restoration of Quarries by Controlled Landfill', Institute of Quarrying, Annual Conference Symposium, 2 October 1989, Bristol, U.K.

UK Dept. of Environment (1992), *Environmental Effects of Surface Mineral Workings,* (Research Report) HMSO, London.

UK Dept. of Environment (1994a), Mineral Planning Guide 6, HMSO, London.

UK Dept. of Environment (1994b), Mineral Planning Guide 7, p. 60123, HMSO, London.

USBM (1994), Mineral Industry Surveys, Annual Review 1994, US Dept. of Interior-Bureau of Mines, Washington, D.C.

USBM (1995), Mineral Industry Surveys, Crushed Stone and Sand and Gravel in the third Quarter of 1995, US Dept. of Interior-Bureau of Mines, Washington, D.C.

US EPA (1994), Memorandum, Fact Sheet FY 1994, Enforcement Highlights, November. Government Printing Office, Washington, D.C.

Vagt, O. (1994), *Cement*, Canadian Minerals Yearbook 1994, Natural Resources Canada, Ottawa, Canada.

Walde, T. (1992), 'Environmental Policies Towards Mining in Developing Countries', Journal of Energy and Natural Resources Law, Vol. **10,** 4, pp. 327–355.

Wilson, D.G., (1977), *Handbook of Solid Waste Management,* MIT, Van Nostrand Reinhold Co., New York.

Wood, R.J. (1995), Book Review Editor, *The Canadian Business Law Journal,* Vol. **24**, 1994–1995, Canadian Law Book Co., Aurora, Ontario.

VITA

Dr. R. (Robert) Lee Aston was born in the former British colony of Virginia, USA. Despite the longevity of his America ancestry, paternally and maternally from the early 17th century, Aston is an Anglophile tracing the Astons to Staffordshire, England, the seat of Aston University where he took his doctoral degree (Ph.D.). His doctoral research was in surface mining law and the reclamation of open-pit mines, comparing the laws of the three Anglo nations, Great Britain, the United States and Canada.

Aston has earned seven university degrees, with master's degrees in three professions, vis., law, mining engineering and geological engineering. He is a multiple graduate of the University of Missouri's (Rolla) School of Mines in engineering and from the College of William & Mary in Virginia.

Aston has practiced as a mining engineer and mining geologist in metallic and industrial minerals in both underground and surface mines for over 40 years and is a professional mining engineer and geologist. He is a member of the Society of Mining Engineers and the Association of Engineering Geologist.

He attended law school in Georgia and has been a practicing trial attorney in civil law, mineral and environmental law for the past eleven years. During his legal career, Aston has been a prolific writer of mining and environmental law articles for mining and legal publications. He is a member of the Energy and Natural Resources Law sections of the International Bar Association and the American Bar Association, and the Australian Mineral and Petroleum Law Association. He is a member of the Georgia, Virginia, Indiana and Montana State Bars and has been admitted to practice before the U.S. Supreme Court, four U.S. Circuit Courts of Appeals and several federal district courts. He has taken appeals to the Supreme Courts of Georgia, Montana and Missouri. In practice, Aston has combined his three professions into one by specialising in mining and environmental law.

Dr. Aston served in the 8th Air Force during World War II. He was stationed in England and flew 35 combat missions as a navigator on a B-24 Liberator, receiving the Distinguished Flying Cross and the Air Medal with five Oak Leaf clusters. He trained as a fighter pilot for a second tour of combat in the Pacific theatre, with the war ending just after he finished his pilot training.

Dr. Aston has been an Adjunct Professor of Mining and Environmental Law at the University of Missouri since 1992. He is nearing completion of another doctoral degree, a Doctor of Engineering, at the University of Missouri. His dissertation will be developed into an environmental law book to be used as a teaching text for engineers and attorneys at the University of Missouri.